The History of Modern Physics

1800 – 1950

Volume 7

The History of Modern Physics, 1800–1950

The History of Modern Physics, 1800–1950

TITLES IN SERIES

VOLUME 1
Alsos
by Samuel A. Goudsmit

VOLUME 2
Project Y: The Los Alamos Story
Part I: Toward Trinity
by David Hawkins
Part II: Beyond Trinity
by Edith C. Truslow and Ralph Carlisle Smith

VOLUME 3
American Physics in Transition: A History of Conceptual Change in the Late Nineteenth Century
by Albert E. Moyer

VOLUME 4
The Question of the Atom: From the Karlsruhe Congress to the First Solvay Conference, 1860–1911
by Mary Jo Nye

VOLUME 5
Physics For a New Century: Papers Presented at the 1904 St. Louis Congress

VOLUME 6
Basic Bethe: Seminal Articles on Nuclear Physics, 1936–1937
by Hans A. Bethe, Robert F. Bacher, and M. Stanley Livingston

VOLUME 7
History of the Theories of Aether and Electricity: Volumes I and II
by Sir Edmund Whittaker

INTRODUCTORY NOTE

The Tomash/American Institute of Physics series in the History of Modern Physics offers the opportunity to follow the evolution of physics from its classical period in the nineteenth century when it emerged as a distinct discipline, through the early decades of the twentieth century when its modern roots were established, into the middle years of this century when physicists continued to develop extraordinary theories and techniques. The one hundred and fifty years covered by the series, 1800 to 1950, were crucial to all mankind not only because profound evolutionary advances occurred but also because some of these led to such applications as the release of nuclear energy. Our primary intent has been to choose a collection of historically important literature which would make this most significant period readily accessible.

We believe that the history of physics is more than just the narrative of the development of theoretical concepts and experimental results: it is also about the physicists individually and as a group—how they pursued their separate tasks, their means of support and avenues of communication, and how they interacted with other elements of their contemporary society. To express these interwoven themes we have identified and selected four types of works: reprints of "classics" no longer readily available; original monographs and works of primary scholarship, some previously only privately circulated, which warrant wider distribution; anthologies of important articles here collected in one place; and dissertations, recently written, revised, and enhanced. Each book is prefaced by an introductory essay written by an acknowledged scholar, which, by placing the material in its historical context, makes the volume more valuable as a reference work.

The books in the series are all noteworthy additions to the literature of the history of physics. They have been selected for their merit, distinction, and uniqueness. We believe that they will be of interest not only to the advanced scholar in the history of physics, but to a much broader, less specialized group of readers who may wish to understand a science that has become a central force in society and an integral part of our twentieth-century culture. Taken in its entirety, the series will bring to the reader a comprehensive picture of this major discipline not readily achieved in any one work. Taken individually, the works selected will surely be enjoyed and valued in themselves.

The History of Modern Physics
1800 – 1950
Volume 7

A History of the Theories of Aether and Electricity

II. The Modern Theories

SIR EDMUND WHITTAKER

Tomash Publishers

American Institute of Physics

Published in 1954 by the Philosophical Library.

Library of Congress Cataloging in Publication Data

Whittaker, E. T. (Edmund Taylor), 1873–1956.
History of the theories of aether and electricity.

(History of modern physics, 1800–1950; v. 7)
Reprint. Originally published: London; New York: Nelson, 1951–1953.
Includes index.
Contents: v. 2. The modern theories.
1. Ether (of space)—History. 2. Electricity—History. 3. Electromagnetic theory—History. I. American Institute of Physics. II. Title. III. Series.
QC177.W63 1987 530′.09 87-1851
ISBN 0-88318-523-7

Preface

THE purpose of this volume is to describe the revolution in physics which took place in the first quarter of the twentieth century, and which included the discoveries of Special Relativity, the older Quantum Theory, General Relativity, Matrix Mechanics and Wave Mechanics.

My original intention was to give an account of the history from 1900 to 1950 in a single volume ; but the wealth of material made this undesirable ; and the period from 1926 to 1950 must be reserved for a third book.

I am greatly indebted to Dr E. T. Copson, Regius Professor of Mathematics in the University of St Andrews, and Dr J. M. Whittaker, Vice-Chancellor of the University of Sheffield, for reading the proofs.

E. T. WHITTAKER

48 George Square
Edinburgh, April 1953

Volume I

Contents

Volume II

Contents

Memorandum on Notation

VECTORS are denoted by letters in black type, as **E**.

The three components of a vector **E** are denoted by E_x, E_y, E_z; and the magnitude of the vector is denoted by E, so that

$$E^2 = E_x^2 + E_y^2 + E_z^2.$$

The *vector product* of two vectors **E** and **H**, which is denoted by [**E** . **H**], is the vector whose components are

$$(E_yH_z - E_zH_y, \quad E_zH_x - E_xH_z, \quad E_xH_y - E_yH_x).$$

Its direction is at right angles to the direction of **E** and **H**, and its magnitude is represented by twice the area of the triangle formed by them.

The *scalar product* of **E** and **H** is $E_xH_x + E_yH_y + E_zH_z$. It is denoted by (**E** . **H**).

The quantity $\frac{\partial E_x}{\partial x} + \frac{\partial E_y}{\partial y} + \frac{\partial E_z}{\partial z}$ is denoted by div **E**.

The vector whose components are

$$\left(\frac{\partial E_z}{\partial y} - \frac{\partial E_y}{\partial z}, \quad \frac{\partial E_x}{\partial z} - \frac{\partial E_z}{\partial x}, \quad \frac{\partial E_y}{\partial x} - \frac{\partial E_x}{\partial y}\right)$$

is denoted by curl **E**.

If V denote a scalar quantity, the vector whose components are $\left(-\frac{\partial V}{\partial x}, \ -\frac{\partial V}{\partial y}, \ -\frac{\partial V}{\partial z}\right)$ is denoted by grad V.

The symbol Δ is used to denote the vector operator whose components are $\frac{\partial}{\partial x}, \frac{\partial}{\partial y}, \frac{\partial}{\partial z}$.

Chapter I

THE AGE OF RUTHERFORD

When Röntgen announced his discovery of the X-rays [1] it was natural to suspect some connection between these rays and the fluorescence (or, as it was generally called at that time, phosphorescence) of the part of the vacuum tube from which they were emitted. Accordingly, a number of workers tried to find whether phosphorescent bodies in general emitted radiations which could pass through opaque bodies and then either affect photographic plates or excite phosphorescence in other bodies.

In particular, Henri Becquerel of Paris (1852–1908) resolved to examine the radiations which are emitted, after exposure to the sun, by the double sulphate of uranium and potassium, a substance which had been shown by his father, Edmond Becquerel (1820–91), to have the property of phosphorescence. The result was communicated to the French Academy on 24 February 1896.[2] 'Let a photographic plate,' he said, 'be wrapped in two sheets of very thick black paper, such that the plate is not affected by exposure to the sun for a day. Outside the paper place a quantity of the phosphorescent substance, and expose the whole to the sun for several hours. When the plate is developed, it displays a silhouette of the phosphorescent substance. So the latter must emit radiations which are capable of passing through paper opaque to ordinary light, and of affecting salts of silver.'

At this time Becquerel supposed the radiation to have been excited by the exposure of the phosphorescent substance to the sun; but a week later he announced [3] that in one experiment the sun had become obscured almost as soon as the exposure was begun, and yet that when the photographic plate was developed, the intensity of the silhouette was as strong as in the other cases: and moreover, he had found that the radiation persisted for an indefinite time after the substance had been removed from the sunlight, and after the luminosity which properly constitutes phosphorescence had died away; and he was thus led to conclude that the activity was spontaneous and permanent. It was soon found that those salts of uranium which do not phosphoresce—that is, the uranous series of salts—and the metal itself, all emit the rays; and it became evident that what Becquerel had discovered was a radically new property, possessed by the element uranium in all its chemical compounds.

[1] cf. Vol. I, p. 357 [2] *Comptes Rendus*, cxxii (1896), p. 420
[3] *Comptes Rendus*, cxxii (2 March 1896), p. 501

Very soon he found[1] that the new rays, like the Röntgen and cathode rays, impart conductivity to gases. The conductivity due to X-rays was at that time being investigated at the Cavendish Laboratory, Cambridge, by J. J. Thomson, who had been joined in the summer of 1895 by a young research student from New Zealand named Ernest Rutherford (1871–1937). They found that the conductivity is due to *ions*, or particles carrying electric charges, which are produced in the gas by the radiation, and which are set in motion when an electric field is applied. Rutherford went on to examine the conductivity produced by the rays from uranium (which, as he showed, is likewise due to ionisation), and the absorption of these rays by matter : he found[2] that the rays are not all of the same kind, but that at least two distinct types are present : one of these, to which he gave the name α-rays, is readily absorbed ; while another, which he named β-radiation, has a penetrating power a hundred times as great as the α-rays.

Early in 1898 two new workers entered the field. Marya Sklodowska, born in Warsaw in 1867 (*d.* 1934), had studied physics in Paris, and in 1895 had married a young French physicist, Pierre Curie (1859–1906). She now resolved to search for other substances having the properties that Becquerel had found in uranium, and showed in April 1898 that these properties were possessed by compounds of thorium,[3] the element which, of the elements known at that time, stood next to uranium in the order of atomic weights ; the same discovery was made simultaneously by G. C. Schmidt[4] in Germany. Madame Curie went on to show that, since the emission of rays by uranium and thorium is unaffected by chemical changes, it must be essentially an *atomic* property.[5] Now the mineral pitchblende, from which the uranium was derived, was found to have an activity much greater than could be accounted for by the uranium contained in it : and from this fact she inferred that the pitchblende must contain yet another 'radio-active' element. Making a systematic chemical analysis, she and her husband in July 1898 discovered a new element which, in honour of her native country, she named *polonium*,[6] and then in December another, having an activity many million times as great as uranium : to this the name *radium* was given.[7] Its spectrum was examined by F. A. Demarçay,[8] and a spectral line was found which was not otherwise identifiable. The next three and a half years were spent chiefly in determining its atomic weight, by a laborious series of successive

[1] *Comptes Rendus*, cxxii (1896), p. 559

[2] This paper was published in *Phil. Mag.*(5) xlvii (1899), p. 109, after Rutherford had left Cambridge for a chair in McGill University.

[3] *Comptes Rendus*, cxxvi (12 April 1898), p. 1101

[4] *Ann. d. Phys.* lxv (19 April 1898), p. 141

[5] Some years later, the Curies described the ideas that had inspired their researches in *Comptes Rendus* cxxxiv (1902), p. 85.

[6] *Comptes Rendus*, cxxvii (1898), p. 175

[7] Ibid. cxxvii (1898), p. 1215

[8] *Comptes Rendus*, cxxvii (1898), p. 1218

fractionations: the value found was 225.[1] Meanwhile another French physicist, André Debierne (*b.* 1874), discovered in the uranium residues yet a further radio-active element,[2] to which he gave the name *actinium.*

Attention was now directed to the α- and β-rays of Rutherford. A few months after their discovery it was shown by Giesel, Becquerel and others, that part of the radiation (the β-rays) was deflected by a magnetic field,[3] while part (the α-rays) was not appreciably deflected.[4] After this Monsieur and Madame Curie [5] found that the deviable rays carry negative electric charges, and Becquerel [6] succeeded in deviating them by an electrostatic field. The deviable or β-rays were thus clearly of the same nature as cathode rays; and when measurements of the electric and magnetic deviations gave for the ratio m/e a value [7] of the order 10^{-7}, the identity of the β-particles with the cathode-ray corpuscles was fully established. They differ only in velocity, the β-rays being very much the swifter.

The α-rays were at this time supposed to be not deviated by a magnetic field: the deviation is in fact small, even when the field is powerful: but in February 1903 Rutherford [8] announced that he had succeeded in deviating them by both magnetic and electrostatic fields. The deviation was in the opposite sense to that of the cathode rays, so the α-radiations must consist of positively charged particles projected with great velocity,[9] and the smallness of the deviation suggested that the expelled particles were massive compared to the electron. A method of observing them was discovered in 1903 by Sir W. Crookes [10] and independently by J. Elster and W. Geitel,[11] who found that when a radio-active substance was brought near a screen of Sidot's hexagonal blende (zinc sulphide), bright scintillations were observed, due to the cleavage of the blende under the bombardment. Rutherford suggested that this property might be used for counting the number of α-particles in the rays.

Meanwhile it had been discovered by P. Villard [12] that in addition to the alpha and beta rays, radium emits a third type of radiation, much more penetrating than either of them, in fact 160 times as penetrating as the beta rays. The thickness of aluminium traversed before the intensity is reduced to one-half is approximately 0·0005 cm. for the α-rays, 0·05 cm. for the β-rays and 8 cm. for the γ-rays, as

[1] Later raised to 226

[2] *Comptes Rendus*, cxxx (2 April 1900), p. 906

[3] F. O. Giesel, *Ann d. Phys.* lxix (1899), p. 834 (working with polonium); Becquerel, *Comptes Rendus*, cxxix (1899), p. 996 (working with radium); S. Meyer and E. v. Schweidler, *Phys. ZS.* (1899), p. 113 (working with polonium and radium)

[4] Becquerel, *Comptes Rendus*, cxxix (1899), p. 1205; cxxx (1900), pp. 206, 372; Curie, ibid., cxxx (1900), p. 73

[5] *Comptes Rendus*, cxxx (1900), p. 647

[6] *Comptes Rendus*, cxxx (1900), p. 809

[7] cf. W. Kaufmann, *Verh. Deutsch Phys. Ges.* ix (1907), p. 667

[8] *Phil. Mag.*(6) v (Feb. 1903), p. 177

[9] This had been conjectured by R. J. Strutt in *Phil. Trans.* cxcvi (1901), p. 507.

[10] *Proc. R.S.* lxxi (30 April 1903), p. 405

[11] *Phys. ZS.* iv (1 May 1903), p. 439

[12] *Comptes Rendus*, cxxx (30 April 1900), p. 1178

Villard's radiation was called. Villard found that the γ-radiation is, like the X-rays, not deviable by magnetic forces.

In 1898 Rutherford was appointed to a chair in McGill University, Montreal, and there with R. B. Owens, the professor of electrical engineering, began an investigation into the radio-activity of the thorium compounds. The conductivity produced by the oxide thoria in the air was found [1] to vary in an unexpected and perplexing manner: it could be altered considerably by slight draughts caused by opening or shutting a door. Eventually Rutherford concluded that thoria emitted [2] very small amounts of some material substance which was itself radio-active, and which could be carried away in an air current: this, to which he gave the name *thorium emanation*, was shown to be a gas belonging to the same chemical family as helium and argon, but of high molecular weight.[3]

Meanwhile in Cambridge C. T. R. Wilson had been developing [4] his cloud-chamber, which was to provide the most powerful of all methods of investigation in atomic physics. In moist air, if a certain degree of supersaturation is exceeded (this can be secured by a sudden expansion of the air) condensation takes place on dust-nuclei, when any are present: if by preliminary operations condensation is made to take place on the dust-nuclei, and the resulting droplets are allowed to settle, the air in the chamber is thereby freed from dust. If now X-rays or radiations from a radio-active substance are passed into the chamber, and if the degree of supersaturation is sufficient, condensation again takes place: this is due to the production of ions by the radiation. Thus the tracks of ionising radiations can be made visible by the sudden expansion of a moist gas, each ion becoming the centre of a visible globule of water. Wilson showed that the ions produced by uranium radiation were identical with those produced by X-rays. J. J. Thomson in July 1899 wrote pointing out the advantages of the Wilson chamber to Rutherford, who henceforth profited immensely by its use. In this way the track of a single atomic projectile or electron could be rendered visible.

An important property, discovered for the first time in connection with thorium emanation, was that the radio-activity connected with it rapidly decreased. This behaviour was found later to be characteristic of all radio-active substances: but in the earliest known cases, uranium and thorium, the half-period (i.e. the time required for the activity to be reduced by one-half) is of the order of millions of years, so the property had not hitherto been noticed. Rutherford found [5] that the intensity of the 'induced radiation' of thorium falls off

[1] Owens, *Phil. Mag.*(5) xlviii (Oct. 1899), p. 360

[2] *Phil. Mag.*(5) xlix (Jan. 1900), p. 1

[3] Soon after this, Friedrich Ernst Dorn of Halle found that radium, like thorium, produced an emanation: *Halle Nat. Ges. Abh.* xxiii (1900).

[4] *Phil. Trans.* clxxxix(A) (1897), p. 265; *Proc. Camb. P.S.* ix (1898), p. 333

[5] *Phil. Mag.* xlix (Feb. 1900), p. 161

exponentially with the time : so that if I_1 is the intensity at any time and I_2 the intensity after the lapse of a time t then

$$I_2 = I_1 e^{-\lambda t}$$

when λ is a constant.

In May 1900 Sir W. Crookes [1] showed that it was possible by chemical means to separate from uranium a small fraction, which he called uranium X, which possessed the whole of the photographic activity of the original substance. He found, moreover, that the activity of the uranium X gradually decayed, while the full activity of the residual uranium was gradually renewed, so that after a sufficient lapse of time it was possible to separate from it a fresh supply of uranium X. These facts had an important share in the formation of the theory.

It was at first supposed that the pure uranium, immediately after the separation, is not radio-active : but F. Soddy (*b.* 1877) observed that though photographically inactive, it is active when tested by the electrical method. Now the α-rays are active electrically but not photographically, whereas the β-rays are active photographically : and the conclusion was drawn [2] that pure uranium emits only α-rays and uranium X only β-rays. Soddy had joined the staff of McGill University in 1900 as Demonstrator in Chemistry, and at once began to assist Rutherford in his work on radio-activity. Further experiments on the thorium emanation involved condensing it by extreme cold, and it was discovered [3] that the emanation was produced not directly by the thorium but by an intermediate substance which, as it had many of the characters of Crookes' uranium X, was named thorium X. This was the first indication that radio-activity involves a chain of transformations of chemical elements.

The work of Rutherford and Soddy on thorium and its radio-active derivatives led them to a general theory of radio-activity, which was published in September 1902–May 1903.[4] The greatest obstacle to a clear understanding of the subject had been, curiously enough, the intense belief of everybody in the principle of conservation of energy : here was an enormous amount of energy being outpoured, and no-one could see where it came from. So long as it was attributed to the absorption of some unknown kind of external radiation, the essence of the matter could not be discovered. Rutherford and Soddy now swept this notion away, and asserted that :

(i) In the radio-active elements radium, thorium and uranium, there is a continuous production of new kinds of matter, which are themselves radio-active.

(ii) When several changes occur together these are not simul-

[1] *Proc. R.S.*(A), lxvi (1900), p. 409

[2] Soddy, *Journ. Chem. Soc.* lxxxi and lxxxii (July 1902), p. 860 ; Rutherford and A. G. Grier, *Phil. Mag.*(6) iv (Sept. 1902), p. 315

[3] *Phil. Mag.*(6) iv (Sept. 1902), p. 370

[4] *Phil. Mag.*(6) iv (Sept. 1902), p. 370 ; ibid. (Nov. 1902), p. 569 ; ibid.(6) v (April 1903), pp. 441, 445 ; ibid. (May 1903), pp. 561, 576

taneous, but successive; thus thorium produces thorium X, the thorium X produces the thorium emanation and the latter produces an excited activity.

(iii) The phenomenon of radio-activity consists in this, that a certain proportion of the atoms undergo spontaneous transformation into atoms of a different nature: these changes are different in character from any changes that have been dealt with before in chemistry, for the energy comes from intra-atomic sources which are not concerned in chemical reactions.

(iv) The number of atoms that disintegrate in unit time is a definite proportion of the atoms that are present and have not yet disintegrated. The proportion is characteristic of the radio-active body, and is constant for that body. This leads at once to an exponential law of decay with the time: thus if n_0 is the initial number of atoms, and n is the number at time t afterwards, then

$$n = n_0 e^{-\lambda t}$$

where λ is the fraction of the total number which disintegrates in unit time, so the average life of an atom is $1/\lambda$.

(v) The α-rays consist of positively charged particles, whose ratio of mass to charge is over 1,000 times as great as for the electrons in cathode rays. If it is assumed that the value of the charge is the same as for the electron, then the α-ray particles must have a mass of the same order as that of the hydrogen atom.

(vi) The rays emitted are an accompaniment of the change of the atom into the one next produced, and there is every reason to suppose, not merely that the expulsion of a charged particle accompanies the change, but that this expulsion actually *is* the change.

The authors remarked (in the paper of November 1902) that in naturally occurring minerals containing radio-elements, the radio-active changes must have been taking place over a very long period, and it was therefore possible that the ultimate products might have accumulated in sufficient quantity to be detected. As helium is usually found in such minerals, it was suggested that helium might be such a product. Several years passed before this was finally established, but its probability was continually increasing. Soddy left Montreal in 1903 to work with Sir William Ramsay at University College, London, and Rutherford, who was in England in the summer of that year, called on Ramsay and Soddy, and with them detected (by its spectrum) the presence of helium in the emanation of radium. It seemed certain, therefore, that helium occupied some place in the sequence of linear descent which begins with radium, and at first the general expectation was that it would prove to be an end-product. Rutherford, however, entertained the idea that it might be formed from the α-particles,[1] which, as we have seen, were known to be of the same order of mass as hydrogen or helium atoms; that the α-particles, in fact, might be positively charged

[1] *Nature,* lxviii (20 Aug. 1903), p. 366

atoms of helium : and for some years this supposition was debated without a definite conclusion being reached. In 1906 Rutherford determined [1] with greater accuracy the ratio e/m of the α-particles from radium C, and found it to be between $5{\cdot}0 \times 10^3$ and $5{\cdot}2 \times 10^3$, which is only half the value of e/m for the hydrogen atom : this, however, left it undecided whether the α-particle is a hydrogen molecule (molecular weight 2) carrying the ionic charge, or a helium atom (atomic weight 4), carrying twice the ionic charge.

In 1904 William Henry Bragg (1862–1942), at that time professor in the University of Adelaide, South Australia, showed [2] that the α-particle, on account of its mass, has only a small probability of being deviated when passing through matter, and that in general it continues in a fixed direction, gradually losing its energy, until it comes to a stop : the distance traversed may be called the *range* of the α-particle. He found in the case of radium definite ranges for four kinds of α-particles, corresponding to emissions from radium, radium emanation, radium A and radium C : the α-particles from any particular kind of atom are all shot out with the same velocity, but this velocity varies from one kind of atom to another, as might be expected from Rutherford's theory. β-particles, on the other hand, are easily deflected from their paths by collisions with gas molecules, and their tracks in a Wilson cloud-chamber are zig-zag : they are scattered by passing through matter, so that a narrow pencil of β-rays, after passing through a metal plate, emerges as an ill-defined beam.

In 1907 Rutherford was translated to the chair of physics in the University of Manchester. Here he found a young graduate of Erlangen, Hans Geiger, with whom he devised [3] an electrical method of counting the α-particles directly, the *Geiger counter* as it has since been generally called. The α-rays were sent through a gas, exposed to an electric field so strong as to be near the breakdown value at which a discharge must pass. When a single α-particle passed and produced a small ionisation, the ions were accelerated by the electric field and the ionisation was magnified by collisions several thousand times. This made possible the passage of a momentary discharge, which could be registered. This counting of atoms one by one was a great achievement : it was found that the number of α-particles emitted by 1 gram of radium in one second is $3{\cdot}4 \times 10^{10}$: when this was combined with the value of the total charge (found in the second paper), it became clear that an α-particle carries double the electron charge, reversed in sign.

The question as to the possible connection of α-rays with helium was finally settled later in the same year. Rutherford placed a

[1] *Phys. Rev.* xxii (Feb. 1906), p. 122 ; *Phil. Mag.* xii (Oct. 1906), p. 348

[2] *Phil. Mag.* viii (Dec. 1904), p. 719 ; W. H. Bragg and R. Kleeman, *Phil. Mag.* x (Sept. 1905), p. 318 ; Paper read before the Royal Society of South Australia, 6 June 1904.

[3] *Proc. R.S.*(A), lxxxi (27 Aug. 1908), pp. 141, 162

quantity of radium emanation in a glass tube, which was so thin that the α-rays generated by the emanation would pass through its walls : they were received on the walls of a surrounding glass tube, and, after diffusing out, were found to give the spectrum of helium. This proved definitely that the α-particles are helium atoms, carrying two unit positive charges,[1] a conclusion which he had also reached [2] a short time before by a different line of reasoning.

An interesting corroboration of Rutherford's account of the emission of α-rays was obtained somewhat later, when he and Geiger investigated the fluctuations [3] in the recorded numbers of particles emitted by a radio-active substance in successive equal intervals of time. H. Bateman had shown that if the emission is a random one, then the probability that n particles will be observed in unit time is

$$\frac{x^n e^{-x}}{n!}$$

where x is the average number per unit time, and n is a whole number ($n=0, 1, 2, \ldots \infty$). Rutherford and Geiger in 1910 verified this formula experimentally.[4] Yet a further completion of the work on α-particles was a measurement, made with Boltwood,[5] of the volume of helium produced by a large quantity of radium. By combining the result now obtained with that of the counting experiment it was possible to evaluate the number of molecules in a quantity of the substance whose weight in grams is equal to the molecular weight of the substance (the Avogadro number).

The determination of this constant had been the object of many researches in the years immediately preceding, beginning with a notable paper by Einstein.[6] Albert Einstein was born at Ulm in Württemberg on 14 March 1879. The circumstances of his father's business compelled the family to leave Germany ; and after receiving a somewhat irregular education in Switzerland, he became an official in the Patent Office in Berne. It was in this situation that he wrote, in six months, four papers, each of which attracted much attention.[7]

The paper now to be considered was really a sequel to two earlier papers [8] on the statistical-kinetic theory of heat, in which, however, Einstein had only obtained independently certain results which had been published a year or two earlier by Willard Gibbs. He now

[1] Rutherford and T. Royds, *Mem. Manchester Lit. and Phil. Soc.* liii (31 Dec. 1908), p. 1 ; *Phil. Mag.* xvii (Feb. 1909), p. 281

[2] *Nature,* lxxix (5 Nov. 1908), p. 12

[3] That the emission of α-particles is a random process, and so subject to the laws of probability, seems to have been first clearly stated by E. von Schweidler, *Premier Cong. Internat. pour l'Étude de la Radiologie*, Liège, 1905.

[4] *Phil. Mag.* xx (Oct. 1910), p. 698

[5] Rutherford and Boltwood, *Phil. Mag.* xxii (Oct. 1911), p. 586

[6] *Ann. d. Phys.*(4) xvii (1905), p. 549 ; continued in *Ann. d. Phys.* xix (1906), p. 371

[7] Two of these will be referred to in Chapter II and one in Chapter III.

[8] *Ann. d. Phys.* ix (1902), p. 417 ; xi (1903), p. 170

applied these results to the motion of very small particles suspended in a liquid. The particles were supposed to be much larger than a molecule, but it was assumed that as a result of collisions with the molecules of the water, they require a random motion, like that of the molecules of a gas. The average velocity of such a suspended particle, even in the case of particles large enough to be seen with a microscope, might be of observable magnitude : but the direction of its motion would change so rapidly, under the bombardment to which it would be exposed, that it would not be directly measurable. However, as a statistical effect of these transient motions, there would be a resultant motion which might be within the range of visibility. Einstein showed that in a finite interval of time t the mean square of the displacement for a spherical particle of radius a is

$$\frac{\mathrm{RT}t}{3\pi a\mu\mathrm{N}}$$

where R is the gas-constant, T the temperature, N is Avogadro's number, and μ is the coefficient of viscosity. Thus by this phenomenon the thermal random motion, hitherto a matter of hypothesis, might actually be made a matter of visible demonstration.

The motion of small particles suspended in liquids had been observed as early as 1828 by Robert Brown [1] (1773–1858), a botanist, after whom it was called the *Brownian motion.* Einstein identified the motion studied by him with the Brownian motion, somewhat tentatively in his first paper, but without hesitation in the second.

The theory of the Brownian motion was investigated almost at the same time by M. von Smoluchowski [2] (1872–1917), and it was confirmed experimentally by Th. Svedberg,[3] M. Seddig,[4] and P. Langevin.[5] Particular mention might be made of the experimental studies made in 1908–9 by Jean-Baptiste Perrin (1870–1942) of Paris.[6] These experiments yielded a value of the mean energy of a particle at a definite temperature, and thus enabled him to deduce the value of Avogadro's number, which is $6{\cdot}06\times10^{23}$.[7] The direct confirmation of the kinetic theory provided by these researches on the Brownian movement was the means of converting to it some notable former opponents, such as Wilhelm Ostwald and Ernst Mach.

The statistical-kinetic theory of heat was confirmed experimentally in a different way in 1911 by L. Dunoyer,[8] who obtained a parallel beam of sodium molecules by allowing the vapour of

[1] *Phil. Mag.* iv (1828), p. 161

[2] *Bull. Acad. Sci. Cracovie,* vii (1906), p. 577 ; *Ann. d. Phys.* xxi (1906), p. 756

[3] *ZS. Elektrochem,* xii (1906), pp. 853, 909 ; *ZS. phys. chem.* lxv (1909), p. 624 ; lxvi (1909), p. 752 ; lxvii (1909), p. 249 ; lxx (1910), p. 571

[4] *Phys. ZS.* ix (1908), p. 465

[5] *Comptes Rendus,* cxlvi (1908), p. 503

[6] *Comptes Rendus,* cxlvi (1908), p. 967 ; cxlvii (1908), pp. 475, 530 ; cxlix (1909), pp. 477, 549 ; *Ann. Chim. Phys.* xvii (1909), p. 5 ; cf. also R. Fürth, *Ann d. Phys.* liii (1917), p. 117

[7] Another method of determining Avogadro's number is to study the diffusion of ions in a gas under the influence of an electric force.

[8] *Comptes Rendus,* clii (1911), p. 592 ; *Le Radium,* viii (1911), p. 142

heated sodium to pass through two diaphragms pierced with small holes in an exhausted tube : the behaviour of the molecular beams was entirely in agreement with the predictions of the kinetic theory of gases.

In his statistical studies, Einstein also recognised that the thermal motion of the carriers of electric charge in a conductor should give rise to random fluctuations of potential difference between the ends of the conductor. The effect was too small to be detected by the means then available, but many years later, after the development of valve amplification, it was observed by J. B. Johnson,[1] and the theory was studied by H. Nyquist.[2] This phenomenon is one of the causes of the disturbance that is called ‘ noise ’ in valve amplifiers.

We must now return to the consideration of the radio-active elements themselves. In 1903–5 Rutherford [3] identified a number of members of the radium sequence later than the emanation : radium A, B and C were known by the summer of 1903, and radium D, E, F were discovered in the next two years. It was suspected that one of these later products was identical with the polonium which had been the first new element found by the Curies, and in fact polonium was shown to be radium F. One of the most remarkable discoveries was that of radium B : for at the time no radiations of any kind could be found accompanying its transformation into radium C, and there was therefore no direct evidence of its existence : the only reason for postulating it was, that to suppose an immediate derivation of radium C from radium A would have violated the laws of radio-active change laid down in 1902–3 ; and it was therefore necessary to assume the reality of an intermediate body.

As soon as the principle that radio-active elements are derived from each other in series had been established in 1902–3, the suspicion was formed that radium, which is found in nature in uranium ores, might be a descendant of uranium ; this conjecture was supported by the facts that uranium is one of the few elements having a higher atomic weight than radium, and that the proportion of radium in pitchblende corresponds roughly with the ratio of activity of radium and uranium. Soddy [4] in 1904 described an experiment which showed that radium is not produced *directly* from uranium : if it is produced at all, it can only be by the agency of intermediate substances. Bertram B. Boltwood (1870–1927), of Yale University,[5] worked on this investigation for several years, and at last in 1907 succeeded in showing that radium is the immediate descendant of

[1] *Nature*, cxix (1927), p. 50 ; *Phys. Rev.* xxix (1927), p. 367 ; xxxii (1928), p. 97
The possibility that under certain conditions the thermal motion of electrons in conductors could create a measurable disturbance in amplifiers had been recognised on theoretic grounds by W. Schottky, *Ann. d. Phys.* lvii (1918), p. 541.

[2] *Phys. Rev.* xxxii (1928), p. 110

[3] *Proc. R.S.*(A), lxxiii (22 June 1904), p. 493 ; *Phil. Trans.* cciv (Nov. 1904), p. 169 (Bakerian lecture) ; *Phil. Mag.* viii (Nov. 1904), p. 636 ; *Nature*, lxxi (Feb. 1905), p. 341

[4] *Nature*, lxx (12 May 1904), p. 30

[5] *Nature*, lxx (26 May 1904), p. 80

a new radio-active element which he named *ionium*, and which is itself descended from uranium.[1]

The great number of different radio-active atoms that had now been discovered raised questions concerning their atomic weights, particularly with regard to their position in what was known as the *periodic table* of the chemical elements. In a paper published in 1864, John A. R. Newlands [2] had pointed out that when the chemical elements are arranged according to the numerical values of their atomic weights, the eighth element starting from any given one is, in regard to its properties, closely akin to the first, 'like the eighth note of an octave in music.' This idea he developed in later papers,[3] calling the relationship the 'Law of Octaves.' [4] He read a paper on the subject before the Chemical Society on 1 March 1866; but it was rejected, on the ground that the Society had 'made it a rule not to publish papers of a purely theoretical nature, since it was likely to lead to correspondence of a controversial character." [5]

Newland's ideas were adopted and developed a few years later by Dmitri Ivanovich Mendeléev (1834–1907),[6] who arranged the elements in a *periodic table*. From gaps in this he inferred the existence and approximate atomic weights of three hitherto unknown elements, to which he gave the names eka-boron, eka-aluminium and eka-silicon; when these were subsequently discovered (they are now known as scandium, gallium and germanium), the importance of the periodic table became universally recognised; and the inert gases helium etc., when they were discovered still later, were found to fit into it perfectly.

As new members of the radio-active sequences were discovered, it was found in some cases that two or more of the atoms in the series had exactly the same chemical properties, so that they belonged to the same place in Newland's and Mendeléev's periodic table. For instance, in 1905, O. Hahn, working with Sir William Ramsay at University College, London, discovered [7] the parent of thorium X, which he called *radio-thorium*. This was found to be not separable chemically from thorium; and Boltwood found that his ionium also was not separable chemically from thorium. It was shown by A. S. Russell and R. Rossi,[8] working in Rutherford's laboratory, that the optical spectrum of ionium is indistinguishable from the

[1] *Nature*, lxxvi (26 Sept. 1907), p. 544; *Amer. Journ. Sci.* xxiv (Oct. 1907), p. 370; xxv (May 1908), p. 365

[2] *Chem. News*, x (20 Aug. 1864), p. 94

[3] *Chem. News*, xii (18 Aug. 1865), p. 83; xii (25 Aug. 1865), p. 94

[4] The group of elements Helium, Neon, Argon, Krypton, Xenon and Niton was not known at the time; when they are introduced into the table, it is the *ninth* element starting from any given one which is akin to the first. We leave aside the complications associated with the rare earths, etc.

[5] J. A. R. Newlands, *The Periodic Law*; London, E. and F. W. Spon, 1884, p. 23

[6] *ZS. f. Chem.* v (1869), p. 405; *Deutsch. Chem. Gesell. Ber.* iv (1871), p. 348; *Ann. d. Chem.*, Supplementband, viii (1873), p. 133

[7] *Proc. R.S.*(A), lxxvi (24 May 1905), p. 115; *Chem. News*, xcii (1 Dec. 1905), p. 251

[8] *Proc. R.S.*(A), lxxxvii (Dec. 1912), p. 478

spectrum of thorium. The radio-active properties of the three substances are, however, totally different, since the half-value period of thorium is of the order of 10^{10} years, that of ionium is of the order of 10^5 years, and that of radio-thorium is 1·9 years: and the atomic weights are different: but chemically they are different forms of the same element.

Curiously enough, the possibility of such a situation had been suggested so far back as 1886 by Sir William Crookes.[1] 'I conceive, therefore,' he said, 'that when we say the atomic weight of, for instance, calcium is 40, we really express the fact that, while the majority of calcium atoms have an actual atomic weight of 40, there are not a few which are represented by 39 or 41, a less number by 38 or 42, and so on.'

As the investigation of the radio-active atoms progressed still further, many other examples became known of atoms which are inseparable by chemical methods but have different radio-active properties and different atomic weights. Attention was drawn to the matter in 1909 by the Swedish chemists D. Strömholm and Th. Svedberg,[2] and in 1910 by Soddy,[3] and much experimental work relating to it was done by Alexander Fleck.[4]

New light on the problem now came from an unexpected quarter. Sir Joseph Thomson (he had been knighted in 1908) took up work on the canal rays,[5] or *positive rays* as he now called them, and devised a method of 'positive-ray analysis' for finding the values of m/e for the positively charged particles which constitute the rays; the method was to shoot the rays through a narrow tube, so as to obtain a small spot on a phosphorescent screen or a photographic plate, and to subject them between the tube and the screen to an electric field and also a magnetic field, so as to deflect the beam of particles, the electrostatic deflection and the magnetic deflection being perpendicular to each other. He showed that all particles having the same value for m/e would be spread out by the two fields so as to strike the screen in points lying on a parabola; thus, particles of different mass would give different parabolas. Parabolas were found corresponding to the atoms and molecules of various gases in the discharge-tube; and the atomic weights of the particles could be at once inferred from measures of the parabolas. On applying this method of positive-ray analysis to the gas neon, he found [6] in addition to a parabola belonging to atomic weight 20, another corresponding to atomic weight 22. These proved to be, both of them, atoms of neon, but of different masses. Thomson had in fact discovered two ordinary non-radio-active atoms having the same chemical behaviour but different physical characteristics. This result, which was immediately

[1] *Brit. Ass. Rep.*, Birmingham, 1886, p. 569
[2] *ZS. f. Anorg. Chem.* lxi (1909), p. 338 and lxiii (1909), p. 197
[3] *Chem. Soc. Ann. Rep.* (1910), p. 285
[4] cf. Fleck, *Brit. Ass. Rep.*, Birmingham, 1913, p. 447
[5] cf. Vol. I, p. 363 [6] *Proc. R.S.*(A), lxxxix (1 Aug. 1913), p. 1

confirmed by Francis William Aston [1] (1877–1945) showed that the phenomenon of a place in the Newlands-Mendeléev table being occupied by more than one element was not confined to the highest places in the table. Elements which are chemically inseparable, but have different atomic weights, were named by Soddy *isotopes*.

A striking example was furnished when the question as to the end-product of radio-active changes was solved. Having become convinced so early as 1905 that the end-product of the radium series was not helium, Rutherford sought for some other element to fill this position ; and both he and Boltwood suggested that it might be lead,[2] since lead appears persistently as a constituent of uranium-radium minerals: it was possible indeed that lead might be radium G. This proved to be correct, and lead was found to be also the final product of the thorium series. The atomic weights of these two kinds of lead are not, however, equal, that of radium lead being 206 and that of thorium lead being 208. (The atomic weight of ordinary lead is 207·20).[3]

Atoms obtained by radio-active disintegrations occupy places in the periodic table which are determined by what are called the *displacement laws*, first enunciated in 1913 by A. S. Russell,[4] K. Fajans [5] (another pupil of Rutherford's) and F. Soddy,[6] which may be stated as follows : a disintegration with emission of an α-particle causes the atom to descend two places in the Newlands-Mendeléev table (i.e. the atomic weight is diminished) ; a disintegration with emission of a β-particle causes the atom to ascend one place in the table, but does not change the atomic weight.

In 1919–20 it was stated by F. W. Aston [7] that within the limits of experimental accuracy the masses of all the isotopes examined by him were expressed by whole numbers when oxygen was taken as 16 : the only exception was hydrogen, whose mass was 1·008.

Possible methods for separating isotopes were indicated in 1919 by F. A. Lindemann and F. W. Aston,[8] but for long no notable success was attained in practice ; in 1932–3, however, two isotopes of hydrogen were successfully separated by electrolytic methods.[9] The isotopes of neon have been separated by repeated diffusion by Gustave Hertz.[10]

[1] *Brit. Ass. Rep.*, Birmingham, 1913, p. 403

[2] Rutherford, *Radioactivity* (second edn., May 1905), p. 484; Boltwood, *Phil. Mag.*(6) ix (April 1905), p. 599

[3] Soddy, *Ann. Rep. Chem. Soc.*, 1913, p. 269 ; *Chem. News*, cvii (28 Feb. 1913), p. 97 ; *Nature*, xci (20 March 1913), p. 57 ; *Nature*, xcviii (15 Feb. 1917), p. 469

[4] *Chem. News*, cvii (31 Jan. 1913), p. 49

[5] *Phys. ZS.* xiv (15 Feb. 1913), pp. 131 and 136

[6] *Chem. News*, cvii (28 Feb. 1913), p. 97

[7] *Nature*, civ (18 Dec. 1919), p. 393 ; cv (4 March 1920), p. 8 ; *Phil. Mag.* xxxix (April 1920), p. 449 ; ibid. (May 1920), p. 611 ; *Nature*, cv (1 July 1920), p. 547

[8] *Phil. Mag.* xxxvii (May 1919), p. 523

[9] E. W. Washburn and H. C. Urey, *Proc. Nat. Acad. Sci.* xviii (July 1932), p. 496 ; E. W. Washburn, E. R. Smith and M. Frandsen, *Bureau of Standards J. of Research*, xi (Oct. 1933), p. 453 ; G. N. Lewis and R. T. Macdonald, *J. Chem. Phys.* i (June 1933), p. 341

[10] *ZS. f. Phys.* lxxix (1932), p. 108

A new phenomenon in radio-activity was described in 1908 by Rutherford,[1] and confirmed later by Fajans [2] and other workers, namely that in some cases (e.g. radium C, thorium C and actinium C) some of the atoms emitted an α-particle, and in the next transformation a β-particle, while the rest of the atoms reversed the order of the transformations, emitting first a β-particle and afterwards an α-particle. This is known as a *branching* of the series. Rutherford in his original paper expressed the belief that in this way uranium might give rise to the actinium family as well as the radium family; a conjecture which was afterwards generally accepted as correct.

An account must now be given of some notable advances concerned with X-rays. Charles Glover Barkla (1877–1944), when a research student under J. J. Thomson at Cambridge, had become interested in X-rays. In 1902 his work was transferred to Liverpool University, and there in 1904 he discovered that the rays may be partly polarised.[3] In the final dispostion [4] of his experiments, a mass of carbon was subjected to a strong primary beam of X-rays, and so became a source of secondary radiation. A beam of this secondary radiation, propagated in a direction at right angles to that of the primary, was studied. In this second beam was placed a second mass of carbon, and the intensities of tertiary radiation proceeding in directions perpendicular to the direction of propagation of the secondary beam were observed. The X-ray tube was turned round the axis of the secondary beam, while the rest of the apparatus was fixed, and the intensities of the tertiary radiations were observed for different positions of the tube. It was found that the intensity of the tertiary radiation was a maximum when the primary and tertiary beams were parallel, and a minimum when they were at right angles to each other, which showed that the secondary radiation was polarised. This result told decidedly in favour of the hypothesis that X-rays were transverse waves.

Continuing his work on X-rays, Barkla resolved to test a suggestion of J. J. Thomson's, that the number of electrons in an atom might be found by observing the amount of the scattering when X-rays fall on the lighter chemical elements, and comparing it with the scattering produced when they fall on a single electron. In 1903 Thomson had already given, in the first edition of his *Conduction of Electricity through Gases*,[5] a theoretical discussion, based on classical electrodynamics, of the scattering of a pulse of electromagnetic force by an electron on which it is incident. He found

[1] *Nature*, lxxvii (5 March 1908), p. 422

[2] *Phys. ZS.* xii (1911), p. 369; xiii (1912), p. 699

[3] *Nature*, lxix (17 March 1904), p. 463; *Proc. R.S.*(A), lxxiv (1905), p. 474; *Phil. Trans.*(A), cciv (1905), p. 467

[4] *Proc. R.S.*(A), lxxvii (1906), p. 247

[5] J. J. Thomson, *Conduction of Electricity through Gases*, 1st edn. (1903), p. 268; 2nd edn. (1906), p. 321; 3rd edn., Vol. II (1933), p. 256. cf. also J. J. Thomson, *Phil. Mag.* xi (1906), p. 769, where he suggested three different methods of determining the number of electrons in an atom, based respectively on (1) the dispersion of light by gases, (2) the scattering of X-rays by gases, (3) the absorption of β-rays.

that the energy radiated by the electron is $8\pi e^4/3m^2$ times the energy passing through unit area of the wave-front of the primary beam (when the charge e is measured in electromagnetic units). Thus if it is assumed that the electrons, in the chemical element exposed to the X-rays, all scatter independently, the value of the mass-scattering coefficient is

$$\frac{8\pi}{3}\frac{e^4 n}{m^2 \rho}$$

where n is the number of electrons per cm^3, and ρ is the density. Now let

N = number of molecules in one gram-molecule
Z = number of free electrons per atom
A = atomic weight.

Then

$$\frac{NZ}{A} = \text{number of electrons in one gram} = \frac{n}{\rho}$$

so the value of the mass-scattering coefficient is

$$\frac{8\pi}{3}\frac{e^4 NZ}{m^2 A}.$$

Barkla [1] found experimentally for the mass-scattering coefficient of the lighter elements (except for hydrogen) a value about 0·2, which would therefore give

$$Z = \frac{3m^2 A}{40\pi e^4 N}.$$

The values accepted at the time for the quantities on the right-hand side of this equation were inaccurate, and the result deduced, namely that there were between 100 and 200 electrons per molecule of air, was replaced by Barkla in 1911 [2] by a much better determination based on Bucherer's value for e/m, Rutherford and Geiger's value for e, and Rutherford's value for N. This gave approximately

$$Z = \tfrac{1}{2}A,$$

i.e. the number of scattering electrons per atom, for the lighter elements, is about half the atomic weight of the element, except in the case of hydrogen, for which Z = 1. These results anticipated later discoveries in a remarkable way.

The secondary X-rays were destined to furnish other contributions to atomic physics. In 1906 Barkla [3] found that in some cases the secondary rays consisted mainly of a radiation which differed altogether in 'hardness,' or penetrating power, from the primary radiation, so that it could not be regarded as the result of 'scattering.'

[1] *Phil. Mag.* vii (May 1904), p. 543. cf. also J. A. Crowther, *Phil. Mag.* xiv (Nov. 1907), p. 653

[2] *Phil. Mag.* xxi (May 1911), p. 648

[3] *Phil. Mag.* xi (June 1906), p. 812

He pursued this matter further, with the help of one of his students, C .A. Sadler, and in 1908 they [1] found that the secondary X-rays emitted by a chemical element exposed to a primary beam of X-rays were of two distinct types :

(i) A scattered radiation, not of great amount, of the same quality as the primary beam.

(ii) A radiation characteristic of the exposed chemical element, and almost, if not quite, *homogeneous*, i.e. all of the same degree of hardness. It was, moreover, emitted uniformly in all directions, unlike the scattered radiation. This characteristic radiation was produced only when the primary X-rays contained a constituent harder than the characteristic radiation that was to be excited. (On this account the characteristic X-rays were often spoken of at the time as 'flourescent.') Barkla found also that the hardness of the characteristic radiation increased as the atomic weight of the emitting chemical element increased.

R. Whiddington [2] found that the primary rays from an X-ray tube can excite the radiation characteristic of an element of atomic weight w only when the velocity of the parent cathode rays exceeds $10^8 w$ cm/sec ; when the velocity of the primary rays is less than this, only a truly 'scattered' radiation is emitted, resembling the primary.

It was found [3] that the characteristic secondary radiations may be divided into several groups, the radiation belonging to each group becoming more penetrating as the atomic weight of the radiating element increases ; in other words, each chemical element emits a line spectrum of X-rays, each line moving to the more penetrating end of the spectrum as the atomic weight of the element increases. Two groups which were described in 1909 received the notation K and L in 1911, and an M-group was found a little later.[4] The K-series, which is the most penetrating, was found together with the L-series for elements from zirconium (atomic wt. 90·6) to silver (atomic wt. 107·88). For elements heavier than silver, the K-series was difficult to excite, since very great velocities would be required in the exciting cathode rays : and for elements lighter than zirconium, the L-series was difficult to observe because it was so easily absorbed.[5]

It was shown by G. W. C. Kaye [6] that the radiation characteristic of a chemical element can be excited not only by exposing it to a

[1] *Phil. Mag.* xiv (Sept. 1907), p. 408 : xvi (Oct. 1908), p. 550

[2] *Proc. R.S.*(A), lxxxv (April 1911), p. 323

[3] *Proc. Camb. Phil. Soc.* xv (1909), p. 257 ; Barkla and J. Nicol, *Nature,* lxxxiv (Aug. 1910), p. 139

[4] Barkla and V. Collier, *Phil. Mag.* xxiii (June 1912), p. 987

[5] E. H. Kürth, *Phys. Rev.* xviii (1921), p. 461, found for the convergence wave-lengths in Ångströms : K-series of carbon, 42·6, oxygen 23·8 : L-series of carbon, 375, oxygen 248, iron 16·3, copper 12·3 : M-series of iron, 54·3, copper 41·6 : N-series of iron 247, copper 116.

[6] *Phil. Trans.*(A), ccix (Nov. 1908), p. 123. Kaye found that the intensity of general X-radiation was nearly proportional to the atomic weight of the element forming the anticathode ; later, W. Duane and T. Schimizu, *Phys. Rev.* xiv (1919), p. 525, showed that the intensity is proportional to the atomic *number.*

beam of primary X-rays, but also by using it as the anticathode in an X-ray tube, so that it is bombarded by cathode rays. It was suggested that this might be an indirect effect, produced by the mediation of non-characteristic X-rays: but this suggestion was disproved by R. T. Beatty,[1] who proved beyond doubt that the characteristic X-rays are excited directly by the impact of cathode rays. In the following year Beatty [2] verified experimentally a result which had been reached theoretically by J. J. Thomson [3] in 1907, namely that the total intensity of general X-radiation is proportional to the fourth power of the velocity of the exciting electrons.

The question as to whether X-rays were corpuscles or waves was still unsettled in 1910. In that year W. H. Bragg published a paper [4] in which, interpreting his experiments by the light of the corpuscular hypothesis, he arrived at conclusions which were in fact true and of great significance. We have seen [5] that when X-rays are passed through a gas they render it a conductor of electricity, and that this property is due to the production of ions in the gas. Bragg now asserted that the X-rays do not ionise the gas directly; they act by ejecting, from a small proportion of the atoms of the gas, electrons (photo-electrons) of high speed, each of which acts as a β-particle and ionises the gas by detaching electrons in a succession of collisions with molecules along its path. The speed of the ejected electron depends only on the hardness or penetrating power of the X-rays (which was later shown to be, in effect, their frequency), and not at all on their intensity, or on the nature of the atom from which the electron is expelled. What Bragg emphasised as specially remarkable was that the energy of the electron was as great as that of an electron in the beam of cathode rays by which the X-rays had been excited originally: the X-ray pulse seemed to have the property of keeping its energy together in a small bundle, without any of the spreading that might have been expected on the wave theory, and to be able to transfer the whole of this energy to a single electron. He enunciated 'the general principle, that if one radiant entity (α-, β-, γ-, X- or cathode-ray) enters an atom, one and only one entity emerges, carrying with it the energy of the entering entity.' '*One* X-ray provides the energy for *one* β-ray, and similarly in the X-ray bulb, *one* β-ray excites *one* X-ray. No energy is lost in the interchange of forms, β- to X-ray and back again; and the speed of the secondary β-ray is independent of the distance that the X-ray has travelled: so the X-ray cannot diffuse its energy as it goes, that is to say, it is a corpuscle.' It was in fact now established that the X-ray behaves in some ways [6] as a wave and in other ways as a corpuscle.

[1] *Proc. R.S.*(A), lxxxvii (Dec. 1912), p. 511
[2] *Proc. R.S.*(A), lxxxix (1913), p. 314 [3] *Phil. Mag.* xiv (1907), p. 226
[4] *Phil. Mag.* xx (Sept. 1910), p. 385; *Brit. Ass. Rep.* 1911, p. 340; W. H. Bragg and H. L. Porter, *Proc. R.S.*(A), lxxxv (1911), p. 349 [5] cf. Vol. I, p. 359
[6] cf. A. Joffé and N. Dobronrawov, *ZS. f. P.*, xxxiv (1925), p. 889

Bragg's conclusions were fully confirmed in 1911–12 by C. T. R. Wilson,[1] using his method of cloud-chamber photographs. The whole of the region traversed by the primary X-ray beam was seen to be filled with minute streaks and patches of cloud : examining the photographs more closely, the cloudlets were seen to be small thread-like objects, consisting of droplets deposited on ions produced along the paths of the β-particles, which were the actually effective ionising agents.

In the early part of the twentieth century, many attempts were made to test the hypothesis that X-rays are waves, by trying to obtain diffraction-effects with them. In 1899 and 1902 H. Haga and C. H. Wind [2] of Groningen observed a broadening of the image of a wedge-shaped slit, and inferred that the wave-length of the vibrations concerned was of the order of one Ångström.[3] However, in 1908, when B. Walter and R. Pohl [4] repeated the experiments, they found that different times of exposure gave different results as regards the image, and concluded that the effect was not confirmed. In 1912 the question was re-opened when P. P. Koch,[5] making a special study of the blackening of photographic plates in general, re-examined Walter and Pohl's images, and decided that there was evidence of genuine diffraction. Thereupon Arnold J. W. Sommerfeld (1868–1951), Professor at Munich, compared the results of theory with Koch's photometric measurements,[6] and deduced a value of 0·3 Ångströms for the wave-length of the X-rays.

At that time a young student, Peter Paul Ewald (*b.* 1888), who had just taken his doctorate at Munich, was interested in the transmission of light through the atomic lattice of a crystal. Some notion of the dimensions of crystal-lattices could by this time be formed ; the Avogadro number (the number of molecules in a number of grams equal to the molecular weight) was known to be approximately 6×10^{23} ; this together with a knowledge of the density and molecular weight of a crystal made it possible to estimate that the distance apart of the atoms in a crystal was of the order of 10^{-8} cm. or one Ångström. A junior lecturer in Munich, Max Laue [7] (*b.* 1879), who was in contact with Sommerfeld and Ewald, saw that if the X-rays had a wave-length of the order suggested by Sommerfeld, then the crystal-lattice had the right dimensions for acting as a three-dimensional diffraction grating, so to speak, for the X-rays. He promptly arranged for an experimental test of this idea, which was carried out by W. Friedrich and P. Knipping ; and a paper

[1] *Proc. R.S.*(A), lxxxv (April 1911), p. 285 ; (A), lxxxvii (Sept. 1912), p. 277

[2] *Proc. Amst. Ac.* (25 March 1899) (English edn. i, p. 420) and 27 Sept. 1902 (English edn. v, p. 247). This work was discussed by Sommerfeld, *Phys. ZS.*, i (1899), p. 105 and ii (1900), p. 55.

[3] cf. Vol. I, p. 367, note 5

[4] *Ann. d. Phys.* xxv (1908), p. 715 ; xxix (1909), p. 331

[5] *Ann. d. Phys.* xxxviii (1912), p. 507

[6] *Ann. d. Phys.* xxxviii (1912), p. 473

[7] About this time Laue's father, who was a general in the German Army, received a title of nobility, so the son was known subsequently as Max von Laue.

was published [1] in June–July 1912 in which it was completely vindicated. A thin pencil of X-rays was allowed to fall on a crystal of zinc sulphide. A photographic plate, placed behind the crystal at right angles to this primary pencil, showed a strong central spot, where it was met by the primary rays, surrounded by a number of other spots, in a regular arrangement: these were situated at the places where the plate was met by diffracted pencils, produced by reflection of the primary X-rays at sets of planes of atoms in the crystal. The positions of the spots in the simplest imaginable case are given by the following rule [2]: Suppose that the atoms are disposed so that the three rectangular co-ordinates of any atom are integral multiples of a length a, every such place being filled, and suppose that the incident rays are parallel to one of the axes. If the distance from an atom A to another atom B is an integral multiple of a, then in the direction AB there will be one of the diffracted pencils that cause the spots. Clearly such directions correspond to all ways of expressing a square as the sum of three squares.

In this case the spots furnish no information regarding the wave-length of the radiation; and indeed the radiation used by Friedrich and Knipping had no definite wave-length, being a heterogeneous mixture of rays whose wave-lengths formed a continuous series.[3] A mathematical theory of the spots in more general cases was given by Laue himself and by other writers.[4]

Laue's discovery was of the first importance, for not only did the diffraction-patterns under suitable conditions serve to determine the wave-length of the X-rays, but the idea was soon developed into a regular method for determining the arrangement of the atoms in crystals; and its merit was fitly recognised by the award to Laue in 1914 of the Nobel Prize for physics. The question as to whether X-rays were corpuscles or waves seemed to be settled in favour of the undulatory hypothesis; it was in fact found that the wave-length of high-frequency X-rays was about one Ångström: but W. H. Bragg wrote [5] 'The problem becomes, it seems to me, not to decide between two theories of X-rays, but to find one theory which possesses the capacity of both'—a remarkable anticipation of the view that was made possible many years later by the discovery of quantum mechanics.

William Lawrence Bragg (*b.* 1890), son of W. H. Bragg, in a paper read before the Cambridge Philosophical Society in the autumn of the same year,[6] introduced considerable simplifications in the theory.

[1] W. Friedrich, P. Knipping and M. Laue, *München Ber.* 8 June, p. 303 and 6 July p. 363, 1912; reprinted *Ann. d. Phys.* xli (1913), p. 971

[2] W. H. Bragg, *Nature*, xc (24 Oct. 1912), p. 219

[3] H. G. J. Moseley and C. G. Darwin, *Phil. Mag.* xxvi (July 1913), p. 210 found a continuous spectrum, with maxima due to characteristic rays from the anticathode.

[4] M. von Laue, loc. cit.: *Ann. d. Phys.* xlii (1913), p. 397; *Phys. ZS.* xiv (1913), p. 1075; H. Moseley and C. G. Darwin, *Nature*, xc (14 Oct. 1912), p. 219; P. P. Ewald, *Phys. ZS.* xiv (1913), p. 1038; xv (1914), p. 399; C. G. Darwin, *Phil. Mag.* xxvii (Feb. 1914), p. 315; ibid. (April 1914), p. 675

[5] *Nature*, xc (28 Nov. 1912), p. 360

[6] *Proc. Camb. Phil. Soc.* xvii (Feb. 1913), p. 43

His leading idea (which replaced Laue's assumption of scattering at the points of a crystal-grating) was that parallel planes in the crystal which are rich in atoms can be regarded, taken together, as a reflecting surface for X-rays ; and experiments with a slip of mica about a millimetre thick whose surface was a cleavage plane showed him that the laws of reflection were obeyed when the rays were incident at nearly glancing angles ; reflection takes place only when the wave-length λ of the rays, the distance d between the parallel planes in the crystal, and the angle of incidence ϕ, are connected by the relation

$$n\lambda = 2d \cos \phi$$

where n is a small whole number. This is known as the *Bragg law.* His father, continuing the work with him,[1] devised an *X-ray spectrometer*, the principle of which is to allow *monochromatic* X-rays to fall in a fixed direction on a crystal, which is made to turn so that each plane can be examined in detail : and with this instrument the arrangement of the atoms in many different crystals was determined. From this point the study of crystal-structure was developed by the Braggs with great success over an immense range : the Nobel Prize for physics was awarded to them in 1915.

The discoveries regarding X-rays led to a better understanding of the γ-rays from radio-active substance. J. A. Gray [2] established the similarity in nature of γ-rays and X-rays by showing that the γ-rays from RaE excite the characteristic X-radiations (K-series) of several elements, just as very penetrating X-rays would : and that the γ-rays behave similarly to X-rays (both qualitatively and quantitatively) in regard to scattering. In 1914 Rutherford and E. N. da C. Andrade,[3] by methods based on the same principle as those used by the Braggs and by Moseley and Darwin, measured the wave-lengths of the γ-rays from radium B and C. The wave-lengths of γ-rays are usually less than those of X-rays, being generally between 0·01 and 0·1 Ångströms.

In the work of Rutherford and Geiger on counting α-particles by the electric method, carried out in 1908,[4] some of the difficulties that had to be overcome were due to the scattering of α-rays in passing through matter. Geiger made a special study of the scattering for small angles of deflection, and in 1909 Rutherford suggested to one of his research students, E. Marsden, an examination of the possibility of scattering through large angles. As a result of this suggestion, experiments were carried out by Geiger and Marsden, which showed[5] that α-particles fired at a thin plate of matter can be scattered inside the material to such an extent that some of them emerge again on the side of the plate at which they entered : and

[1] *Proc. R.S.*(A), lxxxviii (July 1913), p. 428 ; lxxxix (Sept. 1913), pp. 246, 248 ; ibid. (Feb. 1914), p. 468
[2] *Proc. R.S.*(A), lxxxvii (Dec. 1912), p. 489
[3] *Phil. Mag.* xxvii (May 1914), p. 854
[4] cf. p. 7
[5] *Proc. R.S.*(A), lxxxii (July 1909), p. 495 ; *Phil. Mag.* xxv (1913), p. 604

calculation showed that some of the α-particles must have been deflected at single encounters through angles greater than a right angle.

Now at that time the atom was generally pictured in the form suggested by J. J. Thomson in his Silliman lectures of 1903.[1] He was then working out the consequences of supposing that the negative electrons occupy stationary positions in the atom. In order that the atom as a whole may be electrically neutral, there must be also a positive charge : and he saw that this could not be concentrated in positively charged corpuscles, since a mixed assemblage of negative and positive corpuscular charges could not be in stable equilibrium. He therefore assumed that the positive electrification was uniformly distributed throughout a sphere of radius equal to the radius of the atom as inferred from the kinetic theory of gases (about 10^{-8} cm.) : the negative electrons he supposed to be situated inside this sphere, their total charge being equal and opposite to that of the positive electrification.

In attempting to picture the way in which the negative electrons would dispose themselves, Thomson was guided by some experiments with magnets which had been made many years earlier by Alfred Marshall Mayer of the Stevens Institute of Technology, Hoboken.[2] Mayer magnetised a number of sewing-needles with their points of the same polarity, say south. Each needle was run into a small cork, of such a size that it floated the needle in an upright position, the eye end of the needle just coming through the top of the cork. If three of these vertical magnetic needles are floated in a bowl of water, and the north pole of a large magnet is brought down over them, the mutually repellent needles at once approach each other, and finally arrange themselves at the vertices of an equilateral triangle. With four needles a square is obtained, with five either a regular pentagon or (a less stable configuration) a square with one needle at its centre, and so on. The under-water poles of the floating needles, and the upper pole of the large magnet, were regarded as too far away to exert any appreciable influence, so the problem was practically equivalent to that of a number of south poles in presence of a single large north pole.

Thomson examined theoretically the problem of the configurations assumed by a small number of negative electrons inside a sphere of positive electrification, and found that when the number of electrons was small, they disposed themselves in a regular arrangement, all being at the same distance from the centre ; but when the number of electrons was increased, they tended to arrange themselves in rings or spherical shells, and that the model imitated many of the known properties of atoms, particularly the periodic changes with increase of atomic weight which are set forth in the

[1] Published as *Electricity and Matter* in 1904. cf. J. J. Thomson, *Phil. Mag.* vii (1904), p. 237, and for an earlier model, Lord Kelvin, *Phil. Mag.* iii (1902), p. 257.

[2] *Phil. Mag.*(5) v (1878), p. 397 ; (5) vii (1879), p. 98

Newlands-Mendeléev table. If one of the electrons were displaced slightly from its position of equilibrium, it would be acted on by a restitutive force proportional to the displacement. This was a most desirable property, since it was just what was required for an electronic theory of optical dispersion and absorption ; and, moreover, it would explain the monochromatic character of spectral lines : but in no way could Thomson's model be made to give an account of spectral series.[1]

A model atom alternative to Thomson's had been proposed in the same year (1903) by Philipp Lenard [2] (1862–1947) of Kiel, who observed that since cathode-ray particles can penetrate matter, most of the atomic volume must offer no obstacle to their penetration, and who designed his model to exhibit this property. In it there were no electrons and no positive charge separate from the electrons : the atom was constituted entirely of particles which Lenard called *dynamides*, each of which was an electric doublet possessing mass. All the dynamides were supposed to be identical, and an atom contained as many of them as were required to make up its mass. They were distributed throughout the volume of the atom, but their radius was so small ($<0{\cdot}3 \times 10^{-11}$ cm.) compared with the radius of the atom, that most of the atomic volume was actually empty. Lenard's atom, however, never obtained much acceptance, as no evidence could be found for the existence of the dynamides.

The deflection of an α-particle through an angle greater than a right angle was clearly not explicable on the assumption of either Thomson's or Lenard's atom ; and Rutherford in December 1910 came to the conclusion that the phenomenon could be explained only by supposing that an α-particle occasionally (but rarely) passed through a very strong electric field, due to a charged nucleus [3] of very small dimensions in the centre of the atom. This was confirmed a year later by C. T. R. Wilson's photographs [4] of cloud-chamber tracks of α-particles which showed violent sudden deflections at encounters with single atoms.

Thus Rutherford was led to what was perhaps the greatest of all his discoveries, that of the structure of the atom ; the first account of his theory was published in May 1911.[5] He found that if a model atom were imagined with a central charge concentrated within a sphere of less than 3×10^{-12} cm. radius, surrounded by electricity of the opposite sign distributed throughout the rest of the volume of the atom (about 10^{-8} cm. radius), then this atom would satisfy all the known laws of scattering of α- or β-particles, as found by Geiger and Marsden. The central charge necessary would be Ne,

[1] See, however, an attempt by K. F. Herzfeld, *Wien Ber.* cxxi, 2a (1912), p. 593

[2] *Ann. d. Phys.* xii (1903), p. 714, at p. 736

[3] The term *nucleus* for the central charge seems to have been used first in Rutherford's book *Radioactive Substances and their Radiations*, which was published in 1912.

[4] *Proc. R.S.*(A), lxxxvii (Sept. 1912), p. 279

[5] *Phil. Mag.* xxi (May 1911), p. 669 ; xxvii (1914), p. 488. For an anticipation that the atom might prove to be of this type, cf. H. Nagaoka, *Phil. Mag.* vii (1904), p. 445.

where e is the electronic charge, and N is a number equal to about half the atomic weight. This fitted in perfectly with the discovery already made by Barkla,[1] that the number of scattering electrons per atom is (for the lighter elements, except hydrogen) about half the atomic weight: for the positive central charge, and the negative charges on the electrons in the space around it, must exactly neutralise each other.

Thus the Rutherford atom is like the solar system, a small positively charged nucleus in the centre, which contains most of the mass of the atom, being surrounded by negative electrons moving around it like planets, at distances of the order of 10^{-8} cm. Occasionally an α-particle passes near enough to unbind and detach an electron and thus ionise the atom: still more infrequently (only about one α-particle in ten thousand, even in the case of heavy elements) the α-particle may come so close to the nucleus as to experience a violent deflection, due to the electrical repulsion between them. The encounters were studied mathematically by C. G. Darwin[2] (*b.* 1887), who found a satisfactory agreement between theory and experiment, and showed that Geiger and Marsden's results could not be reconciled with any law of force except the electrostatic law of the inverse square, which is obeyed to within 3×10^{-13} cm. of the centre of the atom.

Rutherford now laid down the principle[3] that the positive charge on the nucleus (or the number of negative electrons) is the fundamental constant which determines the chemical properties of the atom: this fact explains the existence of isotopes, which have the same nuclear charge but different nuclear masses, and which have the same chemical properties. He pointed out also that gravitation and radio-activity, being unaffected by chemical changes, must depend on the nucleus. His old discovery, that the α-particle is a doubly ionised atom of helium, was now reformulated in the statement, that the α-particle, at the end of its track, captures two electrons (one at a time), and thus becomes a neutral helium atom; he suggested, moreover, that the nucleus of the hydrogen atom might actually be the 'positive electron.' It was seen that the hydrogen nucleus differed from the negative electron not only in the reversal of sign of its charge, but also in having a much greater mass—in fact, almost all the mass of the hydrogen atom[4]; and at the Cardiff meeting of the British Association in 1920, Rutherford proposed for it the name *proton*, which has been universally accepted.

A proposal for removing the uncertainty which still remained as to the precise amount of the nuclear charge was made in 1913 by

[1] cf. p. 15
[2] *Phil. Mag.* xxvii (March 1914), p. 499
[3] *Nature*, xcii (Dec. 1913), p. 423; *Phil. Mag.* xxvi (Oct. 1913), p. 702; xxvii (March 1914), p. 488
[4] Poincaré in his St. Louis lecture of 1904 had said (*Bull. des Sci. Math.* xxviii (1904), p. 302) 'The mass of a body would be the sum of the masses of its positive electrons, the negative electrons not counting.'

A. van der Broek [1] of Utrecht. He remarked that when α-particles are scattered by a nucleus, the amount of scattering per atom, divided by the square of the charge in the nucleus, must be constant. As Geiger and Marsden had shown, this condition is roughly satisfied if the nuclear charge is assumed to be proportional to the atomic weight; but van der Broek now pointed out that it would be satisfied with far greater accuracy if the nuclear charge were assumed to be proportional to the number representing the place of the element in the Newlands-Mendeléev periodic table. He suggested, therefore, that the nuclear charge should be taken to be Ze, where e is the electronic charge (taken positively) and Z is the ordinal number of the element in the periodic table.

This suggestion received a complete confirmation from experiments performed in Rutherford's laboratory at Manchester by Henry Gwyn-Jeffreys Moseley [2] (*b.* 1887, killed at the Suvla Bay landing in the Dardanelles, 10 August 1915) in continuation of the work which he and Darwin had been carrying on together. Moseley exposed the chemical elements, from calcium to nickel, as anticathodes in an X-ray tube, so that under the bombardment of cathode-rays they emitted their characteristic X-ray spectra, consisting essentially of two strong lines (the K- and L-lines) [3]; and the wave-lengths of these lines were determined by the crystal method. Taking either of these lines and following it from element to element, he found that the square root of its frequency increased by a constant quantity as the transition was made from any element to the next higher element in the periodic table; so that the frequency was expressible in the form $k(N-a)^2$, where k was an absolute constant, N was the 'atomic number' or place in the periodic table, and a was a constant which had different values for the K- and L-lines. So there must be in the atom a fundamental number, which increases by unity as we pass from one element to the next in the periodic table; and, having regard to the results of Rutherford, Geiger, Marsden and van der Broek, this quantity can only be the amount of the nuclear charge, expressed in electron-units. Thus *the number of negative electrons which circulate round the nucleus of an atom of a chemical element is equal to the ordinal number of the element in the periodic table.*

Two incidental results of Moseley's work on X-ray spectra must be mentioned. It now became clear that the atomic numbers of iron, cobalt and nickel must be respectively 26, 27, 28, thus confirming the opinion, already suggested by chemical considerations, that cobalt should have a lower place in the periodic table than

[1] *Phys. ZS.* xiv (1913), p. 32; *Nature*, xcii (27 Nov. 1913), p. 372; xcii (25 Dec. 1913), p. 476; *Phil. Mag.* xxvii (March 1914), p. 455

[2] *Phil. Mag.* xxvi (Dec. 1913), p. 1024; xxvii (April 1914), p. 703. Moseley left Manchester for Oxford at the end of 1913, and completed his work there. cf. his obituary notice in *Proc. R.S.*(A), xciii (1917), p. xxii.

[3] cf. p. 16. Actually each of these lines is a multiplet.

nickel, although it has a higher atomic weight ; and the vexed questions of the number of elements in the group of the rare earths, and of missing elements in the periodic table, could also be settled, since it was now known what the X-ray spectra of these elements must be.[1] Some predictions which had been made by the Danish chemist, Julius Thomsen (1826–1909) were now verified in a remarkable way.

Thus Rutherford, with the help of the young men in his research school—Geiger, Marsden, Moseley and Darwin [2]—created a definite quantitative theory of the atom, lending itself to mathematical treatment, and satisfying every comparison with experiment. It has been the foundation of all later work.

During the years 1914–18 Rutherford was occupied chiefly with matters connected with the war : but in 1919 he made a contribution [3] of the highest importance to atomic physics. It originated from an observation made by Marsden, who had shown [4] that when an α-particle collides with an atom of hydrogen, the hydrogen atom may be set in such swift motion that it travels (nearly in the direction of the impinging particle) four times as far as the colliding α-particle, and that it may be detected by a scintillation produced on a zinc sulphide screen. Rutherford now showed, by measurements of deflections in magnetic and electric fields, that these scintillations were due to hydrogen atoms carrying unit positive charge, in other words, to hydrogen nuclei, or protons as they soon came to be called.

He next bombarded dry air, and nitrogen, with α-particles, and again found scintillations at long range. The similarity in behaviour of the particles obtained from nitrogen to those previously obtained from the hydrogen led him to suspect that they were identical, i.e. that the long-range particles obtained by bombarding nitrogen with α-particles were actually hydrogen nuclei. The general idea now presented itself, that some of the lighter atoms might be actually disintegrated by a collision with a swift α-particle : going beyond the earlier discovery that an α-particle might be deflected through a large angle by a close collision with a nucleus, he now came to the conclusion that on still more rare occasions (say one α-particle in half a million) it might break up the nucleus. The phenomenon was found to occur markedly with nitrogen, but not with dry oxygen.

In the summer of 1919 Rutherford succeeded J. J. Thomson as the Cavendish Professor of Physics at Cambridge. Continuing his experiments there, he succeeded in proving definitely that the

[1] For individual elements the nuclear charges were found directly by J. Chadwick (*Phil. Mag.* xl (1920), p. 734) by experiments on the scattering of pencils of α-rays. His results agreed with those deduced from Moseley's law of X-ray spectra. Practically all of the elements which have been discovered since Moseley's day, and which fill the gaps that then existed in the periodic table, have been identified by the study of their characteristic X-ray spectra.

[2] And Bohr, whose work will be described in a later chapter

[3] *Phil. Mag.* xxxvii (June 1919), p. 537

[4] *Phil. Mag.* xxvii (May 1914), p. 824

nitrogen atom can be disintegrated by bombarding it with α-particles.[1] As P. M. S. Blackett (*b.* 1897) showed, the tracks of the particles could be seen in the Wilson cloud-chamber. Since the nitrogen nucleus, of charge 7 electronic units, captures the α-particle, of charge 2, and expels the proton, of charge 1, the particle obtained by the transformation must have charge 8, that is, it must be the nucleus of an isotope of oxygen. Since the nitrogen nucleus has mass 14, the captured α-particle has mass 4, and the expelled proton has mass 1, it follows that the oxygen isotope must have mass 17.

In 1921 Rutherford and J. Chadwick (*b.* 1891) found[2] that similar transformations could be produced in boron, fluorine, sodium, aluminium and phosphorus: and other elements were later added to the list. In each case the α-particle was captured and a swift proton was ejected, while a new nucleus of mass three units greater and charge one unit higher was formed. Thus the medieval alchemist's dream of the transmutation of matter was realised at last.

Rutherford died at Cambridge on 19 October 1937, and was buried in Westminster Abbey near the graves of Newton and Kelvin. He was survived by his old teacher J. J. Thomson, who had in 1918 been elected Master of the great foundation of which he had been a member uninterruptedly since 1875.

'How fortunate I have been throughout my life!' Thomson wrote, near the end of it, 'I have had good parents, good teachers, good colleagues, good pupils, good friends, great opportunities, good luck and good health.' He lived to be eighty-three, dying at Trinity Lodge on 30 August 1940, and was buried on 4 September in the Abbey.

[1] *Proc. R.S.*(A), xcvii (July 1920), p. 374; *Engineering*, cx (17 Sept. 1920), p. 382 (a paper read to the British Association at its Cardiff meeting); *Proc. Rhys. Soc.* xxxiii (Aug. 1921), p. 389

[2] *Nature*, cvii (10 March 1921), p. 41

Chapter II

THE RELATIVITY THEORY OF POINCARÉ AND LORENTZ

At the end of the nineteenth century, one of the most perplexing unsolved problems of natural philosophy was that of determining the relative motion of the earth and the aether. Let us try to present the matter as it appeared to the physicists of that time.

According to Newton's First Law of Motion, any particle which is free from the action of impressed forces moves, if it moves at all, with uniform velocity in a straight line. But in order that this statement may have a meaning, it is necessary to define the terms *straight line* and *uniform velocity* ; for a particle which is said to be 'moving in a straight line' in a terrestrial laboratory would not appear to be moving in a straight line to an observer on the sun, since he would perceive its motion compounded with the earth's diurnal rotation and her annual revolution in her orbit. We can, however, define a straight line *with reference to a system of axes Oxyz* as the geometrical figure defined by a pair of linear equations between x, y, z ; and we can assert as a fact of experience that certain systems of axes $Oxyz$ exist such that free particles move in straight lines with reference to them. Moreover, we can assert that there exist certain ways of measuring time such that the velocity of free particles along their rectilinear paths is uniform. A set of axes in space and a system of time-measurement, which possess these properties, may be called an *inertial system of reference.*

In Newtonian mechanics, if S is an inertial system of reference, and if S′ is another system such that the axes $O'x'y'z'$ of S′ have any uniform motion of pure translation with respect to the axes $Oxyz$ of S, and if the system of time-measurement is the same in the two cases, then S′ is also an inertial system of reference : the Newtonian laws of motion are valid with respect to S′ just as with respect to S. No one inertial system of reference could be regarded as having a privileged status, in the sense that it could properly be said to be fixed while the others were moving. Newtonian mechanics does not involve the notion of the absolute fixity of a point in space.

The laws of Newtonian dynamics thus presuppose the knowledge of a certain set of systems of reference, which is necessary if the laws are to have any meaning. In the nineteenth century many physicists inquired how this set of systems of reference should be described and defined. When Carl Neumann (1832–1925) was appointed professor of mathematics at Leipzig in 1869, he devoted his inaugural lecture [1] to the question, and introduced the name *The Body Alpha* for these systems of reference collectively. W. Thomson (Kelvin) and P. G. Tait in their *Treatise on Natural Philosophy* [2] suggested as a basis for specifying the Body Alpha that the centre of gravity of all matter in the universe might be considered to be *absolutely at rest*, and that the plane in which the angular momentum of the universe round its centre of gravity is the greatest, might be regarded as *fixed in direction in space*. Other writers proposed that the Body Alpha should be based on the system of the fixed stars, or the aggregate of all the bodies in existence.[3]

In the latter part of the nineteenth century the doctrine of the aether, which was justified by the undulatory theory of light, was generally regarded as involving the concepts of rest and motion relative to the aether, and thus to afford a means of specifying absolute position and defining the Body Alpha. Suppose, for instance, that a disturbance is generated at any point in free aether : this disturbance will spread outwards in the form of a sphere : and the centre of this sphere will for all subsequent time occupy an unchanged position relative to the aether. In this way, or in many other ways, we might hope to determine, by electrical or optical experiments, the velocity of the earth's motion relative to the aether.

In the first years of the twentieth century this problem was provoking a fresh series of experimental investigations. The most interesting of these was due to FitzGerald[4] who, shortly before his death in February 1901, commenced to examine the phenomena exhibited by a charged electrical condenser, as it is carried through space by the terrestrial motion. When the plane of the condenser includes the direction of the aether-drift (the ' longitudinal position '), the moving positive and negative charges on its two plates will be equivalent to currents running tangentially in opposite directions in the plates, so that a magnetic field will be set up in the space between them, and magnetic energy must be stored in this space : but when the plane of the condenser is at right angles to the terrestrial motion (the ' transverse position '), the equivalent currents are in the normal direction, and neutralise each other's magnetic action almost completely. FitzGerald's original idea was that, in order to supply the magnetic energy, there must be a mechanical drag on the condenser at the moment of charging, similar to that which would be produced if the mass of a body at the surface of the earth were suddenly to become greater. Moreover, the co-existence of the electric and magnetic fields in the space between the plates would entail [5] the

[1] Afterwards published as a booklet of 32 pages, *Die Principien der Galilei-Newton'schen Theorie* (Leipzig 1870). He returned to the matter in 1904, in the *Festschrift Boltzmann* (Leipzig, 1904), p. 252.

[2] New edition, Cambridge 1890, Vol. I, p. 241

[3] An account of these suggestions is given by G. Giorgi, *Palermo Rend.*, xxxiv (1912), p. 301.

[4] FitzGerald's *Scientific Writings*, p. 557 ; cf. Larmor, ibid., p. 566

[5] cf. Vol. I, p. 318

existence of an electromagnetic momentum proportional to their vector-product. This momentum is easily seen to be (with sufficient approximation) parallel to the plates, and so would not in general have the same direction as the velocity of the condenser relative to the aether : thus the change in the situation in one second might be represented by the annihilation of the momentum existing at the beginning of the second and the creation of the momentum (equal and parallel to it) existing at the end of the second. But two equal and oppositely-parallel momenta at a distance apart constitute an angular momentum : and we may therefore expect that if the condenser is freely suspended, there will in general be a couple acting on it, proportional to the vector-product of the velocity of the condenser and the electromagnetic momentum. This couple would vanish in either the longitudinal or the transverse orientation, but in intermediate positions would tend to rotate the condenser into the longitudinal position ; the transverse position would be one of unstable equilibrium.

For both effects a search was made by FitzGerald's pupil F. T. Trouton [1] ; in the experiments designed to observe the turning couple, a condenser was suspended in a vertical plane by a fine wire, and charged. The effect to be detected was small : for the magnetic force due to the motion of the charges would be of order (w/c), where w denotes the velocity of the earth : so the magnetic energy of the system, which depends on the square of the force, would be of order $(w/c)^2$: and the couple would likewise be of the second order in (w/c).

No effect of any kind could be detected,[2] a result whose explanation was rightly surmised by P. Langevin [3] to belong to the same order of ideas as FitzGerald's hypothesis of contraction.

It may be remarked that the existence of the couple, had it been observed, would have demonstrated the possibility of drawing on the energy of the earth's motion for purposes of terrestrial utility.

The FitzGerald contraction of matter as it moves through the aether might conceivably be supposed to affect in some way the optical properties of the moving matter : for instance, transparent substances might become doubly refracting. Experiments designed to test this supposition were performed by Lord Rayleigh [4] in 1902 and by D. B. Brace in 1904,[5] but no double refraction comparable with the proportion $(w/c)^2$ of the single refraction could be detected. The FitzGerald contraction of a material body cannot therefore be of the same nature as the contraction which would be produced in the body by pressure, but must be accompanied by such concomitant

[1] *Trans. Roy. Dub. Soc.*, vii (1902), p. 379 ; F. T. Trouton and H. R. Noble, *Phil. Trans.* ccii (1903), p. 165

[2] This negative result was confirmed in 1926 by R. Tomaschek, *Ann. d. Phys.* lxxviii (1926), p. 743 and lxxx (1926), p. 509 ; and by C. T. Chase, *Phys. Rev.* xxviii (1926), p. 378.

[3] *Comptes Rendus*, cxl (1905), p. 1171

[4] *Phil. Mag.* iv (1902), p. 678

[5] *Phil. Mag.* vii (1904), p. 317

changes in the relations of the molecules to the aether, that an isotropic substance does not lose its simply refracting character.

Even before the end of the nineteenth century, the failure of so many promising attempts to measure the velocity of the earth relative to the aether had suggested to the penetrating and original mind of Poincaré a new possibility. In his lectures at the Sorbonne in 1899,[1] after describing the experiments so far made, which had yielded no effects involving either the first or the second powers of the coefficient of aberration (i.e. the ratio of the earth's velocity to the velocity of light), he went on to say,[2] 'I regard it as very probable that optical phenomena depend only on the *relative* motions of the material bodies, luminous sources, and optical apparatus concerned, and that this is true not merely as far as quantities of the order of the square of the aberration, but *rigorously*.' In other words, Poincaré believed in 1899 that *absolute motion is indetectible in principle*, whether by dynamical, optical, or electrical means.

In the following year, at an International Congress of Physics held at Paris, he asserted the same doctrine.[3] 'Our aether,' he said, 'does it really exist? I do not believe that more precise observations could ever reveal anything more than *relative* displacements.' After referring to the circumstance that the explanations then current for the negative results regarding terms of the first order in (w/c) were different from the explanations regarding the second order terms, he went on, 'It is necessary to find the *same* explanation for the negative results obtained regarding terms of these two orders: and there is every reason to suppose that this explanation will then apply equally to terms of higher orders, and that the mutual destruction of the terms will be rigorous and absolute.' A new principle would thus be introduced into physics, which would resemble the Second Law of Thermodynamics in as much as it asserted the *impossibility of doing something*: in this case, the impossibility of determining the velocity of the earth relative to the aether.[4]

In a lecture to a Congress of Arts and Science at St Louis, U.S.A., on 24 September 1904, Poincaré gave to a generalised form of this principle the name, *The Principle of Relativity*.[5] 'According to the Principle of Relativity,' he said, 'the laws of physical phenomena must be the same for a "fixed" observer as for an observer who has a uniform motion of translation relative to him: so that we have not, and cannot possibly have, any means of discerning whether we are, or are not, carried along in such a motion.' After examining the records of observation in the light of this principle, he declared,

[1] Edited by E. Néculcéa, and printed in 1901 under the title *Electricité et Optique*, Paris, Carré et Naud. [2] loc. cit., p. 536

[3] *Rapports présentés au Congrès International de Physique réuni à Paris en 1900* (Paris, Gauthier-Villars, 1900), Tome I, p. 1, at pp. 21, 22

[4] In April 1904 Lorentz asserted the same general principle: cf. *Versl. Kon. Akad. v. Wet.*, Amsterdam, Dl. xii (1904), p. 986; English edn. (*Amst. Proc.*), vi (1904), p. 809.

[5] This address appeared in *Bull. des Sc. Math.*(2) xxviii (1904), p. 302; an English translation by G. B. Halsted was published in *The Monist* for January 1905.

'From all these results there must arise an entirely new kind of dynamics, *which will be characterised above all by the rule, that no velocity can exceed the velocity of light.*'

We have now to see how an analytical scheme was devised which enabled the whole science of physics to be reformulated in accordance with Poincaré's Principle of Relativity.

That Principle, as its author had pointed out, required that observers who have uniform motions of translation relative to each other should express the laws of nature in the same form. Let us consider in particular the laws of the electromagnetic field.

Lorentz, as we have seen,[1] had obtained the equations of a moving electric system by applying a transformation to the fundamental equations of the aether. In the original form of this transformation, quantities of order higher than the first in (w/c) were neglected. But in 1900 Larmor [2] extended the analysis so as to include quantities of the second order. Lorentz in 1903 went further still,[3] and obtained the transformation in a form which is exact to all orders of the small quantity (w/c). In this form we shall now consider it.

The fundamental equations of the aether in empty space are

$$\operatorname{div} \mathbf{d} = 0, \qquad c \operatorname{curl} \mathbf{d} = -\frac{\partial \mathbf{h}}{\partial t}$$

$$\operatorname{div} \mathbf{h} = 0, \qquad c \operatorname{curl} \mathbf{h} = \frac{\partial \mathbf{d}}{\partial t}.$$

It is desired to find a transformation from the variables t, x, y, z, $\mathbf{d}$, $\mathbf{h}$, to new variables t_1, x_1, y_1, z_1, $\mathbf{d}_1$, $\mathbf{h}_1$, such that the equations in terms of these new variables may take the same form as the original equations, namely

$$\operatorname{div}_1 \mathbf{d}_1 = 0, \qquad c \operatorname{curl}_1 \mathbf{d}_1 = -\frac{\partial \mathbf{h}_1}{\partial t_1}$$

$$\operatorname{div}_1 \mathbf{h}_1 = 0, \qquad c \operatorname{curl}_1 \mathbf{h}_1 = \frac{\partial \mathbf{d}_1}{\partial t_1}.$$

Evidently one particular class of such transformation is that which corresponds to rotations of the axes of co-ordinates about the origin. These may be described as the linear homogeneous transformations of determinant unity which transform the expression $(x^2+y^2+z^2)$ into itself. It had, however, already become clear from Lorentz's earlier work that some of the transformations must involve not only

[1] cf. Vol. I, p. 406. cf. also Lorentz, *Proc. Amst. Acad.* (English edn.), i (1899), p. 427

[2] Larmor, *Aether and Matter* (1900), p. 173

[3] *Proc. Amst. Acad.* (English edn.), vi (1903), p. 809

x, y, z, but also the variable t.[1] So (guided by the approximate formulae already obtained) he now replaced the condition of transforming $(x^2+y^2+z^2)$ into itself, by the condition of transforming the expression $(x^2+y^2+z^2-c^2t^2)$ into itself; and, as we shall now show, he succeeded in proving that the transformations so obtained have the property of transforming the differential equations of the aether in the manner required.

We shall first consider a transformation of this class in which the variables y and z are unchanged. The equations of this transformation may easily be derived by considering that the equation of the rectangular hyperbola

$$x^2-(ct)^2=1$$

(in the plane of the variable x, ct) is unaltered when any pair of conjugate diameters are taken as new axes, and a new unit of length is taken proportional to the length of either of these diameters. The equations of transformation thus obtained are

$$\begin{aligned} ct &= ct_1 \cosh \alpha + x_1 \sinh \alpha \\ x &= x_1 \cosh \alpha + ct_1 \sinh \alpha \\ y &= y_1 \\ z &= z_1 \end{aligned} \qquad (1)$$

where α denotes a constant parameter. The simpler equations previously given by Lorentz[2] may evidently be derived from these by writing $w=c \tanh \alpha$, and neglecting powers of (w/c) above the first. It will be observed that not only is the system of measuring the abscissa x changed, but also the system of measuring the time t: the necessity for this had been recognised in Lorentz's original memoir by his introduction of 'local time.'

Let us find the physical interpretation of this transformation (1). If we consider the point in the (t_1, x_1, y_1, z_1) system for which x_1, y_1, z_1 are all zero, its co-ordinates in the other system are given by the equations

$$t=t_1 \cosh \alpha, \qquad x=ct_1 \sinh \alpha, \qquad y=0, \qquad z=0,$$

so

$$x=ct \tanh \alpha, \qquad y=0, \qquad z=0.$$

Thus if we regard the axes of (x_1, y_1, z_1) and the axes of (x, y, z) as two rectangular co-ordinate systems in space, then the origin of the (x_1, y_1, z_1) system has the co-ordinates $(ct \tanh \alpha, 0, 0)$, that is to

[1] Larmor, *Aether and Matter* (1900), in commenting on the FitzGerald contraction, had recognised that clocks, as well as rods, are affected by motion: a clock moving with velocity v relative to the aether must run slower, in the ratio

$$\sqrt{\left(1-\frac{v^2}{c^2}\right)} : 1.$$

[2] cf. Vol. I, p. 407

say, the origin of the (x_1, y_1, z_1) system moves with a uniform velocity $c \tanh \alpha$ along the x-axis of the (x, y, z) system. Thus if w is the relative velocity of the two systems, we have

$$\cosh \alpha = \left(1 - \frac{w^2}{c^2}\right)^{-\frac{1}{2}}, \qquad \sinh \alpha = \frac{w}{c}\left(1 - \frac{w^2}{c^2}\right)^{-\frac{1}{2}},$$

and Lorentz's transformation between their co-ordinates may be written

$$t = \frac{t_1 - \dfrac{wx_1}{c^2}}{\sqrt{\left(1 - \dfrac{w^2}{c^2}\right)}}, \qquad x = \frac{x_1 + wt_1}{\sqrt{\left(1 - \dfrac{w^2}{c^2}\right)}}, \qquad y = y_1, \qquad z = z_1.$$

In this transformation the variable x plays a privileged part, as compared with y or z. We can of course at once write down similar transformations in which y or z plays the privileged part; and we can combine any number of these transformations by performing them in succession. The aggregate of all the transformations so obtained, combined with the aggregate of all the rotations in ordinary space, constitutes a *group*, to which Poincaré [1] gave the name the group of *Lorentz transformations*.

By a natural extension of the equations formerly given by Lorentz for the electric and magnetic forces, it is seen that the equations for transforming these, when (t, x, v, z) are transformed by equations (1), are

$$\begin{aligned} d_x &= d_{x1} & h_x &= h_{x1} \\ d_y &= d_{y1} \cosh \alpha + h_{z1} \sinh \alpha & h_y &= h_{y1} \cosh \alpha - d_{z1} \sinh \alpha \\ d_z &= d_{z1} \cosh \alpha - h_{y1} \sinh \alpha & h_z &= h_{z1} \cosh \alpha + d_{y1} \sinh \alpha. \end{aligned} \tag{2}$$

When the original variables are by direct substitution replaced by the new variables defined by (1) and (2) in the fundamental differential equations of the aether, the latter take the form

$$\operatorname{div}_1 \mathbf{d}_1 = 0, \qquad c \operatorname{curl}_1 \mathbf{d}_1 = -\frac{\partial \mathbf{h}_1}{\partial t_1}$$

$$\operatorname{div}_1 \mathbf{h}_1 = 0, \qquad c \operatorname{curl}_1 \mathbf{h}_1 = \frac{\partial \mathbf{d}_1}{\partial t_1}$$

that is to say, *the fundamental equations of the aether retain their form unaltered, when the variables (t, x, y, z) are subjected to the Lorentz*

[1] *Comptes Rendus*, cxl (5 June 1905), p. 1504. It should be added that these transformations had been applied to the equation of vibratory motions many years before by W. Voigt, *Gött. Nach.* (1887), p. 41.

transformation (1), *and at the same time the electric and magnetic intensities are subjected ot the transformation* (2).

The fact that the electric and magnetic intensities undergo the transformation (2) when the co-ordinates undergo the transformation (1), raises the question as to whether the transformation (2) is familiar to us in other connections. That this is so may be seen as follows.

In 1868–9 J. Plücker and A. Cayley introduced into geometry the notion of *line co-ordinates*; if (x_0, x_1, x_2, x_3) and (y_0, y_1, y_2, y_3) are the tetrahedral co-ordinates of two points of a straight line p, and if we write

$$x_m y_n - x_n y_m = p_{mn},$$

then the six quantities

$$p_{01}, \quad p_{02}, \quad p_{03}, \quad p_{23}, \quad p_{31}, \quad p_{12}$$

are called the *line-co-ordinates* of p.

Now suppose that the transformation

$$\begin{aligned} x_0 &= x'_0 \cosh a + x'_1 \sinh a \\ x_1 &= x'_1 \cosh a + x'_0 \sinh a \\ x_2 &= x'_2 \\ x_3 &= x'_3 \end{aligned} \qquad (3)$$

is performed on the co-ordinates (x_0, x_1, x_2, x_3), and the same transformation is performed on the co-ordinates (y_0, y_1, y_2, y_3). Then we have

$$\begin{aligned} p_{01} &= x_0 y_1 - x_1 y_0 \\ &= (x'_0 \cosh a + x'_1 \sinh a)(y'_1 \cosh a + y'_0 \sinh a) \\ &\qquad - (x'_1 \cosh a + x'_0 \sinh a)(y'_0 \cosh a + y'_1 \sinh a) \\ &= x'_0 y'_1 - x'_1 y'_0 \\ &= p'_{01} \end{aligned}$$

and in the same way we find

$$\begin{aligned} p_{02} &= p'_{02} \cosh a + p'_{12} \sinh a \\ p_{03} &= p'_{03} \cosh a - p'_{31} \sinh a \\ p_{23} &= p'_{23} \\ p_{31} &= p'_{31} \cosh a - p'_{03} \sinh a \\ p_{12} &= p'_{12} \cosh a + p'_{02} \sinh a. \end{aligned}$$

But these equations of transformation of the p's are precisely the same as the equations of transformation (2) of the electric and magnetic intensities, provided we write

$$p_{01} = d_x, \quad p_{02} = d_y, \quad p_{03} = d_z, \quad p_{23} = h_x, \quad p_{31} = h_y, \quad p_{12} = h_z.$$

The line-co-ordinates of a line have this property of transforming like the six components of the electric and magnetic intensities not only for the particular Lorentz transformation (1) but for the *most general* Lorentz transformation. A set of six quantities which transform like the line-co-ordinates of a line when the co-ordinates are subjected to any Lorentz transformation whatever, is called a *six-vector*. Thus we may say that the *quantities* (d_x, d_y, d_z, h_x, h_y, h_z) *constitute a six-vector*.[1] In the older physics, **d** was regarded as a vector, and **h** as a distinct vector: but if an electrostatic system (in which **d** exists but **h** is zero) is referred to axes which are in motion with respect to it, then the magnetic force with respect to these axes will not be zero. The six-vector transformation takes account of this fact, and furnishes the value of the magnetic force which thus appears.

We see, therefore, that in electromagnetic theory, as in Newtonian dynamics, there are *inertial systems* of co-ordinate axes with associated systems of measurement of time, such that the path of a free material particle relative to an inertial system is a straight line described with uniform velocity, and *also* that the equations of the electromagnetic field relative to the inertial system are Maxwell's equations, and *any system of axes which moves with a uniform motion of translation, relative to any given inertial system of axes, is itself an inertial system of axes, the measurement of time and distance in the two systems being connected by a Lorentz transformation. All the laws of nature have the same form in the co-ordinates belonging to one inertial system as in the co-ordinates belonging to any other inertial system.* No inertial system of reference can be regarded as having a privileged status, in the sense that it should be regarded as fixed while the others are moving: the notion of absolute fixity in space, which in the latter part of the nineteenth century was thought to be required by the theory of aether and electrons was shown in 1900–4 by the Poincaré-Lorentz theory of relativity to be without foundation.

Suppose that an inertial system of reference (t, x, y, z) is known on earth: and imagine a distant star which is moving with a uniform velocity relative to this framework (t, x, y, z). The theorem of relativity shows that there exists another framework (t_1, x_1, y_1, z_1) with respect to which the star is at rest, and in which, moreover, a luminous disturbance generated at time t_1 at any point (x_1, y_1, z_1) will spread outwards in the form of a sphere

$$(X_1 - x_1)^2 + (Y_1 - y_1)^2 + (Z_1 - z_1)^2 = c^2(T_1 - t_1)^2,$$

the centre of this sphere occupying for all subsequent time an unchanged position in the co-ordinate system (x_1, y_1, z_1). This framework is peculiarly fitted for the representation of phenomena which happen on the star, whose inhabitants would therefore naturally adopt it as their system of space and time. Beings, on the other hand, who dwell on a body which is at rest with respect to the axes (t, x, y, z), would prefer to use the latter system; and from the point of view of the universe at large, either of these systems is as good as the other. The electromagnetic equations are the same with respect to both sets of co-ordinates, and therefore neither can

[1] *Raum-Zeit-Vektor II Art* of H. Minkowski, *Gött. Nach.* 1908, p. 53

claim to possess the only property which could confer a primacy—namely, a special relation to the aether.

Some of the consequences of the new theory seemed to contemporary physicists very strange. Suppose, for example, that two inertial sets of axes A and B are in motion relative to each other, and that at a certain instant their origins coincide: and suppose that at this instant a flash of light is generated at the common origin. Then, by what has been said in the subsequent propagation, the wave-fronts of the light, as observed in A and in B, are spheres whose centres are the origins of A and B respectively, and therefore *different* spheres. How can this be?

The paradox is explained when it is remembered that a wave-front is defined to be the locus of points which are *simultaneously* in the same phase of disturbance. Now events taking place at different points, which are simultaneous according to A's system of measuring time, are not in general simultaneous according to B's way of measuring: and therefore what A calls a wave-front is not the same thing as what B calls a wave-front. Moreover, since the system of measuring space is different in the two inertial systems, what A calls a sphere is not the same thing as what B calls a sphere. Thus there is no contradiction in the statement that the wave-fronts for A are spheres with A's origin as centre, while the wave-fronts for B are spheres with B's origin as centre.

In common language we speak of events which happen at different points of space as happening 'at the same instant of time,' and we also speak of events which happen at different instants of time as happening 'at the same point of space.' We now see that such expressions can have a meaning only by virtue of artificial conventions; they do not correspond to any essential physical realities.

It is usual to regard Poincaré as primarily a mathematician, and Lorentz as primarily a theoretical physicist: but as regards their contributions to relativity theory, the positions were reversed: it was Poincaré who proposed the general physical principle, and Lorentz who supplied much of the mathematical embodiment. Indeed, Lorentz was for many years doubtful about the physical theory: in a lecture which he gave in October 1910 [1] he spoke of 'die Vorstellung (die auch Redner nur ungern aufgeben würde), dass Raum und Zeit etwas vollig Verschiedenes seien und dass es eine "wahre Zeit" gebe (die Gleichzeitigkeit würde denn unabhängig vom Orte bestehen).' [2]

A distinguished physicist who visited Lorentz in Holland shortly before his death found that his opinions on this question were unchanged.

We are now in a position to show the connection between the

[1] Printed in *Phys. ZS.* xi (1910), p. 1234

[2] 'The concept (which the present author would dislike to abandon) that space and time are something completely distinct and that a "true time" exists (simultaneity would then have a meaning independent of position).'

Lorentz transformation and FitzGerald's hypothesis of contraction; this connection was first established by Larmor [1] for his approximate form of the Lorentz transformation, which is accurate only to the second order in (w/c), but the extension to the full Lorentz transformation is easy.

Suppose that a rod is moving along the axis of x with uniform velocity w; let the co-ordinates of its ends at the instant t be x_1 and x_2. Take a system of axes $O'x'y'z'$ which move with the rod, the axis $O'x'$ being in the same line as the axis Ox, and the axes $O'y'$ and $O'z'$ being constantly parallel to the axes Oy and Oz respectively. In this system the length of the rod will be $x'_2-x'_1$, where, of course, x'_2 and x'_1 do not vary with the time. The Lorentz transformation gives

$$x'_2=x_2 \cosh a-ct \sinh a$$
$$x'_1=x_1 \cosh a-ct \sinh a$$

where $\tanh a=(w/c)$. Subtracting, we have

$$x'_2-x'_1=(x_2-x_1) \cosh a=(x_2-x_1)\{1-(w/c)^2\}^{-\frac{1}{2}}$$

or

$$x_2-x_1=\sqrt{1-\frac{w^2}{c^2}}\,(x'_2-x'_1).$$

This equation shows that the distance between the ends of the rod, in the system of measurement furnished by the original axes, with reference to which the rod is moving with velocity w, bears the ratio $(1-w^2/c^2)^{\frac{1}{2}}:1$ to their distance in the system of measurement furnished by the transformed axes, with reference to which the rod is at rest: and this is precisely FitzGerald's hypothesis of contraction. The hypothesis of FitzGerald may evidently be expressed by the statement, that *the equations of the figures of material bodies are covariant with respect to those transformations for which the fundamental equations of the aether are covariant:* that is, for all Lorentz transformations.

Now let us look into Poincaré's remark [2] that the Principle of Relativity requires the creation of a new mechanics in which no velocity can exceed the velocity of light.

Suppose that an inertial system B is being translated relative to an inertial system A with velocity w along the axis of x. Let a point P moving along the axis of x have the co-ordinates $(t, x, 0, 0)$ in system A and $(t', x', 0, 0)$ in system B. Denote the components of velocity dx/dt and dx'/dt' by v_x, v'_x, respectively, and let $w=c \tanh a$. Then the Lorentz transformation gives at once

$$v_x=\frac{dx}{dt}=\frac{c(dx' \cosh a+cdt' \sinh a)}{c\,dt' \cosh a+dx' \sinh a}=\frac{v'_x+w}{1+\frac{v_x w}{c^2}}.$$

[1] *Aether and Matter* (1900), p. 173 [2] cf. p. 31

Now, v_x being the velocity of P relative to A, v'_x the velocity of P relative to B, and w the velocity of B relative to A, in Newtonian kinematics we should have $v_x = v'_x + w$. The denominator $(1 + v_x w/c^2)$ in the relativist formula expresses the difference between Newtonian theory and relativity theory, so far as concerns the composition of velocities. We see that if $v'_x = c$, then $v_x = c$; that is to say, *any velocity compounded with c gives as the resultant c over again*, and therefore that no velocity can exceed the velocity of light.

This result enables us to solve a problem which had perplexed many generations of physicists. It had been supposed that if the correct theory of light is the corpuscular theory, then the corpuscles emitted by a moving star should have a velocity which is compounded of the velocity of the star and the velocity of light relative to a source at rest, just as an object thrown from a carriage window in a moving railway train has a velocity which is obtained by compounding its velocity relative to the carriage with the velocity of the train (the *ballistic* theory); whereas, if the correct theory of light is the wave-theory, the velocity of the light emitted by the star should be unaffected by the velocity of the star, just as the waves created by throwing a stone into a pond move outwards from the point where the stone entered the water, without being affected by the velocity of the stone. The new relativist theory led to the surprising conclusion that the velocity of light would be unaffected by the velocity of its source *even on the corpuscular theory.*

An attempt to explain the Michelson-Morley experiment, and the other evidence which had given rise to relativity theory, without assuming that the velocity of light is independent of the velocity of its source, was made in 1908 by W. Ritz,[1] who postulated that the velocity of light and the velocity of the source are additive, as in the old physics. It is, however, now known certainly that the velocity of light is independent of the motion of the source. The astronomical evidence for this statement has been marshalled by several writers,[2] and further confirmation has been furnished by Majorana by direct experiment.[3] It should be remarked that since in purely terrestrial experiments the light rays always describe closed paths, the results to be expected from 'ballistic' and non-ballistic theories can differ only by quantities of the second order,[4] but the performance of the

[1] *Ann. de chim. et phys.* xiii (1908), p. 145; *Arch. de Génève,* xxvi (1908), p. 232; cf. a careful discussion of it by R. C. Tolman, *Phys. Rev.* xxxv (1912), p. 136

[2] Particularly by R. C. Tolman, *Phys. Rev.* xxxi (1910), p. 26; W. de Sitter, *Amsterdam Proc.* xv (1913), p. 1297; xvi (1913), p. 395; *Phys. ZS.* xiv (1913), pp. 429, 1267; *Bull. of the Astron. Inst. of the Netherlands,* ii (1924), pp. 121, 163; R. S. Capon, *Month. Not. R.A.S.* lxxiv (1914), pp. 507, 658; H. C. Plummer, ibid., p. 660; H. Thirring, *ZS. f. P.* xxxi (1925), p. 133; G. Wataghin, *ZS. f. P.* xl (1926), p. 378

[3] *Comptes Rendus,* clxv (1917), p. 424; clxvii (1918), p. 71; clxix (1919), p. 719; *Phys. Rev.* xi (1918), p. 411; *Phil. Mag.* xxxvii (1919), p. 145; xxxix (1920), p. 488; cf. also Jeans, *Nature,* cvii (1921), pp. 42, 169

[4] cf. P. Ehrenfest, *Phys. ZS.* xiii (1912), p. 317; F. Michaud, *Comptes Rendus,* clxviii (1919), p. 507

Michelson-Morley experiment with light from astronomical sources by R. Tomaschek [1] in 1924 definitely disproved the ballistic hypothesis.

A further result in harmony with the new theory was obtained when Michelson [2] showed experimentally that the velocity of a moving mirror is without influence on the velocity of light reflected at its surface.

It was now recognised that these observational findings, which in the nineteenth century might have been supposed to tell in favour of the wave-theory, were actually without significance one way or the other in the dispute between the wave and corpuscular theories of light. For, according to relativity theory, even on the corpuscular hypothesis, a corpuscle which had a velocity c relative to its source would have the same velocity relative to any observer, whether he shared in the motion of the source or not.

In 1905 Poincaré [3] completed the theorem of Lorentz [4] on the covariance of Maxwell's equations with respect to the Lorentz transformation, by obtaining the formulae of transformation of the electric density ρ and current $\rho\mathbf{v}$. The fundamental equations are

$$\operatorname{div} \mathbf{d} = 4\pi\rho\,; \qquad c \operatorname{curl} \mathbf{d} = -\frac{\partial \mathbf{h}}{\partial t}$$

$$\operatorname{div} \mathbf{h} = 0\,; \qquad c \operatorname{curl} \mathbf{h} = \frac{\partial \mathbf{d}}{\partial t} + 4\pi\rho\mathbf{v}$$

and it is desired to find a transformation from the variable t, x, y, z, ρ, $\mathbf{d}$, $\mathbf{h}$, $\mathbf{v}$ to new variables t_1, x_1, y_1, z_1, ρ_1, $\mathbf{d}_1$, $\mathbf{h}_1$, $\mathbf{v}_1$, such that the equations in terms of these new variables may have the same form as the original equations. The transformations of t, x, y, z, $\mathbf{d}$, $\mathbf{h}$ have already been found. Poincaré now showed that

$$\begin{aligned} \rho &= \rho_1 \cosh \alpha + (\rho_1 v_{x_1}/c) \sinh \alpha \\ \rho v_x &= \rho_1 v_{x_1} \cosh \alpha + c\rho_1 \sinh \alpha \\ \rho v_y &= \rho_1 v_{y_1} \\ \rho v_z &= \rho_1 v_{z_1}. \end{aligned}$$

When the original variables are by direct substitution replaced by the new variables in the differential equations, the latter take the form

$$\operatorname{div}_1 \mathbf{d}_1 = 4\pi\rho_1, \qquad c \operatorname{curl}_1 \mathbf{d}_1 = -\frac{\partial \mathbf{h}_1}{\partial t_1}$$

$$\operatorname{div}_1 \mathbf{h}_1 = 0, \qquad c \operatorname{curl}_1 \mathbf{h}_1 = \frac{\partial \mathbf{d}_1}{\partial t_1} + 4\pi\rho_1\mathbf{v}_1$$

[1] *Ann. d. Phys.* lxxiii (1924), p. 105

[2] *Astrophys. J.* xxxvii (1913), p. 190

[3] *Comptes Rendus*, cxl (June 1905), p. 1504

[4] cf. p. 33

that is to say, the fundamental equations of aether and electrons retain their form unaltered, when the variables are subjected to the transformation which has been specified.

In the autumn of the same year, in the same volume of the *Annalen der Physik* as his paper on the Brownian motion,[1] Einstein published a paper which set forth the relativity theory of Poincaré and Lorentz with some amplifications, and which attracted much attention. He asserted as a fundamental principle the *constancy of the velocity of light*, i.e. that the velocity of light *in vacuo* is the same in all systems of reference which are moving relatively to each other : an assertion which at the time was widely accepted, but has been severely criticised by later writers.[2] In this paper Einstein gave the modifications which must now be introduced into the formulae for aberration and the Doppler effect.[3]

Consider a star, which is observed from the earth on two occasions. The distance of the star is assumed to be so great that its apparent proper motion in the interval between the observations is negligible. Denote an inertial system of axes at the earth at the time of the first observation by K, and an inertial system of axes at the earth at the time of the second observation by K′ : and choose these axes so that the x-axis has the direction of the velocity w $(=c \tanh \alpha)$ of K′ relative to K. Let ψ be the angle which the ray of light arriving at the earth from the star makes with the x-axis as measured in K, and ψ' the corresponding angle in the system K′. Then the Lorentz transformation gives for the co-ordinates of the star in the two systems

$$\begin{cases} ct' = ct \cosh \alpha - x \sinh \alpha \\ x' = x \cosh \alpha - ct \sinh \alpha \\ y' = y \end{cases}$$

(taking the plane of xy to contain the star) : and since light is propagated with velocity c in both systems, we have $ct = \sqrt{(x^2+y^2)}$, $ct' = \sqrt{(x'^2+y'^2)}$. Thus

$$\cos \psi' = \frac{x'}{\sqrt{(x'^2+y'^2)}} = \frac{x'}{ct'} = \frac{x \cosh \alpha - ct \sinh \alpha}{ct \cosh \alpha - x \sinh \alpha} = \frac{\cos \psi \cosh \alpha - \sinh \alpha}{\cosh \alpha - \cos \psi \sinh \alpha}$$

or

$$\cos \psi' = \frac{c \cos \psi - w}{c - w \cos \psi}.$$

This is the relativist formula for aberration : it may be written

$$\sin \frac{\psi' - \psi}{2} = \tanh \frac{\alpha}{2} \sin \frac{\psi' + \psi}{2}.$$

[1] *Ann. d. Phys.* xvii (Sept. 1905), p. 891

[2] e.g. H. E. Ives, *Proc. Amer. Phil. Soc.* xcv (1951), p. 125 ; *Sc. Proc. R.D.S.* xxvi (1952), p. 9, at pp. 21–2

[3] *cf.* Vol. I, pp. 368 and 389

When powers of (w/c) above the first are neglected, this gives

$$\psi' - \psi = \frac{w}{c} \sin \psi,$$

which is the aberration-formula of classical physics.

To find the relativist formula for the Doppler effect, we suppose that K′ is an inertial system with respect to which the star is at rest, and K is an inertial system in which the earth is at rest: and choose the axes so that the system K′ is moving with velocity w $(=c \tanh \alpha)$ parallel to the axis of x in the system K. Let ψ be the angle which the line joining the star to the observer makes with the x-axis in the system K, and let ψ' be the corresponding angle in the system K′. Then the phase in the system K is determined by

$$\nu \left(t + \frac{x \cos \psi + y \sin \psi}{c}\right)$$

where ν is the frequency of the light as observed by the terrestrial observer; and as the phase is a physical invariant, we must have

$$\nu \left(t + \frac{x \cos \psi + y \sin \psi}{c}\right) = \nu' \left(t' + \frac{x' \cos \psi' + y' \sin \psi'}{c}\right)$$

when ν' is the frequency of the light as measured by an observer on the star. Thus

$$\nu \left\{t' \cosh \alpha + \frac{x'}{c} \sinh \alpha + \frac{1}{c}\left[(x' \cosh \alpha + ct' \sinh \alpha) \cos \psi + y' \sin \psi\right]\right\}$$

$$= \nu' \left(t' + \frac{x' \cos \psi' + y' \sin \psi'}{c}\right).$$

Equating coefficients of t', we have

$$\nu\,(\cosh \alpha + \sinh \alpha \cos \psi) = \nu'$$

or

$$\frac{\nu'}{\nu} = \frac{1 + \frac{w}{c} \cos \psi}{\sqrt{\left(1 - \frac{w^2}{c^2}\right)}}.$$

This is the relativist formula for the Doppler effect. When only first-order terms in (w/c) are retained, it gives

$$\nu = \nu' \left(1 - \frac{w_r}{c}\right)$$

where w_r is the radial component of w : which is the older formula for the Doppler effect.

It will be noticed that the relativist formula differs from the older formula by the presence of the factor $\sqrt{(1-w^2/c^2)}$. Now if an observer moving with velocity w relative to an inertial system passes a place P where a clock belonging to the inertial system reads t_1, and if he afterwards passes a place Q where the clock in the inertial system reads t_2, and if t' is the interval of time registered by the observer's clock between the positions P and Q, then it follows at once from the equations of the Lorentz transformation that

$$\frac{t'}{t_2-t_1}=\sqrt{\left(1-\frac{w^2}{c^2}\right)},$$

so that we can (somewhat loosely) speak of the factor $\sqrt{(1-w^2/c^2)}$ as representing the slower rate at which the observer's clock is running as compared with clocks that are at rest on the star. It is obvious that this factor must occur in the relativist formula.

It will be observed that in the relativist formula, the Doppler effect is not zero even when the relative motion of the source and observer is at right angles to the direction of propagation of the light; in this case ($\psi=\frac{1}{2}\pi$) we have

$$\nu=\nu'\left(1-\frac{w^2}{c^2}\right)^{\frac{1}{2}}$$

or in the first approximation

$$\frac{\nu-\nu'}{\nu'}=-\frac{w^2}{2c^2}.$$

This is called the *transverse Doppler effect.* In 1907 Einstein suggested [1] that it might be observed by examining the light emitted by canal rays [2] in hydrogen, on which J. Stark [3] had published a paper in 1906. Stark's experimental results, however, did not seem to confirm the theoretical formula : and it was not until more than thirty years later that H. E. Ives and G. R. Stillwell [4] succeeded in carrying out this experiment with any degree of success.

It is clear, from the history set forth in the present chapter, that the theory of relativity had its origin in the theory of aether and electrons. When relativity had become recognised as a doctrine covering the whole operation of physical nature, efforts were made to present it in a form free from any special association with electro-

[1] *Ann. d. Phys.* xxiii (1907), p. 197
[2] cf. Vol. I, p 363
[3] *Ann. d. Phys.* xxi (1906), p. 401
[4] *J. Opt. Soc. Amer.* xxviii (1938), p. 215; xxxi (1941), p. 369

magnetic theory, and deducible logically from a definite set of axioms of greater or less plausibility.[1]

It should be mentioned also that when relativity theory had become generally accepted, the Michelson-Morley experiment was rediscussed with a much more complete understanding and exactitude.[2]

An account may be given here of some experiments performed long after the time with which we are at present mainly concerned, which confirmed in a striking way the predictions of relativity theory. In one of them, due to A. B. Wood, G. A. Tomlinson and L. Essen,[3] a rod in longitudinal vibration was rotated in a horizontal plane, so that its length varied periodically by reason of the FitzGerald contraction. Accurate measurements were made of the vibration frequency, which would have varied with the length, if the length only had been affected. According to relativity theory, however, there should be a complete compensation of the contraction in length, by a modification of the elasticity of the rod according to its orientation with respect to the direction of its motion, so that no change of frequency should be observed. The experiment was carried out with two similar longitudinal piezo-electric quartz oscillators, one rotating and the other stationary, the relative frequency being measured. The experiment yielded a null result within narrow limits of uncertainty of about ± 4 parts in 10^{11}, thus fully confirming the prediction of the Poincaré-Lorentz theory of relativity.

Still later, a prediction of the theory was verified in a striking

[1] Papers on axiomatics are many. Attention may be directed specially to the following: P. Frank and H. Rothe, *Ann. d. Phys.* xxxiv (1911), p. 825; E. V. Huntington, *Phil. Mag.* xxiii (1912), p. 494; L. A. Pars, *Phil. Mag.* xlii (1921), p. 249; C. Carathéodory, *Berlin Sitz.* v (1924), p. 12; V. V. Narliker, *Proc. Camb. Phil. Soc.* xxviii (1932), p. 460; G. J. Whitrow, *Quart. J. Math.* iv (1933), p. 161; L. R. Gomes, *Lincei Rend.* xxi (1935), p. 433; N. R. Sen, *Indian J. of Phys.* x (1936), p. 341; F. Severi, *Proc. Phys.-Math. Soc. Japan*, xviii (1936), p. 257; E. Esclangon, *Comptes Rendus*, ccii (1936), p. 708; *Bull. Astron.* x (1937), p. 1; J. Meurers, *ZS. f. P.* cii (1936), p. 611; V. Lalan, *Comptes Rendus*, ciii (1936), p. 1491; *Bull. Soc. Math. France*, lxv (1937), p. 83; G. Temple, *Quart. J. Math.* ix (1938), p. 283; H. E. Ives, *Proc. Amer. Phil. Soc.* xcv (1951), p. 125. A valuable paper by H. P. Robertson, *Rev. Mod. Phys.* xxi (1949), p. 378, is in a somewhat different category. Robertson discusses the justification of the axioms on the ground of experimental results, and shows that most of the axioms can be based securely on (i) the Michelson-Morley experiment, (ii) the experiment of Ives and Stilwell on the transverse Doppler effect (cf. p. 42), and (iii) an experiment performed in 1932 by R. J. Kennedy and E. M. Thorndike [*Phys. Rev.* xlii (1932), p. 400]; in this, a pencil of homogeneous light was split at a half-reflecting surface into two beams, which, after traversing paths of different lengths, were brought together again and made to interfere; the positions of the fringes in the interference pattern were observed when the velocity of the system was varied owing to the motions of rotation and revolution of the earth. The predictions of relativity theory were verified. An interesting experiment with a rotating interferometer was performed by G. Sagnac in 1913; *Comptes Rendus* clvii (1913), pp. 708, 1410; *J. Phys. Rad.* iv (1914), p. 177; cf. A. Metz, *J. Phys. Rad.* xiii (1952), p. 224.

[2] cf. E. Kohl, *Ann. d. Phys.* xxviii (1909), pp. 259, 662; E. Budde, *Phys. ZS.* xii (1911), p. 979; M. von Laue, *Ann. d. Phys.* xxxiii (1910), p. 186; *Phys. ZS.* xiii (1912), p. 501; A. Right, *Le Radium*, xi (1919), p. 321; *N. Cimento*, xviii (1919), p. 91; J. Villey, *Comptes Rendus*, clxx (1920), p. 1175; clxxi (1920), p. 298; E. H. Kennard and D. E. Richmond, *Phys. Rev.* xix (1922), p. 572; J. L. Synge, *Sci. Proc. Roy. Dub. Soc.* xxvi (1952), p. 45; *Nature* clxx (1952), p. 244

[3] *Proc. R.S.*(A), clviii (1937), p. 606

way. If two events (1) and (2) are considered, and if in an inertial system A these events happen at different points of space, whereas in an inertial system B (moving with velocity w relative to A), the two events happen at the same point of space, then the Lorentz transformation gives

$$t_1^{\mathrm{A}} = \frac{t_1^{\mathrm{B}} - \dfrac{wx_1^{\mathrm{B}}}{c^2}}{\sqrt{\left(1 - \dfrac{w^2}{c^2}\right)}}, \qquad t_2^{\mathrm{A}} = \frac{t_2^{\mathrm{B}} - \dfrac{wx_2^{\mathrm{B}}}{c^2}}{\sqrt{\left(1 - \dfrac{w^2}{c^2}\right)}}.$$

Since $x_1^{\mathrm{B}} = x_2^{\mathrm{B}}$, these equations give

$$(t_2 - t_1)^{\mathrm{A}} = \frac{(t_2 - t_1)^{\mathrm{B}}}{\sqrt{\left(1 - \dfrac{w^2}{c^2}\right)}}$$

so the time between the events, measured in system A, is $(1 - w^2/c^2)^{-\frac{1}{2}}$ times greater than the time between the events, measured in system B.

Now certain particles called *cosmic-ray mesons*, discovered observationally in 1937, disintegrate spontaneously; and it may be assumed that the rate of disintegration depends on time as measured by an observer travelling with the meson. Thus to an observer who is stationary with respect to the earth, the rate of disintegration should appear to be slower, the faster the meson is moving. This was found in 1941 to be actually the case.[1]

The study of relativist dynamics was begun in 1906, when Max Planck[2] found the equations which, according to the new theory, should replace the Newtonian equations of motion of a material particle. Considering first the one-dimensional case, let a particle of mass m and charge e be moving along the axis of x with velocity $w(=c \tanh \alpha)$ in the system $Oxyz$, in a field of electric force parallel to Ox. Let $O'x'y'z'$ be axes parallel to these, whose origin O' moves with the particle. The relations between (t, x, y, z) and (t', x', y', z') are

$$\begin{cases} ct' = ct \cosh \alpha - x \sinh \alpha \\ x' = x \cosh \alpha - ct \sinh \alpha \\ y' = y \\ z' = z. \end{cases}$$

The Newtonian equation of motion is assumed to be valid with respect to the axes $O'x'y'z'$, so the equation of motion of the particle is

$$m\frac{d^2x'}{dt'^2} = ed'_{x'} = ed_x$$

[1] B. Rossi and D. B. Hall, *Phys. Rev.* lix (1941), p. 223
[2] *Verh. d. Deutsch, Phys. Ges.* viii (1906), p. 136

where $d'_{x'}$ and d_x denote the electric force in the two systems.[1] Now

$$\frac{dx'}{c\,dt'}=\frac{dx/dt\ \cosh\alpha-c\sinh\alpha}{c\cosh\alpha-dx/dt\ \sinh\alpha}$$

so

$$\frac{d^2x'}{c^2dt'^2}=\frac{\dfrac{d}{dt}\left\{\dfrac{(dx/dt)\cosh\alpha-c\sinh\alpha}{c\cosh\alpha-(dx/dt)\sinh\alpha}\right\}}{c\cosh\alpha-\dfrac{dx}{dt}\sinh\alpha}=\frac{\dfrac{d^2x}{dt^2}\cosh\alpha}{\left\{c\cosh\alpha-\dfrac{dx}{dt}\sinh\alpha\right\}^2},$$

remembering that $\frac{dx}{dt}\cosh\alpha-\sinh\alpha=0$. But

$$c\cosh\alpha-\frac{dx}{dt}\sinh\alpha=c\cosh\alpha-\frac{c\sinh^2\alpha}{\cosh\alpha}=\frac{c}{\cosh\alpha}$$

and therefore

$$\frac{d^2x'}{dt'^2}=\frac{d^2x}{dt^2}\cosh^3\alpha=\left(1-\frac{w^2}{c^2}\right)^{-\frac{3}{2}}\frac{dw}{dt}=\frac{d}{dt}\left\{\frac{w}{\sqrt{(1-w^2/c^2)}}\right\}.$$

Thus the equation of motion is (writing X for the moving force on the particle, namely ed_x),

$$\frac{d}{dt}\left\{\frac{mw}{\sqrt{(1-w^2/c^2)}}\right\}=X\,;$$

and extending the investigation to three dimensions, we can show that if the components of velocity are dx/dt, dy/dt, dz/dt, and if their resultant is w, then *the general equations of motion of a particle acted on by a force* (X, Y, Z) are

$$\frac{d}{dt}\left\{\frac{m\,dx/dt}{\sqrt{(1-w^2/c^2)}}\right\}=X,$$

$$\frac{d}{dt}\left\{\frac{m\,dy/dt}{\sqrt{(1-w^2/c^2)}}\right\}=Y, \qquad (1)$$

$$\frac{d}{dt}\left\{\frac{m\,dz/dt}{\sqrt{(1-w^2/c^2)}}\right\}=Z.$$

When $c\to\infty$, these evidently reduce to the Newtonian equations

$$m\frac{d^2x}{dt^2}=X,\qquad m\frac{d^2y}{dt^2}=Y,\qquad m\frac{d^2z}{dt^2}=Z.$$

[1] cf. p. 33

To obtain the law of conservation of energy, multiply the equations (1) by dx/dt, dy/dt, dz/dt respectively, and add. Thus

$$X\frac{dx}{dt}+Y\frac{dy}{dt}+Z\frac{dz}{dt}$$

$$=\frac{dx}{dt}\frac{d}{dt}\left\{\frac{m\,dx/dt}{\sqrt{(1-w^2/c^2)}}\right\}+\frac{dy}{dt}\frac{d}{dt}\left\{\frac{m\,dy/dt}{\sqrt{(1-w^2/c^2)}}\right\}+\frac{dz}{dt}\frac{d}{dt}\left\{\frac{m\,dz/dt}{\sqrt{(1-w^2/c^2)}}\right\}$$

$$=w^2\frac{d}{dt}\left\{m\left(1-\frac{w^2}{c^2}\right)^{-\frac{1}{2}}\right\}+m\left(1-\frac{w^2}{c^2}\right)^{-\frac{1}{2}}\left(\frac{dx}{dt}\frac{d^2x}{dt^2}+\frac{dy}{dt}\frac{d^2y}{dt^2}+\frac{dz}{dt}\frac{d^2z}{dt^2}\right)$$

$$=\frac{mw^3}{c^2}\left(1-\frac{w^2}{c^2}\right)^{-\frac{3}{2}}\frac{dw}{dt}+m\left(1-\frac{w^2}{c^2}\right)^{-\frac{1}{2}}w\frac{dw}{dt}$$

$$=mw\left(1-\frac{w^2}{c^2}\right)^{-\frac{3}{2}}\frac{dw}{dt}.$$

So that

$$X\frac{dx}{dt}+Y\frac{dy}{dt}+Z\frac{dz}{dt}=\frac{d}{dt}\left\{mc^2\left(1-\frac{w^2}{c^2}\right)^{-\frac{1}{2}}\right\}.$$

The left-hand side of this equation is evidently the rate at which work is being done on the particle, so the right-hand side must represent the rate of increase of the kinetic energy of the particle; that is, the kinetic energy of the particle is

$$\frac{mc^2}{\sqrt{\left(1-\frac{w^2}{c^2}\right)}}+C$$

where C denotes a constant; or, expanding the radical by the binomial theorem,

$$mc^2\left(1+\frac{w^2}{2c^2}+\text{higher powers of }\frac{w^2}{c^2}\right)+C.$$

In order that this may agree with the Newtonian value of the kinetic energy, namely, $\frac{1}{2}mw^2$, when the higher powers of w^2/c^2 are neglected, we must have $C=-mc^2$. Thus *the kinetic energy of the particle is*

$$\frac{mc^2}{\sqrt{\left(1-\frac{w^2}{c^2}\right)}}-mc^2. \tag{2}$$

It is easily seen that the equations (1) may be written

$$\frac{d}{dt}\left\{\frac{\partial L}{\partial\left(\frac{dx}{dt}\right)}\right\}=X, \qquad \frac{d}{dt}\left\{\frac{\partial L}{\partial\left(\frac{dy}{dt}\right)}\right\}=Y, \qquad \frac{d}{dt}\left\{\frac{\partial L}{\partial\left(\frac{dz}{dt}\right)}\right\}=Z$$

where

$$L=-mc^2\sqrt{\left(1-\frac{w^2}{c^2}\right)},$$

so L is the Lagrangean function or kinetic potential. Moreover, if we introduce

$$p_x=\frac{\partial L}{\partial\left(\frac{dx}{dt}\right)}=\frac{m}{\sqrt{\left(1-\frac{w^2}{c^2}\right)}}\frac{dx}{dt} \tag{3}$$

and similar expressions for p_y and p_z, and if we write

$$H=mc^2\sqrt{\left(1+\frac{p_x^2+p_y^2+p_z^2}{m^2c^2}\right)},$$

then the equations of motion may be written

$$\frac{dp_x}{dt}=X, \qquad \frac{dp_y}{dt}=Y, \qquad \frac{dp_z}{dt}=Z,$$

$$\frac{dx}{dt}=\frac{\partial H}{\partial p_x}, \qquad \frac{dy}{dt}=\frac{\partial H}{\partial p_y}, \qquad \frac{dz}{dt}=\frac{\partial H}{\partial p_z},$$

which is the Hamiltonian form.

Remembering that the moving force is the time-rate of the momentum, it is evident from equations (1) that the components of momentum of the particle are

$$\frac{m}{\sqrt{\left(1-\frac{w^2}{c^2}\right)}}\frac{dx}{dt}, \qquad \frac{m}{\sqrt{\left(1-\frac{w^2}{c^2}\right)}}\frac{dy}{dt}, \qquad \frac{m}{\sqrt{\left(1-\frac{w^2}{c^2}\right)}}\frac{dz}{dt} \tag{4}$$

which reduce to the Newtonian expressions $m\,dx/dt$, $m\,dy/dt$, $m\,dz/dt$, when $c\longrightarrow\infty$. The same result is obtained from equations (3) when we remember that the components of momentum are the derivates of the Lagrangean function with respect to the components of velocity: and it fits in with a remark which Laplace had made more than a century earlier,[1] namely, that if the momentum of a particle, instead of being mw were $m\phi(w)$, then the kinetic energy must be $\int m\phi'(w)w\,dw$.

[1] *Mécanique céleste*, première partie, Livre I (An vii)

For from (4) we have in this case

$$\phi(w) = \frac{w}{\sqrt{\left(1-\frac{w^2}{c^2}\right)}}, \qquad \phi'(w) = \left(1-\frac{w^2}{c^2}\right)^{-\frac{3}{2}},$$

and the kinetic energy $= \int \frac{mw\,dw}{\left(1-\frac{w^2}{c^2}\right)^{\frac{3}{2}}} = \frac{mc^2}{\sqrt{\left(1-\frac{w^2}{c^2}\right)}} + \text{Constant}$

in agreement with (2).

Equations (2) and (4) fulfil the prediction made by Poincaré in his St Louis lecture of 24 September 1904, that there would be 'a new mechanics, where, the inertia increasing with the velocity, the velocity of light would become a limit that could not be exceeded.'

The arguments by which Planck derived his expressions for the kinetic energy and momentum of a material particle in relativity theory were felt to be perhaps not completely cogent. However, three years afterwards, Gilbert N. Lewis (1875–1946) and Richard C. Tolman (1881–1948)[1] gave a proof of a very different character.

Consider two systems of reference (A) and (B), in relative motion with velocity w parallel to the axes of x and x'. Let a ball P have components of velocity $(0, -u, 0)$ in (A), and let an exactly similar ball Q have components of velocity $(0, u, 0)$ in (B). Let the balls be smooth and perfectly elastic. The experiment is so planned that the balls collide and rebound. From the relativist formulae

$$v_x = \frac{v'_x + w}{1+\frac{v'_x w}{c^2}}, \qquad v_y = \frac{v'_y\,(1-w^2/c^2)^{\frac{1}{2}}}{1+\frac{v'_x w}{c^2}}, \qquad v_z = \frac{v'_z\,(1-w^2/c^2)^{\frac{1}{2}}}{1+\frac{v'_x w}{c^2}},$$

we see that the velocity of Q as estimated by (A) before the collision is

$$\left(w,\ u\sqrt{\left(1-\frac{w^2}{c^2}\right)},\ 0\right).$$

The collision is perfectly symmetrical. But as estimated by (A), the y-component of Q's velocity changes from $u\sqrt{(1-w^2/c^2)}$ to $-u\sqrt{(1-w^2/c^2)}$, and the y-component of P's velocity changes from $-u$ to u.

We assume that there exists a vector quantity called the *momentum* depending on the mass and velocity, which is such that the momentum gained by one of the spheres in a collision is equal to the momentum lost by the sphere which collides with it. We assume further that this momentum approximates to the ordinary Newtonian

[1] *Phil. Mag.* xviii (1909), p. 517

momentum when the velocity is very small compared with that of light. So the components of momentum may be written

$$f(v)v_x, \qquad f(v)v_y, \qquad f(v)v_z,$$

where $v=(v^2_x+v^2_y+v^2_z)^{\frac{1}{2}}$, and the function $f(v)$ reduces to the mass m when $v\to 0$. From the law of conservation of momentum, (A) assumes that the ball P experiences the same change of momentum as the ball Q. Therefore

$$f(v_Q)u\left(1-\frac{w^2}{c^2}\right)^{\frac{1}{2}}=f(v_P)u$$

where v_Q and v_P are the total velocities of Q and P in (A)'s system. Divide by u.

$$f(v_Q)\left(1-\frac{w^2}{c^2}\right)^{\frac{1}{2}}=f(v_P).$$

Now make u tend to zero. Thus

$$f(w)\left(1-\frac{w^2}{c^2}\right)^{\frac{1}{2}}=f(0)=m$$

or

$$f(w)=\frac{m}{\sqrt{\left(1-\frac{w^2}{c^2}\right)}}$$

so the momentum of a particle whose mass is m, and which is moving with velocity (v_x, v_y, v_z) *is*

$$\left\{\frac{mv_x}{\sqrt{\left(1-\frac{v^2}{c^2}\right)}},\ \frac{mv_y}{\sqrt{\left(1-\frac{v^2}{c^2}\right)}},\ \frac{mv_z}{\sqrt{\left(1-\frac{v^2}{c^2}\right)}}\right\},$$

where $v^2=v_x^2+v_y^2+v_z^2$.

Next consider a collision between two elastic spheres, whose masses are m_1 and m_2 respectively, and which are moving along the axis of x with velocities (u_1, u_2) before the collision, and with velocities (u'_1, u'_2), after the collision. The condition of conservation of momentum gives the equation:

$$\frac{m_1u_1}{\sqrt{\left(1-\frac{u_1^2}{c^2}\right)}}+\frac{m_2u_2}{\sqrt{\left(1-\frac{u_2^2}{c^2}\right)}}=\frac{m_1u_1'}{\sqrt{\left(1-\frac{u'^2_1}{c^2}\right)}}+\frac{m_2u_2'}{\sqrt{\left(1-\frac{u'^2_2}{c^2}\right)}}. \qquad (1)$$

Now consider another set of axes, which are moving relatively to the first set with velocity $c \tanh \alpha$ parallel to the axis of x. Let the velocities relative to this second set of axes be denoted by grave accents placed over the letters, so that for any one of the u's we have

$$\grave{u} = \frac{u \cosh \alpha - c \sinh \alpha}{\cosh \alpha - (u/c) \sinh \alpha},$$

$$\sqrt{\left(1 - \frac{\grave{u}^2}{c^2}\right)} = \frac{1}{\cosh \alpha - (u/c) \sinh \alpha} \sqrt{\left(1 - \frac{u^2}{c^2}\right)}. \qquad (2)$$

Substituting from (2) in the equation

$$\frac{m_1 \grave{u}_1}{\sqrt{\left(1 - \frac{\grave{u}_1^2}{c^2}\right)}} + \frac{m_2 \grave{u}_2}{\sqrt{\left(1 - \frac{\grave{u}_2^2}{c^2}\right)}} = \frac{m_1 \grave{u}'_1}{\sqrt{\left(1 - \frac{\grave{u}'^2_1}{c^2}\right)}} + \frac{m_2 \grave{u}'_2}{\sqrt{\left(1 - \frac{\grave{u}'^2_2}{c^2}\right)}},$$

we obtain

$$\frac{m_1(u_1 \cosh \alpha - c \sinh \alpha)}{\sqrt{\left(1 - \frac{u_1^2}{c^2}\right)}} + \frac{m_2(u_2 \cosh \alpha - c \sinh \alpha)}{\sqrt{\left(1 - \frac{u_2^2}{c^2}\right)}}$$

$$= \frac{m_1(u'_1 \cosh \alpha - c \sinh \alpha)}{\sqrt{\left(1 - \frac{u'^2_1}{c^2}\right)}} + \frac{m_2(u'_2 \cosh \alpha - c \sinh \alpha)}{\sqrt{\left(1 - \frac{u'^2_2}{c^2}\right)}}.$$

Subtracting this equation from equation (1) multiplied by cosh α, and dividing the resulting equation by c sinh α, we have

$$\frac{m_1}{\sqrt{\left(1 - \frac{u_1^2}{c^2}\right)}} + \frac{m_2}{\sqrt{\left(1 - \frac{u_2^2}{c^2}\right)}} = \frac{m_1}{\sqrt{\left(1 - \frac{u'^2_1}{c^2}\right)}} + \frac{m_2}{\sqrt{\left(1 - \frac{u'^2_2}{c^2}\right)}}. \qquad (3)$$

This equation shows that if the quantity $m\,(1 - u^2/c^2)^{-\frac{1}{2}}$ be calculated for each of the colliding spheres, then the sum of these quantities for the two spheres is unaltered by the impact. We have therefore obtained a new invariant property. Let us see what corresponds to this in Newtonian dynamics. Supposing that (u_1/c) and (u_2/c) are small, and expanding by the binomial theorem, we have

$$m_1\left(1 + \tfrac{1}{2}\frac{u_1^2}{c^2} + \tfrac{3}{8}\frac{u_1^4}{c^4} + \ldots\right) + m_2\left(1 + \tfrac{1}{2}\frac{u_2^2}{c^2} + \tfrac{3}{8}\frac{u_2^4}{c^4} + \ldots\right)$$

$$= m_1\left(1 + \tfrac{1}{2}\frac{u'^2_1}{c^2} + \tfrac{3}{8}\frac{u'^4_1}{c^4} + \ldots\right) + m_2\left(1 + \tfrac{1}{2}\frac{u'^2_2}{c^2} + \tfrac{3}{8}\frac{u'^4_2}{c^4} + \ldots\right)$$

or

$$\tfrac{1}{2}m_1u_1^2+\tfrac{3}{8}m_1\frac{u_1^4}{c^2}+\ldots+\tfrac{1}{2}m_2u_2^2+\tfrac{3}{8}m_2\frac{u_2^4}{c^2}+\ldots$$

$$=\tfrac{1}{2}m_1u'^2_1+\tfrac{3}{8}m_1\frac{u'^4_1}{c^2}+\ldots+\tfrac{1}{2}m_2u'^2_2+\tfrac{3}{8}m_2\frac{u'^4_2}{c^2}+\ldots$$

When $c\to\infty$, this equation becomes the ordinary equation of conservation of kinetic energy in the collision. We therefore describe (3) as the *equation of conservation of energy* in the relativist theory of the impact, and we call

$$\frac{mc^2}{\sqrt{\left(1-\frac{v^2}{c^2}\right)}}$$

(save for an additive constant) the *kinetic energy* of a particle, whose mass at rest is m, which is moving with velocity v. The c^2 is inserted in the numerator in order to make the expansion in ascending powers of (u/c) begin with the terms [Constant $+\tfrac{1}{2}mu^2$] and thus be assimilated to the Newtonian kinetic energy.

Thus Planck's expressions for the momentum and kinetic energy of a material particle were verified. The quantity m is called the *proper mass.*

We have now to trace the gradual emergence of one of the greatest discoveries of the twentieth century, namely, the connection of mass with energy.

As we have seen,[1] J. J. Thomson in 1881 arrived at the result that a charged spherical conductor moving in a straight line behaves as if it had an additional mass of amount $(4/3c^2)$ times the energy of its electrostatic field.[2] In 1900 Poincaré,[3] referring to the fact that in free aether the electromagnetic momentum is $(1/c^2)$ times the Poynting flux of energy, suggested that electromagnetic energy might possess mass density equal to $(1/c^2)$ times the energy density: that is to say, $E=mc^2$ where E is energy and m is mass: and he remarked that if this were so, then a Hertz oscillator, which sends out electromagnetic energy preponderantly in one direction, should recoil as a gun does when it is fired. In 1904 F. Hasenöhrl[4] (1874–1915) considered a hollow box with perfectly reflecting walls filled with radiation, and found that when it is in motion there is an

[1] Vol. I, pp. 306-310

[2] It was shown long afterwards by E. Fermi, *Lincei Rend.* xxxi_1 (1922), pp. 184, 306, that the transport of the stress system set up in the material of the sphere should be taken into account, and that when this is done, Thomson's result becomes

$$\text{Additional mass} = \frac{1}{c^2}\times \text{Energy of field.}$$

The same result was obtained in a different way by W. Wilson, *Proc. Phys. Soc.* xlviii (1936), p. 736.

[3] *Archives Néerland.* v (1900), p. 252

[4] *Ann. d. Phys.* xv (1904), p. 344; *Wien Sitz.* cxiii, 2a (1904), p. 1039

apparent addition to its mass, of amount $(8/3c^2)$ times the energy possessed by the radiation when the box is at rest: in the following year [1] he corrected this to $(4/3c^2)$ times the energy possessed by the radiation when the box is at rest [2]; that is, he agreed with J. J. Thomson's $E=\frac{3}{4}mc^2$ rather than with Poincaré's $E=mc^2$. In 1905 A. Einstein [3] asserted that when a body is losing energy in the form of radiation its mass is diminished approximately (i.e. neglecting quantities of the fourth order) by $(1/c^2)$ times the amount of energy lost. He remarked that it is not essential that the energy lost by the body should consist of radiation, and suggested the general conclusion, in agreement with Poincaré, that the mass of a body is a measure of its energy content: if the energy changes by E ergs, the mass changes in the same sense by (E/c^2) grams. In the following year he claimed [4] that this law is the necessary and sufficient condition that the law of conservation of motion of the centre of gravity should be valid for systems in which electromagnetic as well as mechanical processes are taking place.

In 1908 G. N. Lewis [5] proved, by means of the theory of radiation-pressure, that a body which absorbs radiant energy increases in mass according to the equation

$$dE = c^2 dm$$

and affirmed that the mass of a body is a direct measure of its total energy, according to the equation [6]

$$E = mc^2.$$

As we have seen, Poincaré had suggested this equation but had given practically no proof, while Einstein, who had also suggested it, had given a proof (which, however, was put forward only as approximate) for a particular case: Lewis regarded it as an exact equation, but his proof also was not of a general character. Lewis, however, pointed out that if this principle is accepted, then in Planck's equation of 1906

$$\left(\begin{array}{l}\text{Kinetic energy of a particle whose}\\ \text{mass when at rest is } m\end{array}\right) = \frac{mc^2}{\sqrt{\left(1-\frac{w^2}{c^2}\right)}} - mc^2$$

[1] *Ann. d. Phys.* xvi (1905), p. 589

[2] The moving hollow box filled with radiation was discussed further by K. von Mosengeil (a pupil of Planck), *Ann. d. Phys.* xxii (1907), p. 867, and M. Planck, *Berlin Sitz.* (1907), p. 542, whose formulae essentially involve the general law $E=mc^2$.

[3] *Ann. d. Phys.* xviii (1905), p. 639; his reasoning has, however, been criticised; cf. H. E. Ives, *J. Opt. Soc. Amer.* xlii (1952), p. 540

[4] *Ann. d. Phys.* xx (1906), p. 627; cf. a further paper in *Ann. d. Phys.* xxiii (1907), p. 371

[5] *Phil. Mag.* xvi (1908), p. 705; cf. however the above note on Planck's paper of 1907

[6] A little earlier D. F. Comstock, *Phil. Mag.* xv (1908), p. 1, had obtained $E=\frac{3}{4}mc^2$ in accordance with the formulae of J. J. Thomson and Hasenöhrl, and had remarked that 'assuming the loss of mass accompanying the dissipation of energy, the sun's mass must have decreased steadily through millions of years.'

the last term, mc^2, must be interpreted to mean the energy of the particle when at rest, whereas the difference

$$\frac{mc^2}{\sqrt{\left(1-\frac{w^2}{c^2}\right)}} - mc^2$$

represents the additional energy which it possesses when in motion; and therefore the total energy of the particle when in motion must be simply [1]

$$\frac{mc^2}{\sqrt{\left(1-\frac{w^2}{c^2}\right)}}$$

For confirmation of this, Lewis referred to experiments by W. Kaufmann [2] and A. H. Bucherer,[3] who studied the magnetic and electric deviations of the β-rays for radio-active substances. The original experiments of Kaufmann [4] showed only that for great velocities the 'mass' of the electron increases with its velocity in general qualitative agreement with the formula $m/\sqrt{(1-w^2/c^2)}$: but Bucherer showed that the formula is accurate to a high degree of precision for values of (w/c) ranging from 0·38 to 0·69.

The mass of a system can therefore be calculated from its total energy by the equation

$$m = \frac{1}{c^2}\,\mathrm{E}:$$

and the researches that have been described show that in calculating E, we must include energy resident in the aether. In 1911 Lorentz [5] showed that *every* kind of energy must be included—masses, stretched

[1] Lorentz in 1904 (*Amst. Proc.* vi (1904), p. 809] had given the formula

$$m = \frac{m_0}{\sqrt{(1-(w/c)^2)}}$$

for the mass of an electron whose mass when at rest is m_0, and which is moving with velocity w, on the assumption that electrons in their motion experience the FitzGerald contraction.

[2] *Gött. Nach.* (1901), p. 143 ; (1902), p. 291 ; (1903), p. 90 ; *Phys. ZS.* iv (1902), p. 54 ; *Berlin Sitz.* xlv (1905), p. 949 ; *Ann. d. Phys.* xix (1906), p. 487 ; cf. also Planck, *Verh. d. Deutsch, Phys. Ges.* ix (1907), p. 301 ; Kaufmann, ibid. p. 607 ; Stark, ibid. x (1908), p. 14

[3] *Berl. Phys. Ges.* vi (1908), p. 688 ; *Ann. d. Phys.* xxviii (1909), p. 513 ; *Phys. ZS.* ix (1908), p. 755 ; cf. also C. Schaefer and G. Neumann, *Phys. ZS.* xiv (1913), p. 1117 ; G. Neumann, *Ann. d. Phys.* xlv (1914), p. 529 ; C. Guye and Ch. Lavanchy, *Arch. des Sc.* (Geneva) xlii (1916), p. 286

[4] It may be mentioned that in pre-relativity days the interpretation placed on Kaufmann's experiments was that by means of them it would be possible to find for the electron the proportion of proper mass (which was independent of velocity) to electromagnetic mass (which increased with velocity).

[5] *Amst. Versl.* xx (1911), p. 87

strings, light rays, etc. For example, if a system consisting of two electrically charged spheres, of charges e_1 and e_2, at distance a apart, is considered, then when we calculate the value of E for the system, we do not obtain simply the sum of the values of E for the two spheres separately (as calculated when they are infinitely remote from each other), but we must include also a term representing the electrostatic *mutual* potential energy of the two charges, namely, (e_1e_2/a) : and therefore the mass of the system must include a term [1] (e_1e_2/c^2a).

Similarly, the mass of a system of gravitating bodies is not the sum of their masses taken separately, but includes a term representing $(1/c^2)$ times their mutual potential energy.[2] Thus, if two Newtonian gravitating particles m_1 and m_2 are at rest at a distance a apart, their mass is

$$m_1 + m_2 - \frac{\gamma m_1 m_2}{c^2 a}$$

where γ is the Newtonian constant of gravitation.

The equivalence of mass and energy was expressed by Planck in 1908 [3] in the form of a unified definition of momentum. The flux of energy, he said, is a vector, which when divided by c^2 is the density of momentum. This had long been known in the case of electromagnetic energy, by the relation between the Poynting vector and the momentum density resident in the aether. But Planck now asserted that it was universally true, e.g. in the cases of radiation, or of conduction or convection of heat. In the case of a single particle of proper-mass m and velocity v, the energy is $mc^2/\sqrt{(1-v^2/c^2)}$, the streaming of energy is $mc^2\mathbf{v}/\sqrt{(1-v^2/c^2)}$, and this divided by c^2 is $m\mathbf{v}/\sqrt{(1-v^2/c^2)}$, which is precisely the momentum of the particle. The unified definition of momentum is a more general expression of the equivalence of mass and energy than the equation $E = mc^2$, for the concept of mass becomes more difficult to define when, e.g. momentum and velocity are no longer parallel to each other.

Planck's new conception of momentum was soon found to be capable of explaining some paradoxical consequences which could

[1] A value not agreeing with this was found by L. Silberstein in 1911 [*Phys. ZS.* xii (1911), p. 87], but an error in his method was pointed out by E. Fermi [*Rend. Lincei*, xxxi$_1$ (1922), pp. 184, 306], whose work led to the correct value.

[2] On this problem cf. A. S. Eddington and G. L. Clark, *Proc. R.S.*(A), clxvi (1938), p. 465 ; Eddington [*Proc. R.S.*(A), clxxiv (1940), p. 16] proposed to define the mass of a system to be that of a point-particle which would produce the same gravitational field as the system at very great distances. This and other definitions were discussed by G. L. Clark, *Proc. R.S. E.* lxii (1949), p. 412. It was shown by Josephine M. Gilloch and W. H. McCrea, *Proc. Camb. Ph. Soc.* xlvii (1951), p. 190, that in the case of a cylinder rotating freely on its axis, the gravitational mass is (to a first approximation) the sum of the proper-mass and $(1/c^2)$ times the kinetic energy ; as was to be expected according to the general principle of the equivalence of mass and energy.

[3] *Verh. d. Deutsch. Phys. Ges.* x (1908), p. 728 ; *Phys. ZS.* ix (1908), p. 828. This statement had been to some extent anticipated (in connection with the moving box containing radiation) by Planck, *Berlin Sitz.* (1907), p. 542, and by F. Hasenöhrl, *Wien Sitz.* cxvi 2a (1907), p. 1391.

apparently be deduced from the theory of relativity. One of these, due to Lewis and Tolman,[1] may be described as follows. Consider a rigid bent lever *abc* at rest, pivoted at *b*, whose arms *ba* and *bc* are equal and perpendicular, and suppose that forces F_x and F_y, each equal to F_0, are applied at *a* and *c* in directions parallel to *bc* and *ba* respectively. The system is thus in equilibrium.

Now let the whole system be referred to axes with respect to which it is moving with velocity *w* in the direction *bc*. Obviously it will still be in equilibrium. But according to the theory of relativity, with reference to the new axes the arm *bc* should experience the FitzGerald contraction, and so should be shortened in the ratio $\sqrt{(1-w^2/c^2)}$ to 1, while *ab* has the same length as at rest. Moreover, if force is defined as the rate of communication of momentum with respect to the time used in the inertial system concerned, we can show that the values of the forces referred to the new axes are

$$F_x = F_0, \qquad F_y = F_0\sqrt{\left(1-\frac{w^2}{c^2}\right)}.$$

Thus the forces produce a moment

$$F_0 \,.\, ba - F_0\sqrt{\left(1-\frac{w^2}{c^2}\right)}\sqrt{\left(1-\frac{w^2}{c^2}\right)} \,.\, bc$$

or

$$\frac{1}{c^2} F_0 w^2 \,.\, ba$$

tending to turn the system round *b*; so apparently it would not be in equilibrium.

The paradox is resolved by the following explanation, which is due to Sommerfeld and Laue.[2] At the point *a* the force F_x furnishes the work at the rate wF_0. An energy current of this strength enters the lever at *a*, travels to *b* and then passes into the axis of the lever, since the axis does work at the rate $-wF_0$ on the lever. Corresponding to this flux of energy there is, by Planck's principle, a momentum parallel to *ab*, of amount $(1/c^2)$ times the volume integral of the energy flux, or $(1/c^2) \,.\, ab \,.\, wF_0$. Due to the existence of this momentum there is an angular momentum about a fixed origin *O*, lying in the prolongation of *ab*, of amount $(1/c^2) \,.\, ab \,.\, Ob \,.\, wF_0$, and its rate of increase with respect to the time is

$$\frac{1}{c^2} \,.\, ab \,.\, \frac{d(Ob)}{dt} \,.\, wF_0 \quad \text{or} \quad (1/c^2) \,.\, F_0 w^2 \,.\, ab.$$

Thus we see that the couple $(1/c^2) \,.\, F_0 w^2 \,.\, ba$, produced by the two forces F_x and F_y, is needed in order to account for the rate of increase

[1] *Phil. Mag.* xviii (1909), p. 510. Relativity statics is treated fully by P. S. Epstein, *Ann. d. Phys.* xxxvi (1911), p. 779.

[2] Laue, *Verh. Deutsch, Phys. Gesells.* (1911), p. 513

$(1/c^2) \, . \, F_0 w^2 \, . \, ab$ of the angular momentum of the lever, and the difficulty is satisfactorily explained.

It may be remarked that if the lever is contained in a case, which supports the axis *b* of the lever, and also (e.g. by elastic strings attached to points of the case) provides the forces F_x and F_y which act at *a* and *c*, then the energy current after leaving the lever at *b* enters the case there, and after travelling in the case re-enters the lever by the elastic string which is attached to *a*. The energy current is therefore closed, so the system consisting of case and lever together has not a variable angular momentum. The case and lever in fact exert equal and opposite couples on each other.

This may be regarded as a model of the Trouton-Noble experiment,[1] the electric field being compared to the lever and the material condenser to the case. Neither the elctromagnetic momentum of the field nor the mechanical momentum of the condenser is parallel to the velocity, and both therefore need couples in order to preserve their orientation in translatory motion, but these couples are equal and opposite, and the system condenser plus field requires no couple.

Not long after the publication of Planck's paper of 1906 writers on the theory of relativity began to take advantage of some developments in pure mathematics, of which an account must now be given.

It was Felix Klein (1849–1925) in his famous *Erlanger Programm* of 1872 [2] who first clearly indicated the essential nature of a *vector*. Let (p, q, r) be the components of a vector with respect to the rectangular axes $O\,x\,y\,z$. Then $px+qy+rz$ is the product of the lengths of the vectors (p, q, r) and (x, y, z) into the cosine of the angle between them, and is therefore invariant if the axes of reference are changed by a rotation about the origin to any other set of rectangular axes. Klein regarded all geometry as the invariant theory of some definite group, and following him, we can take the property just mentioned as the *definition* of a vector : that is, a set of three numbers (p, q, r) will be called a *vector* if $px+qy+rz$ is invariant under the group of rotations of orthogonal axes. This definition suffices to furnish the laws according to which (p, q, r) are transformed when the axes of reference are changed. Since $\{(x \, . \, x) + (y \, . \, y) + (z \, . \, z)\}$ or $(x^2+y^2+z^2)$ is invariant under a rotation of the axes, we see that (x, y, z) is a particular vector. And since all vectors are transformed in the same way, we may say that (p, q, r) is a vector if its components (p, q, r) are transformed like (x, y, z).

Vectors are not the only physical quantities that are related to direction: another class is represented by *elastic stresses*. If we denote by (X_x, Y_x, Z_x) the components of traction across the *yz*-plane at a given point P, by (X_y, Y_y, Z_y) the components of traction across the *zx*-plane at P, and by (X_z, Y_z, Z_z) the components of traction

[1] cf. p. 29

[2] *Programm zum Eintritt in die philosophische Fakultät d. Univ. zu Erlangen*, Erlangen, A. Deichert, 1872. Reprinted in 1893 in *Math. Ann.* xliii, and in Klein's *Ges. Math. Abhandl.* i, p. 460.

across the xy-plane at P, then, as is known, we have $Z_y = Y_z$, $X_z = Z_x$, $Y_x = X_y$, so we can write

$$X_x = a, \quad Y_y = b, \quad Z_z = c, \quad Z_y = Y_y = f, \quad X_z = Z_x = g, \quad Y_x = X_y = h,$$

and the stress can be represented by the six numbers (a, h, c, f, g, h). Now let the axes of reference be changed by any rotation about the origin. Then, as is known, if the components of stress at P with respect to the new axes $Ox'y'z'$ are denoted by (a', b', c', f', g', h'), the expression

$$ax^2 + by^2 + cz^2 + 2fyz + 2gzx + 2hxy$$

is transformed into the expression

$$a'x'^2 + b'y'^2 + a'z'^2 + 2f'y'z' + 2g'z'x' + 2h'x'y'.$$

Any set of six quantities (a, b, c, f, g, h) which, when the axes are changed by a rotation about the origin, changes in this way, that is, in the same way as the coefficients of a quadric surface, is said to constitute a *symmetrical tensor*[1] *of rank* 2. The analogy with the definition of a vector is obvious, and a vector may be called a *tensor of rank* 1. A quantity which is invariant under all rotations of the axes of co-ordinates is called a *scalar* or *tensor of rank zero*.

Since

$$x^2 \,.\, x^2 + y^2 \,.\, y^2 + z^2 \,.\, z^2 + 2yz \,.\, yz + 2zx \,.\, zx + 2xy \,.\, xv = (x^2 + y^2 + z^2)^2$$

is an invariant for rotations of the system of co-ordinate axes, it follows that

$$(x^2, \quad y^2, \quad z^2, \quad yz, \quad zx, \quad xy)$$

is a particular symmetric tensor of rank 2, and since all symmetric tensors of rank 2 are transformed in the same way, we see that *a set of* 6 *quantities* (a, b, c, f, g, h) *constitutes a symmetric tensor of rank* 2, *if* (a, b, c, f, g, h) *are transformed in the same way as* $(x^2, y^2, z^2, yz, zx, xy)$. It is easily shown, for example, that if A, B, C, F, G, H denote the moments and products of inertia of a system of masses with respect to the co-ordinate axes, then (A, B, C, − F, − G, − H) is a symmetric tensor of rank 2.

The definition just given can be generalised, so as to furnish a definition of a tensor of rank 2 which is not necessarily symmetrical. Let (p_1, q_1, r_1) and (p_2, q_2, r_2) be two different vectors. Then a *set of nine numbers*

$$t_{11}, \quad t_{22}, \quad t_{33}, \quad t_{23}, \quad t_{32}, \quad t_{31}, \quad t_{13}, \quad t_{12}, \quad t_{21},$$

[1] Attention was drawn to the properties of sets of quantities obeying these laws of transformation by C. Niven, *Trans. R. S. E.* xxvii (1874), p. 473 ; cf. also W. Thomson (Kelvin), *Phil. Trans.* cxlvi (1856), p. 481 and W. J. M. Rankine, ibid. p. 261. The name *tensor* (with this meaning) is due to J. Willard Gibbs, *Vector Analysis*, New Haven (1881–4), p. 57.

will be called a tensor of rank 2, *if they transform in the same way as*

$$p_1p_2,\ q_1q_2,\ r_1r_2,\ q_1r_2,\ r_1q_2,\ r_1p_2,\ p_1r_2,\ p_1q_2,\ q_1p_2.$$

So far we have considered only tensors which have invariant properties with respect to the rotations of a system of orthogonal co-ordinate axes in three-dimensional space. This theory was generalised into a tensor-calculus applicable to transformations in curved space of any number of dimensions by Gregorio Ricci-Curbastro (1853–1925) of Padua, from 1887 onwards: it first became widely known when a celebrated memoir describing it was published in 1900 by Ricci and Levi-Civita.[1]

Let $x_1, x_2, \ldots x_n$ be any 'generalised co-ordinates' specifying the position of a point in space of n dimensions. Let n new variables $\bar{x}_1, \bar{x}_2, \ldots \bar{x}_n$ be introduced by arbitrary equations

$$\bar{x}_r = f_r(x_1, x_2, \ldots x_n) \qquad (r=1, 2, \ldots n). \qquad (1)$$

Then the differentials of the co-ordinates are transformed according to the equations

$$d\bar{x}_r = \sum_{k=1}^{n} \frac{\partial \bar{x}_r}{\partial x_k} dx_k \qquad (r=1, 2, \ldots n).$$

At a point P of the n-dimensional space we can consider various types of quantities analogous to the scalars, vectors and tensors that we have already considered.

Firstly, there may be a function of position whose value is unchanged when we perform the transformation (1). Such a function is called a *scalar*, or *tensor of rank zero*.

Secondly, we consider a set of n numbers $(V^1, V^2, \ldots V^n)$, which are defined with respect to all co-ordinate systems and which, when we perform the transformation (1), are transformed in the same way as the dx_r, so that

$$\bar{V}^r = \sum_{k=1}^{n} \frac{\partial \bar{x}_r}{\partial x_k} V^k \qquad (r=1, 2, \ldots n)$$

whence

$$V^r = \sum_{k=1}^{n} \frac{\partial x_r}{\partial \bar{x}_k} \bar{V}^k \qquad (r=1, 2, \ldots n).$$

Such a set of n numbers is called a *contravariant tensor of rank* 1, or *contravariant vector*, and the numbers are called its *components*.

Next, consider sets of n numbers $(X_1, X_2, \ldots X_n)$, which are such that if $(V^1, V^2, \ldots V^n)$ is any contravariant tensor of rank 1, the sum $X_1V^1 + X_2V^2 + \ldots + X_nV^n$ is a scalar. Such

[1] *Math. Ann.* liv (1900), p. 125; cf. J. A. Schouten, *Jahresb. d. Deutsch. Math.-Verein.* xxxii (1923), p. 91

a set of n numbers is called a *covariant tensor of rank* 1 or *covariant vector.*

Since

$$\sum_k X_k V^k = \sum_r \overline{X}_r \overline{V}^r = \sum_r \overline{X}_r \sum_k \frac{\partial \overline{x}_r}{\partial x_k} V^k,$$

we have

$$X_k = \sum_r \frac{\partial \overline{x}_r}{\partial x_k} \overline{X}_r, \quad \text{whence } \overline{X}_k = \sum_r \frac{\partial x_r}{\partial \overline{x}_k} X_r.$$

The covariant or contravariant character is indicated by placing the index in the lower or upper position respectively. In Euclidean space, for rotations of rectangular axes, there is no distinction between contravariant and covariant tensors.

If, at the point P of the n-dimensional space, we have n^2 numbers $(V^{11}, V^{12}, \ldots V^{nn})$ which, when we perform the transformation of co-ordinates, are transformed like $(P^1Q^1, P^1Q^2, \ldots P^nQ^n)$, where $(P^1, \ldots P^n)$ and $(Q^1, \ldots Q^n)$ are two different contravariant tensors of rank 1, then $(V^{11}, V^{12}, \ldots V^{nn})$ are said to be the components of a *contravariant tensor of rank* 2. Similarly n^2 numbers $(X_{11}, X_{12}, \ldots X_{nn})$ which transform like $(X_1Y_1, X_1Y_2, \ldots X_nY_n)$ where $(X_1 \ldots X_n)$ and $(Y_1, \ldots Y_n)$ are two different covariant tensors of rank 1, are said to be the components of a *covariant tensor of rank* 2; while n^2 numbers $(W^1{}_1, W^1{}_2, W^2{}_1, \ldots W^n)$ which transform like $(P^1X_1, P^1X_2, P^2X_1, \ldots P^nX_n)$ where $(P^1, P^2, \ldots P^n)$ is a contravariant tensor of rank 1 and $(X_1, X_2, \ldots X_n)$ is a covariant tensor of rank 1, is called a *mixed tensor of rank* 2. Tensors of rank greater than 2 are defined in a similar way. A tensor whose typical component is, say, $X^{\,p}_{rs}$, is often denoted by $(X^{\,p}_{rs})$.

Consider a tensor such that any two of its components, which may be obtained from each other by a simple interchange of two indices, are equal to each other; thus, $V^{pq} = V^{qp}$. If this property holds for any one system of co-ordinates, it will still hold after any change of the co-ordinate system, as is evident from the equations of transformation. Such a tensor is said to be *symmetric.* If a tensor is such that two components which may be derived from each other by a simple interchange of two indices are equal in magnitude but opposite in sign, thus $V^{pq} = -V^{qp}$, the tensor is said to be *skew.* This property also holds in all systems of co-ordinates, provided it holds in any one system.

Two tensors of the same kind (contravariant, covariant or mixed) and of the same rank, are said to be *equal* if their corresponding components are equal in all co-ordinate systems. This is the case if the corresponding components are equal in any one co-ordinate system.

Consider the transformation of tensors when the co-ordinates are subjected to the particular Lorentz transformation (writing $ct=x_0, \quad x=x_1, \quad y=x_2, \quad z=x_3$)

$$dx_0=d\bar{x}_0 \cosh a+d\bar{x}_1 \sinh a, \quad dx_1=d\bar{x}_0 \sinh a+d\bar{x}_1 \cosh a,$$
$$d\bar{x}_2=dx_2, \quad dx_3=d\bar{x}_3.$$

It is found at once that :

for any contravariant vector :

$$J^0=\bar{J}^0 \cosh a+\bar{J}^1 \sinh a, \; J^1=\bar{J}^0 \sinh a+\bar{J}^1 \cosh a, \; J^2=\bar{J}^2, \; J^3=\bar{J}^3.$$

for any covariant vector :

$$J_0=\bar{J}_0 \cosh a-\bar{J}_1 \sinh a, \; J_1=-\bar{J}_0 \sinh a+\bar{J}_1 \cosh a, \; J_2=\bar{J}_2, \; J_3=\bar{J}_3.$$

for any covariant symmetric tensor of rank 2 :

$$\bar{X}_{00}=X_{00} \cosh^2 a+2X_{01} \cosh a \sinh a+X_{11} \sinh^2 a$$
$$\bar{X}_{11}=X_{00} \sinh^2 a+2X_{01} \sinh a \cosh a+X_{11} \cosh^2 a$$
$$\bar{X}_{22}=X_{22}, \quad \bar{X}_{33}=X_{33}, \quad \bar{X}_{32}=\bar{X}_{23}=X_{23}$$
$$\bar{X}_{10}=\bar{X}_{01}=X_{00} \cosh a \sinh a+X_{01} (\cosh^2 a+\sinh^2 a)+X_{11} \sinh a \cosh a$$
$$\bar{X}_{20}=\bar{X}_{02}=X_{02} \cosh a+X_{12} \sinh a$$
$$\bar{X}_{30}=\bar{X}_{03}=X_{03} \cosh a+X_{13} \sinh a$$
$$\bar{X}_{12}=\bar{X}_{21}=X_{02} \sinh a+X_{12} \cosh a$$
$$\bar{X}_{13}=\bar{X}_{31}=X_{03} \sinh a+X_{13} \cosh a$$

for any covariant skew tensor of rank 2 :

$$\bar{X}_{01}=X_{01}, \quad \bar{X}_{02}=X_{02} \cosh a+X_{12} \sinh a, \quad \bar{X}_{03}=X_{03} \cosh a+X_{13} \sinh a$$
$$\bar{X}_{23}=X_{23}, \quad \bar{X}_{31}=X_{31} \cosh a+X_{30} \sinh a, \quad \bar{X}_{12}=X_{12} \cosh a+X_{02} \sinh a$$

It is evident from these last equations that *a six-vector,*[1] such as is constituted by the electric and magnetic intensities *in vacuo, is a skew tensor of rank* 2. We can write

$$X_{10}=d_x, \quad X_{20}=d_y, \quad X_{30}=d_z, \quad X_{23}=h_x, \quad X_{31}=h_y, \quad X_{12}=h_z$$

From the definition of a tensor, it is evident that if two tensors of the same type are taken, say (X^p_{rs}), and (Y^p_{rs}), then the quantities formed by adding corresponding components of these tensors

$$Z^p_{rs}=X^p_{rs}+Y^p_{rs}$$

are the components of a tensor of the same type, which is called the *sum* of the tensors (X^p_{rs}) and (Y^p_{rs}).

[1] cf. p. 35

Moreover it is evident from the definitions that if two tensors, say of rank λ and rank μ, are given in n-dimensional space, and if we multiply each of the n^λ components of one by each of the n^μ components of the other, then the $n^{\lambda+\mu}$ products so formed are the components of a new tensor of rank $(\lambda+\mu)$, thus :

$$X_{ij}\ Y^p_{rs} = U^p_{ijrs}.$$

The tensor $\left(U^p_{ijrs}\right)$ is called the *outer product* of the tensors (X_{ij}) and $\left(Y^p_{rs}\right)$. It may properly be called a product, since the distributive law

$$X(Y+Z) = XY + XZ$$

holds. We can form in this way the outer product of any number of tensors.

An arbitrary tensor cannot in general be expressed as an outer product of tensors of rank 1, since there would not be enough quantities at our disposal to satisfy all the conditions. Thus, if a tensor $\left(X^p_{rs}\right)$ is given, we cannot in general find tensors of rank 1 (Y_r), (Z_s), and (V^p), such that

$$X^p_{rs} = Y_r\, Z_s\, V^p \qquad (p,\ r,\ s = 1,\ 2\ .\ .\ .\ n):$$

but the sum of any number of outer products of this type will be a tensor of the type $\left(X^p_{rs}\right)$; and by taking the number of such products sufficiently great, we shall have enough quantities at our disposal to represent any tensor $\left(X^p_{rs}\right)$ in the form

$$X^p_{rs} = Y_r\, Z_s\, V^p + H_r\, K_s\, L^p + E_r\, F_s\, G^p + \ .\ .\ .$$

Next consider a tensor which has both contravariant and covariant indices, e.g. $\left(X^{lk}_{pqr}\right)$. Make one of the upper or contravariant indices identical with one of the lower or covariant indices, and sum with respect to this index, thus :

$$\sum_{p=1}^{n}\ X^{pk}_{pqr}.$$

Then we can show that *the numbers thus obtained*, when $k,\ q,\ r = 1,\ 2 \ldots n$, *are the components of a new tensor* $\left(Y^k_{qr}\right)$, which is two units lower in rank than $\left(X^{lk}_{pqr}\right)$. To prove this, we remark that $\left(X^{lk}_{pqr}\right)$ can be expressed as a sum of outer products of tensors of rank 1, and the theorem will therefore evidently be true in general if it is true for the case when (X^{lk}_{pqr}) is a *single* outer product of tensors of rank 1, say

$$X^{lk}_{pqr} = Y_p\, Z_q\, T_r\, U^l\, V^k.$$

Then we have

$$\sum_{p=1}^{n} X^{pk}_{pqr} = \sum_{p=1}^{n} \left(Y_p U^p\right) Z_q T_r V^k :$$

and since $\sum_{p=1}^{n} T_p U^p$ is a scalar, these quantities are the components of a tensor of type (Y^{k}_{qr}) ; which establishes the theorem. This process is called *contraction.*

By forming the outer product of any number of tensors, and then contracting (once or oftener) the tensor thereby obtained, we obtain results such as

$$\sum_{abc, \ldots \alpha\beta\gamma \ldots} \left(X^{abc \ldots rst \ldots}_{\alpha\beta\gamma \ldots \rho\sigma\tau \ldots} \; Y^{\alpha\beta\gamma \ldots lmn \ldots}_{abc \ldots jhk \ldots} \right) = Z^{rst \ldots lmn \ldots}_{\rho\sigma\tau \ldots jhk \ldots}$$

This process is called *transvection.*

The spaces we consider will generally be supposed each to possess a *metric*, that is to say, there will be an equation expressing an element ds of arc-length at any point of the space in terms of the infinitesimal differences of the co-ordinates between the ends of the arc-element : thus in ordinary Euclidean three-dimensional space with rectangular co-ordinates (x, y, z), we have

$$(ds)^2 = (dx)^2 + (dy)^2 + (dz)^2,$$

and with spherical-polar co-ordinates (r, θ, ϕ), we have

$$(ds)^2 = (dr)^2 + r^2(d\theta)^2 + r^2 \sin^2\theta (d\phi)^2.$$

We assume generally that the square of the line-element ds is a homogeneous quadratic form in the differentials of the co-ordinates. These differentials will be written $(dx^1, dx^2, \ldots dx^n)$, the index being placed above since $(dx^1, \ldots dx^n)$ is a contravariant vector : thus

$$(ds)^2 = \sum_{p,q} g_{pq} dx^p dx^q.$$

Since $(ds)^2$ is a scalar, it is obvious from this equation that the numbers g_{pq} $(p, q = 1, 2, \ldots n)$ must be the components of a covariant symmetric tensor of rank 2, (g_{pq}) ; this is called the *covariant fundamental tensor.*

Let g denote the determinant $||g_{pq}||$ of the coefficients g_{pq}, and let g^{pq} denote $(1/g)$ times the co-factor of g_{pq} in g, so that $\sum_{p=1} g_{pr}g^{pq} = \delta_r^q$, where δ_r^q is equal to 1 or 0 according as q is equal to, or different from, r. Then

$$\sum_{p,q} g_{qs}g_{pr}g^{pq} = \sum_{q} g_{qs}\delta_r^{\ q} = g_{rs}.$$

Now if X_p and Y_p are two arbitrary covariant vectors, and if

$$X_p = \sum_r g_{pr}X^r$$

so that X^r is a contravariant vector, we have

$$\sum_{pq} g^{pq}X_pY_q = \sum_{pqrs} g_{pr}g_{qs}g^{pq}X^rY^s$$

$$= \sum_{rs} g_{rs}X^rY^s$$

which is a scalar: and therefore the g^{pq} are the components of a contravariant tensor of rank 2. It is called the *contravariant fundamental tensor.*

Moreover, if U^p is a contravariant vector, and X_q is any covariant vector, we have

$$\sum_{pq} \delta_q^{\ p}U^pX_q = \sum_p U^pX_p = \text{a scalar}$$

and therefore (δ_p^q) is a tensor of rank 2, covariant with respect to the index p and contravariant with respect to the index q. It is called the *mixed fundamental tensor.*

By aid of the fundamental tensor (g^{pq}) we can derive from any covariant tensor $(X_{p_1 p_2 \ldots p_m})$ a contravariant tensor of the same rank by writing

$$X^{q_1 q_2 \ldots q_m} = \sum_{p_1, p_2, \ldots p_m} g^{p_1 q_1} g^{p_2 q_2} \ldots g^{p_m q_m} X_{p_1 p_2 \ldots p_m}.$$

It is easily shown that this equation is equivalent to

$$X_{p_1 p_2 \ldots p_m} = \sum_{q_1, q_2, \ldots q_m} g_{p_1 q_1} g_{p_2 q_2} \ldots g_{p_m q_m} X^{q_1 q_2 \ldots q_m}$$

Thus to every contravariant tensor we can correlate a definite covariant tensor; and we may say that the distinction between covariant and contravariant tensors loses most of its importance when the fundamental tensor is given, i.e. in a *metrical* space, since it is not the tensors that are essentially different, but only their mode of expression, i.e. their components. For example, we regard (g^{pq}), (g_{pq}), and $(\delta_p{}^q)$, as essentially the *same* tensor.

If two vectors (X) and (Y) are such that when (X) is expressed in covariant form (X_p) and the other in contravariant form (Y^q), we have

$$\sum_p X_p Y^p = 0,$$

then the two vectors are said to be *orthogonal*.

After this rather long excursus on Ricci's tensor calculus, we can return to physics. A contribution of great importance to relativity theory was made in 1908 by Hermann Minkowski (1864–1909).[1] Its ostensible purpose, as indicated in its title, was to show that the differential equations of the electromagnetic field in moving ponderable bodies under the most general conditions (e.g. of magnetisation) can be derived from the differential equations for the same system of bodies at rest, by the principle of relativity : and to criticise some of the formulae that had been given by Lorentz. But these were not actually the most important elements in the paper ; the great advances made by Minkowski [2] were connected with his formulation of physics in terms of a four-dimensional manifold, the use of tensors in this manifold, and the discovery of some of the more important of these tensors.[3]

The phenomena studied in natural philosophy take place each at a definite location at a definite moment, the whole constituting a four-dimensional world of space and time. The theory of relativity had now made it clear that the separation of this four-dimensional world into a three-dimensional world of space and an independent one-dimensional world of time may be effected in an infinite number of ways, each of which is distinguished from the others only by characteristics that are merely arbitrary and accidental. In order to represent natural phenomena without introducing this contingent element, it is necessary to abandon the customary three-dimensional system of co-ordinates, and to operate in four dimensions.

If (t_1, x_1, y_1, z_1) and (t_2, x_2, y_2, z_2) are the time-and-space co-

[1] *Gött. Nach.* (1908), p. 53 ; cf. also *Math. Ann.* lxviii (1910), p. 472

[2] Minkowski had been to some extent anticipated by Poincaré, who had substantially introduced the metric

$$ds^2 = c^2dt^2 - dx^2 - dy^2 - dz^2 = -\sum_{r=1}^{4} dx_r{}^2$$

(where $x_1 = x$, $x_2 = y$, $x_3 = z$, $x_4 = ct\sqrt{-1}$) in *Rend. circ. Palermo*, xxi (1906), p. 129.

[3] The principle of treating the time co-ordinate on the same level as the other co-ordinates was introduced and developed simultaneously with Minkowski's paper by R. Hargreaves, [*Camb. Phil. Trans.* xxi (1908), p. 107] : his work suggests the use of space-time vectors just as Minkowski's does. For comments on this point, cf. H. Bateman, *Phys. Rev.* xii (1918), p. 459.

ordinates of two point-events referred to an inertial system, then, as we have seen, the expression

$$(t_2-t_1)^2-\frac{1}{c^2}\left\{(x_2-x_1)^2+(y_2-y_1)^2+(z_2-z_1)^2\right\}$$

is invariant under all Lorentz transformations, and therefore has the same value *whatever be the inertial framework of reference.* This quantity is therefore an invariant of the two point-events, which is the same for all observers : and we can make our four-dimensional space-time suited to describe nature when we impose a metric on it, which we do by taking the *interval* (the four-dimensional analogue of length) between the two events (t_1, x_1, y_1, z_1) and (t_2, x_2, y_2, z_2) to be [1]

$$\left[(t_2-t_1)^2-\frac{1}{c^2}\left\{(x_2-x_1)^2+(y_2-y_1)^2+(z_2-z_1)^2\right\}\right]^{\frac{1}{2}}.$$

Taking any point in the four-dimensional manifold as origin, the cone

$$x^2+y^2+z^2-c^2t^2=0$$

which is called the *null cone*, partitions space-time into two regions, of which one is defined by the inequality

$$c^2t^2 < x^2+y^2+z^2$$

and includes the hyperplane $t=0$: the directions at the origin satisfying this inequality are said to be *spatial* : directions at the origin in the other region are said to be *temporal.* Lorentz transformations are simply the rotations and translations in this manifold.

Now consider tensors in the manifold.

Minkowski had not properly assimilated the Ricci tensor-calculus as applied to non-Euclidean manifolds, and in order to be able to work with a space of Euclidean type, he used the device of writing x_4 for $ct\sqrt{-1}$ (the space-co-ordinates being denoted by x_1, x_2, x_3), so that the expression

$$(dx)^2+(dy)^2+(dz)^2-c^2(dt)^2$$

which is invariant under all Lorentz transformations, became

$$(dx_1)^2+(dx_2)^2+(dx_3)^2+(dx_4)^2 \ :$$

[1] The metric of space-time thus introduced is that of a four-dimensional Cayley-Klein manifold which has for absolute (in homogeneous co-ordinates)

$$\left.\begin{aligned} x^2+y^2+z^2-c^2t^2 &= 0 \\ w^2 &= 0 \end{aligned}\right\}$$

a double hyperplane at infinity containing a quadric hypersurface, which is real but with imaginary generators, like an ordinary sphere.

this enabled him to take as his metric

$$(ds)^2 = (dx_1)^2 + (dx_2)^2 + (dx_3)^2 + (dx_4)^2$$

which defines a four-dimensional Euclidean manifold.[1]

It is, however, simpler to work with the real value of the time, and to express Minkowski's results in terms of tensors which exist in the non-Euclidean four-dimensional manifold we have introduced, whose metric is specified by

$$(ds)^2 = c^2(dt)^2 - (dx)^2 - (dy)^2 - (dz)^2$$

which we may write

$$(ds)^2 = (dx^0)^2 - (dx^1)^2 - (dx^2)^2 - (dx^3)^2$$

so

$$g_{00} = 1, \quad g_{11} = g_{22} = g_{33} = -1,$$
$$g^{00} = 1, \quad g^{11} = g^{22} = g^{33} = -1.$$

His greatest discovery [2] was that at any point in the electromagnetic field *in vacuo* there exists a tensor of rank 2 of outstanding physical importance, which in its mixed form (E_p^q) may be defined by the equation

$$E_p^q = \frac{1}{16\pi} \delta_p^q \sum_{\alpha\beta} X_{\alpha\beta} X^{\alpha\beta} - \frac{1}{4\pi} \sum_t X_{pt} X^{qt}$$

where X_{pq} is the electromagnetic six-vector, that is to say, if (d_x, d_y, d_z) and (h_x, h_y, h_z) are the electric and magnetic intensities respectively, then [3]

$$d_x = X^{01} = -X_{01}, \quad d_y = X^{02} = -X_{02}, \quad d_z = X^{03} = -X_{03}$$
$$h_x = X^{23} = X_{23}, \quad h_y = X^{31} = X_{31}, \quad h_z = X^{12} = X_{12}.$$

Substituting in the equation which defines E_p^q, we find the values of the components of this tensor, namely,

$$E_0^0 = \frac{1}{8\pi} (d_x^2 + d_y^2 + d_z^2 + h_x^2 + h_y^2 + h^2_z) :$$

this represents the density of electromagnetic energy, discovered by W. Thomson (Kelvin) in 1853 [4];

$$E_0^1 = \frac{1}{4\pi}(d_y h_z - d_z h_y), \quad E_0^2 = \frac{1}{4\pi}(d_z h_x - d_x h_z), \quad E_0^3 = \frac{1}{4\pi}(d_x h_y - d_y h_x) :$$

[1] Minkowski's use of $x_4 = ct\sqrt{-1}$ led some philosophers to an outpouring of metaphysical nonsense about time being an imaginary fourth dimension of space.

[2] loc. cit., equation (74)

[3] X_{pq} is immediately derived from X^{pq} by the formula

$$X_{pq} = \sum_{ts} g_{ps}\, g_{qt}\, X^{st}.$$

[4] cf. Vol. I, pp. 222, 224

(E_0^1, E_0^2, E_0^3) represents $(1/c)$ times the flux of electromagnetic energy, discovered by Poynting and Heaviside in 1884[1];

$$E_1^0 = -\frac{1}{4\pi}(d_y h_z - d_z h_y),\ E_2^0 = -\frac{1}{4\pi}(d_z h_x - d_x h_z),\ E_3^0 = -\frac{1}{4\pi}(d_x h_y - d_y h_x):$$

$(-E_1^0, \ -E_2^0, \ -E_3^0)$ represents c times the density of electromagnetic momentum, discovered by J. J. Thomson in 1893[2];

$$E_1^1 = \frac{1}{8\pi}(d_x^2 - d_y^2 - d_z^2 + h_x^2 - h_y^2 - h_z^2),$$

and similarly for E_2^2 and E_3^3;

$$E_2^3 = E_3^2 = \frac{1}{4\pi}(d_y d_z + h_y h_z),$$

and similarly for E_3^1, E_1^3, E_1^2, E_2^1.

The nine quantities

$$\begin{matrix} E_1^1 & E_2^1 & E_3^1 \\ E_1^2 & E_2^2 & E_3^2 \\ E_1^3 & E_2^3 & E_3^3 \end{matrix}$$

represent the components of stress in the aether, discovered by Maxwell in 1873.[3] Thus, *each component of the tensor* E_p^q *has a physical interpretation*, which in every case had been discovered many years before Minkowski showed that these 16 components constitute a tensor of rank 2. The tensor E_p^q is called the *energy tensor* of the electromagnetic field.

Since $E_{qp} = \sum_r g_{pr} E_q^r = g_{pp} E_q^p$ for this metric, we have

$$E_{0p} = E_0^p,\ E_{1p} = -E_1^p,\ E_{2p} = -E_2^p,\ E_{3p} = -E_3^p,$$

and hence we find $E_{01} = E_{10}$ and generally $E_{pq} = E_{qp}$, that is, E_{pq} *is a symmetric tensor.*

Moreover we can show that if ρ is the density of electricity and v its velocity, then

$$\begin{aligned}
\frac{\partial E_0^0}{\partial x^0} + \frac{\partial E_0^1}{\partial x^1} + \frac{\partial E_0^2}{\partial x^2} + \frac{\partial E_0^3}{\partial x^3} &= -\frac{\rho}{c}(v_x d_x + v_y d_y + v_z d_z) \\
\frac{\partial E_1^0}{\partial x^0} + \frac{\partial E_1^1}{\partial x^1} + \frac{\partial E_1^2}{\partial x^2} + \frac{\partial E_1^3}{\partial x^3} &= \frac{\rho}{c}(d_x + v_y h_z - v_z h_y) \\
\frac{\partial E_2^0}{\partial x^0} + \frac{\partial E_2^1}{\partial x^1} + \frac{\partial E_2^2}{\partial x^2} + \frac{\partial E_2^3}{\partial x^3} &= \frac{\rho}{c}(d_y + v_z h_x - v_x h_z) \qquad \text{(A)} \\
\frac{\partial E_3^0}{\partial x^0} + \frac{\partial E_3^1}{\partial x^1} + \frac{\partial E_3^2}{\partial x^2} + \frac{\partial E_3^3}{\partial x^3} &= \frac{\rho}{c}(d_z + v_x h_y - v_y h_x).
\end{aligned}$$

[1] cf. Vol. I, pp. 313-4 [2] cf. Vol. I, p. 317 [3] cf. Vol. I, pp. 271-2

The first of these equations is

$$\frac{\partial}{c\partial t}\frac{1}{8\pi}(d_x^2+d_y^2+d_z^2+h_x^2+h_y^2+h_z^2)+\frac{\partial}{\partial x}\frac{1}{4\pi}(d_yh_z-d_zh_y)$$
$$+\frac{\partial}{\partial y}\frac{1}{4\pi}(d_zh_x-d_xh_z)+\frac{\partial}{\partial z}\frac{1}{4\pi}(d_xh_y-d_yh_x)$$

$$=-\frac{\rho}{c}(v_xd_x+v_yd_y+v_zd_z)$$

or

$\partial/\partial t$ (density of electromagnetic energy) + $\partial/\partial x$ (x-component of flux of electromagnetic energy + $\partial/\partial y$ (y-component of flux of electromagnetic energy) + $\partial/\partial z$ (z-component of flux of electromagnetic energy)

$$=-\rho(v_xd_x+v_yd_y+v_zd_z)$$

or

rate of increase of electromagnetic energy in unit volume + rate at which energy is leaving unit volume
= – (work done by the electromagnetic forces on electric charges within the unit volume)
and this is clearly nothing but the *equation of conservation of energy*. Similarly the other three of the equations (A) are the equations of conservation of x-momentum, y-momentum and z-momentum respectively.[1]

In an appendix to his paper,[2] Minkowski threw a new light on the equations of the relativistic dynamics of a material particle, which had been discovered by Planck two years earlier.[3] Denoting by (x, y, z) the co-ordinates of the particle at the instant t, he introduced the notion of the *proper-time* τ of the particle, whose differential is defined by the equation

$$(d\tau)^2=(dt)^2-\frac{1}{c^2}\left\{(dx)^2+(dy)^2+(dz)^2\right\}.$$

It is evident from this equation that $d\tau$ is invariant under all Lorentz transformations of (t, x, y, z), i.e. it is, in the language of the tensor-calculus, a scalar. Now writing $x^0=ct$, $x^1=x$, $x^2=y$, $x^3=z$, we know that

$$(dx^0,\quad dx^1,\quad dx^2,\quad dx^3)$$

[1] On the energy tensor cf. also A. Sommerfeld, *Ann. d. Phys.* xxxii (1910), p. 749; xxxiii (1910), p. 649; and M. Abraham, *Palermo Rend.* xxx (1910), p. 33

[2] loc. cit. [3] cf. p. 44

is a contravariant vector, and therefore

$$\left(\frac{dx^0}{d\tau}, \frac{dx^1}{d\tau}, \frac{dx^2}{d\tau}, \frac{dx^3}{d\tau}\right)$$

is a contravariant vector. But

$$\frac{d\tau}{dx^0} = \frac{1}{c}\left(1 - \frac{v^2}{c^2}\right)^{\frac{1}{2}},$$

where v denotes the velocity of the particle. Therefore

$$\frac{c}{\sqrt{\left(1-\frac{v^2}{c^2}\right)}}, \quad \frac{v_x}{\sqrt{\left(1-\frac{v^2}{c^2}\right)}}, \quad \frac{v_y}{\sqrt{\left(1-\frac{v^2}{c^2}\right)}}, \quad \frac{v_z}{\sqrt{\left(1-\frac{v^2}{c^2}\right)}}$$

are the components of a contravariant vector.

Now Planck had shown that if m is the mass of the particle, its energy E is $mc^2(1 - v^2/c^2)^{-\frac{1}{2}}$, and its components of momentum are

$$p_x = \frac{mv_x}{\sqrt{\left(1-\frac{v^2}{c^2}\right)}}, \quad p_y = \frac{mv_y}{\sqrt{\left(1-\frac{v^2}{c^2}\right)}}, \quad p_z = \frac{mv_z}{\sqrt{\left(1-\frac{v^2}{c^2}\right)}}.$$

Thus

$$(E/c, \quad p_x, \quad p_y, \quad p_z)$$

is a contravariant vector. This is called the *energy-momentum vector*.

The Newtonian and relativist definitions of *force* may be compared as follows. In Newtonian physics the momentum (p_x, p_y, p_z) is a vector and the time t is a scalar, so dp_x/dt, dp_y/dt, dp_z/dt is a vector, namely, the Newtonian *force*. In relativist physics, as we have seen, instead of a momentum vector (p_x, p_y, p_z), we have the contravariant energy-momentum vector $(E/c, p_x, p_y, p_z)$, or in its covariant form $(E/c, -p_x, -p_y, -p_z)$, and the scalar which takes the place of the time is the interval of proper-time,

$$d\tau = \left[(dt^2) - \frac{1}{c^2}\left\{(dx)^2 + (dy)^2 + (dz)^2\right\}\right]^{\frac{1}{2}}.$$

Thus it is natural to represent a *force* in relativity by the covariant vector

$$(F_k) = \left(-\frac{1}{c}\frac{dE}{d\tau}, \frac{dp_x}{d\tau}, \frac{dp_y}{d\tau}, \frac{dp_z}{d\tau}\right).$$

Now the equations of motion of a particle, as found by Planck, were

$$m\frac{d}{dt}\left\{\left(1-\frac{v^2}{c^2}\right)^{-\frac{1}{2}}\frac{dx}{dt}\right\}=X, \qquad m\frac{d}{dt}\left\{\left(1-\frac{v^2}{c^2}\right)^{-\frac{1}{2}}\frac{dy}{dt}\right\}=Y,$$

$$m\frac{d}{dt}\left\{\left(1-\frac{v^2}{c^2}\right)^{-\frac{1}{2}}\frac{dz}{dt}\right\}=Z,$$

or

$$\frac{dp_x}{dt}=X, \qquad \frac{dp_y}{dt}=Y, \qquad \frac{dp_z}{dt}=Z.$$

Comparing these results, we see that we must take the last three components of the relativist force to be

$$F_1=\left(1-\frac{v^2}{c^2}\right)^{-\frac{1}{2}}X, \quad F_2=\left(1-\frac{v^2}{c^2}\right)^{-\frac{1}{2}}Y, \quad F_3=\left(1-\frac{v^2}{c^2}\right)^{-\frac{1}{2}}Z,$$

and then the last three relativist equations of motion will be

$$m\frac{d^2x}{d\tau^2}=F_1, \qquad m\frac{d^2y}{d\tau^2}=F_2, \qquad m\frac{d^2z}{d\tau^2}=F_3.$$

Since

$$\left(c\frac{d^2t}{d\tau^2}, -\frac{d^2x}{d\tau^2}, -\frac{d^2y}{d\tau^2}, -\frac{d^2z}{d\tau^2}\right)$$

is a covariant vector, the first relativist equation of motion must evidently be

$$-mc\frac{d^2t}{d\tau^2}=F_0.$$

This completes Minkowski's set of equations of motion. The last equation may be written

$$mc^2\frac{d^2t}{d\tau^2}=\frac{dE}{dt},$$

where E is the energy; which is evidently true, since $E=mc^2\, dt/d\tau$.

Since

$$dE=X\,dx+Y\,dy+Z\,dz,$$

we have

$$\left(1-\frac{v^2}{c^2}\right)^{-\frac{1}{2}}dE=F_1dx+F_2dy+F_3dz$$

or

$$-c\,d\tau\left(1-\frac{v^2}{c^2}\right)^{-\frac{1}{2}}F_0=F_1dx+F_2dy+F_3dz$$

or

$$F_0 c dt + F_1 dx + F_2 dy + F_3 dz = 0$$

an equation which may be expressed geometrically by the statement that *the vector* (F_0, F_1, F_2, F_3) is *orthogonal to the vector which represents the velocity of the particle, namely,*

$$\left(c\frac{dt}{d\tau}, \frac{dx}{d\tau}, \frac{dy}{d\tau}, \frac{dz}{d\tau}\right).$$

We can now obtain a simple expression for the ponderomotive force on a particle of charge e and velocity $\mathbf{v}$ in electromagnetic theory. In Newtonian physics the three components are

$$e\left(d_x+\frac{h_z}{c}\frac{dy}{dt}-\frac{h_y}{c}\frac{dz}{dt}\right), \quad e\left(d_y+\frac{h_x}{c}\frac{dz}{dt}-\frac{h_z}{c}\frac{dx}{dt}\right), \quad e\left(d_z+\frac{h_y}{c}\frac{dx}{dt}-\frac{h_x}{c}\frac{dy}{dt}\right).$$

The corresponding force in relativity theory will have for its last three components these quantities multiplied by $dt/d\tau$. So if the relativist force is (F_k), we have

$$F_1=\frac{e}{c}\left(d_x\frac{cdt}{d\tau}+h_z\frac{dy}{d\tau}-h_y\frac{dz}{d\tau}\right)=\frac{e}{c}(X_{10}V^0+X_{12}V^2+X_{13}V^3)$$

where
$$(V^q)=\left(c\frac{dt}{d\tau}, \frac{dx}{d\tau}, \frac{dy}{d\tau}, \frac{dz}{d\tau}\right)$$

is the contravariant vector representing the relativist velocity of the particle. This may be written

$$F_1=\frac{e}{c}\sum_q X_{1q}V^q$$

and similarly we have

$$F_2=\frac{e}{c}\sum_q X_{2q}V^q, \qquad F_3=\frac{e}{c}\sum_q X_{3q}V^q.$$

These are the last three components of the covariant vector which is obtained by transvecting the electromagnetic six-vector X_{pq} with the particle's velocity V_q. Clearly the first component of the force must be the first component of this transvectant: and therefore *the relativist force on a particle of charge e and velocity* V_q *is*

$$F_k=\frac{e}{c}\sum_q X_{kp}V^q.$$

The fact that energy-density occurs as the component E_0^0 of Minkowski's energy-tensor, while energy occurs as the first component of his energy-momentum vector, leads naturally to an inquiry into the connection between these two tensors. This can be investigated as follows.

Consider a system occupying a finite volume and involving energy of any kind (e.g. electromagnetic energy, or stress-energy, or gravitational energy) for which we can define an energy tensor T_p^q such that T_0^0 is the energy-density, (T_0^1, T_0^2, T_0^3) is $(1/c)$ times the flux of the energy, (T_1^0, T_2^0, T_3^0) is $(-c)$ times the density of momentum, and $(T_1^1, T_2^2, T_3^3, T_2^3, T_3^2, T_1^3, T_3^1, T_2^1, T_1^2)$ are the components of flux of momentum, just as in the case of Minkowski's energy-tensor of the electromagnetic field *in vacuo*: and suppose that the following conditions are satisfied:

(i) the system is rigidly-connected, and is considered in the first place as being at rest:

(ii) its state does no vary with the time:

(iii) there is conservation of momentum, so

$$\frac{\partial T_r^1}{\partial x}+\frac{\partial T_r^2}{\partial y}+\frac{\partial T_r^3}{\partial z}=0 \quad (r=1,\ 2,\ 3) \tag{1}$$

(iv) there is no flux of energy in the state of rest, so

$$T_r^0=0, \qquad T_0^r=0 \quad (r=1,\ 2,\ 3). \tag{2}$$

From (1) we have

$$\iiint T_1^1 dx\,dy\,dz=\iiint\left\{\frac{\partial}{\partial x}(xT_1^1)+\frac{\partial}{\partial y}(xT_1^2)+\frac{\partial}{\partial z}(xT_1^3)\right\}dx\,dy\,dz$$

where the integration is taken over the whole volume occupied by the system and therefore

$$\iiint T_1^1 dx\,dy\,dz=\iint x(lT_1^1+mT_1^2+nT_1^3)\,dS$$

where the last integral is taken over a surface S enclosing the whole system, and (l, m, n) are the direction-cosines of the outward-drawn normal to S.

If we suppose the surface S so large that it includes the whole of the space in which there are any sensible effects due to the system, then T_p^q is zero on S, and therefore the last integral vanishes: so we have

$$\iiint T_1^1 dx\,dy\,dz=0. \tag{3}$$

Now suppose that (t, x, y, z) is the frame of reference relative to which the system is at rest, and let $(\bar{t}, \bar{x}, \bar{y}, \bar{z})$ be a frame of reference such that relative to it, the system is in motion parallel to the axis of $\bar{x}$ with velocity $w = c \tanh \alpha$. The axes of $\bar{x}, \bar{y}, \bar{z}$, are taken to be parallel to the axes of x, y, z respectively. Then the two sets of co-ordinates are connected by the equations

$$t = \bar{t} \cosh \alpha - (\bar{x}/c) \sinh \alpha$$
$$x = \bar{x} \cosh \alpha - c\bar{t} \sinh \alpha$$
$$y = \bar{y}, \qquad z = \bar{z},$$

and the equations of transformation of the mixed tensor $T_p{}^q$ give

$$\bar{T}_0{}^0 = \cosh^2\alpha\, T_0{}^0 - c \sinh\alpha \cosh\alpha\, T_1{}^0 + \frac{1}{c} \cosh\alpha \sinh\alpha\, T_0{}^1 - \sinh^2\alpha\, T_1{}^1$$
$$= \cosh^2\alpha\, T_0{}^0 - \sinh^2\alpha\, T_1{}^1$$

by (2). Thus

$$\iiint \bar{T}_0{}^0\, d\bar{x}\, d\bar{y}\, d\bar{z} = \iiint (\cosh^2\alpha\, T_0{}^0 - \sinh^2\alpha\, T_1{}^1) d\bar{x}\, d\bar{y}\, d\bar{z}$$
$$= \iiint (\cosh^2\alpha\, T_0{}^0 - \sinh^2\alpha\, T_1{}^1) \operatorname{sech} \alpha\, dx\, dy\, dz,$$

since $\partial(x, y, z)/\partial(\bar{x}, \bar{y}, \bar{z}) = \cosh \alpha$, it being understood that $\bar{x}, \bar{y}, \bar{z}$ are measured over the field at a constant value of $\bar{t}$. So by (3),

$$\iiint \bar{T}_0{}^0\, d\bar{x}\, d\bar{y}\, d\bar{z} = \cosh \alpha \iiint T_0{}^0\, dx\, dy\, dz.$$

Now let $U = \iiint T_0{}^0\, dx\, dy\, dz$, so U represents the total energy of the system when at rest. Then since $\cosh \alpha = (1 - w^2/c^2)^{-\frac{1}{2}}$, the result now becomes :

Total energy associated with the moving system $= U\ (1 - w^2/c^2)^{-\frac{1}{2}}$.

This may be regarded as an extension, to systems of finite size, of the formula that in relativity theory the energy of a particle of proper-mass m, moving with velocity w, is

$$\frac{mc^2}{\sqrt{\left(1 - \frac{w^2}{c^2}\right)}}.$$

Now consider the momentum. The equation of transformation of a mixed tensor of rank 2 gives

$$\bar{T}_1{}^0 = -\frac{1}{c} \sinh\alpha \cosh\alpha\, T_0{}^0 + \frac{1}{c} \cosh\alpha \sinh\alpha\, T_1{}^1,$$

since T_0^1 and T_1^0 are zero. Therefore

$$\iiint \overline{T}_1^0 \, d\bar{x} \, d\bar{y} \, d\bar{z} = -\frac{1}{c} \sinh \alpha \cosh \alpha \iiint (T_0^0 - T_1^1) \, d\bar{x} \, d\bar{y} \, d\bar{z}$$

$$= -\frac{1}{c} \sinh \alpha \iiint (T_0^0 - T_1^1) \, dx \, dy \, dz$$

$$= -\frac{1}{c} \sinh \alpha \iiint T_0^0 \, dx \, dy \, dz, \quad \text{by (3)}$$

$$= - \frac{wU}{c^2 \sqrt{\left(1 - \frac{w^2}{c^2}\right)}}.$$

Now $-\overline{T}_1^0$ represents the density of $\bar{x}$-momentum. Therefore the total momentum of the moving system parallel to the x-axis is

$$\frac{w}{\sqrt{\left(1 - \frac{w^2}{c^2}\right)}} \frac{U}{c^2}.$$

This may be regarded as an extension, to systems of finite size, of the formula for the x-component of momentum of a particle.

The above analysis shows how the components of the energy-momentum vector (now no longer restricted by the condition that it is to apply only to a single particle) can be derived from those of the energy-tensor. It is evident that whereas the energy-tensor is *localised* (i.e. each of its components is a function of position in space), the *energy-momentum vector is not localised*.[1]

Before the discovery of relativity theory, physicists were accustomed to think of energy not as a component of a tensor, but as a scalar: and indeed even in relativity theory, energy *as observed by a particular observer* is a scalar. For let an observer be moving along the axis of x with velocity v, so that the covariant vector representing his velocity, namely,

$$\left(c \frac{dt}{d\tau}, -\frac{dx}{d\tau}, -\frac{dy}{d\tau}, -\frac{dz}{d\tau}\right) \quad \text{or} \quad (\xi_0, \xi_1, \xi_2, \xi_3)$$

is given by

$$\xi_0 = c \left(1 - \frac{v^2}{c^2}\right)^{-\frac{1}{2}}, \quad \xi_1 = -v \left(1 - \frac{v^2}{c^2}\right)^{-\frac{1}{2}}, \quad \xi_2 = 0, \quad \xi_3 = 0.$$

[1] On the relation of the energy-momentum vector of a particle to the energy-tensor of a continuous field, see further H. P. Robertson, *Proc. Edin. Math. Soc.*(2) v (1937), p. 63, and M. Mathisson, *Proc. Camb. Phil. Soc.* xxxvi (1940), p. 331.

Let a particle of proper-mass m be moving in the same straight line with velocity w, so that its contravariant energy-momentum vector is

$$\eta^0 = mc\left(1-\frac{w^2}{c^2}\right)^{-\frac{1}{2}}, \quad \eta^1 = mw\left(1-\frac{w^2}{c^2}\right)^{-\frac{1}{2}}, \quad \eta^2 = 0, \quad \eta^3 = 0.$$

The transvectant of these vectors, namely, $\xi_0\eta^0+\xi_1\eta^1+\xi_2\eta^2+\xi_3\eta^3$, is a scalar : its value is

$$m(c^2 - vw)\left(1-\frac{v^2}{c^2}\right)^{-\frac{1}{2}}\left(1-\frac{w^2}{c^2}\right)^{-\frac{1}{2}}. \qquad \text{(A)}$$

Now the relative velocity of the particle and the observer is, by the relativist formula,

$$\frac{v-w}{1-\dfrac{vw}{c^2}}$$

and the energy of a particle moving with this velocity relative to the axes of reference is

$$mc^2\left\{1-\frac{(v-w)^2}{c^2(1-vw/c^2)^2}\right\}^{-\frac{1}{2}}$$

which reduces at once to the expression (A). Thus we see that *the energy of an observed particle may properly be regarded as a scalar, being the transvectant of the particle's energy-momentum vector and the observer's velocity.*

A vector which is of importance in electromagnetic theory may be introduced in the following way. We have seen that the electric intensity (d_x, d_y, d_z) and the magnetic intensity (h_x, h_y, h_z), at a point in free aether, are parts of a six-vector

$$d_x = X^{01}, \quad d_y = X^{02}, \quad d_z = X^{03}, \quad h_x = X^{23}, \quad h_y = X^{31}, \quad h_z = X^{12}.$$

Now if ϕ is the electric potential, and (a_x, a_y, a_z) the vector-potential, we have

$$d_x = -\frac{\partial\phi}{\partial x} - \frac{\partial a_x}{c\partial t}, \qquad h_x = \frac{\partial a_z}{\partial y} - \frac{\partial a_y}{\partial z}$$

and four similar equations. The question therefore suggests itself, what is the character of the potential ϕ, a_x, a_y, a_z, from the point of view of the tensor-calculus ? The answer, which is easily verified by examining the effects of Lorentz transformations, is that *if*

$$\phi_0 = \phi, \qquad \phi_1 = -a_x, \qquad \phi_2 = -a_y, \qquad \phi_3 = -a_z,$$

then $(\phi_0, \phi_1, \phi_2, \phi_3)$ *is a covariant vector* : its connection with the six-vector is given by the equations

$$X_{pq} = \frac{\partial \phi_p}{\partial x^q} - \frac{\partial \phi_q}{\partial x^p}.$$

It was discovered in 1915 by D. Hilbert [1] that the energy-tensor of a system can be expressed in terms of the Lagrangean function of the system. This theorem was developed further by E. Schrödinger [2] and H. Bateman [3] in 1927 : the rule was given by Schrödinger as follows :

Let (a_0, a_1, a_2, a_3) *be one of the four-vectors on which the Lagrangean* L *depends* (as e.g. the Lagrangean in electromagnetic theory depends on the electromagnetic potential-vector), *and let* a_{pq} *denote the derivative of* a_p *with respect to the co-ordinate* x_q : *then the components of the energy-tensor are given by*

$$E_p{}^q = \sum \left(\sum_{t=0}^{3} a_{tp} \frac{\partial L}{\partial a_{tq}} + \sum_{t=0}^{3} a_{pt} \frac{\partial L}{\partial a_{qt}} + a_p \frac{\partial L}{\partial a_q} \right) - \delta_p{}^q L,$$

where $\delta_p{}^q = 0$ *or* 1 *according as q is, or is not, different from p, and the summation is taken over all the four-vectors a.*

For example, consider the electromagnetic field in free aether, for which the Lagrangean function is

$$L = \frac{1}{8\pi} \left(d_x^2 + d_y^2 + d_z^2 - h_x^2 - h_y^2 - h_z^2 \right),$$

or, if (a_0, a_1, a_2, a_3) denotes the covariant electromagnetic potential vector,

$$L = \frac{1}{8\pi} \left[\begin{array}{l} \left(-\frac{\partial a_0}{\partial x_1} + \frac{\partial a_1}{\partial x_0} \right)^2 + \left(-\frac{\partial a_0}{\partial x_2} + \frac{\partial a_2}{\partial x_0} \right)^2 + \left(-\frac{\partial a_0}{\partial x_3} + \frac{\partial a_3}{\partial x_0} \right)^2 \\ - \left(-\frac{\partial a_3}{\partial x_2} + \frac{\partial a_2}{\partial x_3} \right)^2 - \left(-\frac{\partial a_1}{\partial x_3} + \frac{\partial a_3}{\partial x_1} \right)^2 - \left(-\frac{\partial a_2}{\partial x_1} + \frac{\partial a_1}{\partial x_2} \right)^2 \end{array} \right];$$

from this we have

$$\frac{\partial L}{\partial a_{kn}} = \frac{1}{4\pi} X^{nk}$$

[1] *Gött. Nach.* (1915), p. 395 ; cf. also F. Klein, *Gött. Nach.* (1917), p. 469 and Hilbert, *Math. Ann.* xcii (1924), p. 1

[2] *Ann. d. Phys.* lxxxii (1927), p. 265

[3] *Proc. Nat. Acad. Sci.* xiii (1927), p. 326

Thus Schrödinger's formula gives

$$E_p{}^q = \frac{1}{4\pi} \sum_{t=0}^{3} \frac{\partial a_t}{\partial x_p} X^{qt} + \frac{1}{4\pi} \sum_{t=0}^{3} \frac{\partial a_p}{\partial x_t} X^{tq} - \delta_p{}^q L$$

$$= -\frac{1}{4\pi} \sum_t \left(\frac{\partial a_p}{\partial x_t} - \frac{\partial a_t}{\partial x_p} \right) X^{qt} - \delta_p{}^q L$$

$$= -\frac{1}{4\pi} \sum_t X_{pt} X^{qt} + \frac{1}{16\pi} \delta_p{}^q \sum_{\alpha, \beta} X_{\alpha\beta} X^{\alpha\beta}$$

which is the usual formula for Minkowski's energy tensor.

Chapter III

THE BEGINNINGS OF QUANTUM THEORY

At the end of the nineteenth century the theory of radiation was in a most unsatisfactory state. For the energy per cm.^{3} of pure-temperature or black-body radiation, in the range of wave-lengths from λ to $\lambda + d\lambda$, two different formulae had been proposed. Firstly, that of Wien,[1]

$$E = C\lambda^{-5}e^{-b/\lambda T}d\lambda$$

where λ is wave-length, T is absolute temperature, and b and C are constants. This formula is asymptotically correct in the region of short waves (more precisely, when λT is small) ; but, as O. Lummer and E. Pringsheim showed,[2] is irreconcilable with the observational results for long waves. Secondly, that of Rayleigh and Jeans [3]

$$E = 8\pi kT\lambda^{-4}d\lambda$$

where k is Boltzmann's constant ; which, as shown by the experiments of Rubens and Kurlbaum,[4] is asymptotically correct for the long waves, but is inapplicable at the other end of the spectrum. What was wanted was a formula which for the extreme limits $\lambda \to 0$ and $\lambda \to \infty$ would tend asymptotically to Wien's and Rayleigh's formulae respectively, and which would agree with the experimental values over the whole range of wave-lengths.

In the spring and summer of 1900 attempts were made to construct such a formula empirically by M. Thiesen,[5] by O. Lummer and E. Jahnke,[6] and by O. Lummer and E. Pringsheim.[7] These formulae were of the type

$$E = CT^{5-\mu}\lambda^{-\mu}e^{-b/(\lambda T)^{\nu}}$$

which for $\mu = 5$, $\nu = 1$, gives Wien's law, and for $\mu = 4$, $\nu = 1$, $b = 0$, gives Rayleigh's.

The correct law was first given by Max Karl Ernst Ludwig Planck (1858–1947) in a communication which was read on 19 October 1900 before the German Physical Society.[8] Planck was the son of

[1] Vol. I, p. 381
[2] *Verh. d. deutsch phys. Ges.* i (1899), p. 215 ; ii (1900), p. 163
[3] Vol. I, p. 384
[4] loc. cit.
[5] *Verh. d. deutsch. phys. Ges.* ii (1900), p. 65
[6] *Ann. d. Phys.* iii (1900), p. 283
[7] *Verh. d. deutsch. phys. Ges.* ii (1900), p. 163
[8] *Verh. d. deutsch. phys. Ges.* ii (1900), p. 202

a professor of law at Kiel, later translated to Munich; he was educated at the University of Munich, but for one year attended the lectures of Helmholtz and Kirchhoff at Berlin. After four years as *professor extraordinarius* at Kiel, he was called in 1889 to succeed Kirchhoff at Berlin, where the rest of his academic life was spent.

In the study of pure-temperature radiation, his starting-point was the known fact that in a hollow chamber at a given temperature, the distribution of radiant energy among wave-lengths is altogether independent of the material of which the chamber is composed; and he was therefore free to suppose the walls of the chamber to have any constitution which was convenient for the calculations, so long as they were capable of absorbing and emitting radiation, and thereby making possible the exchange of energy between matter and aether. He chose them to be of the simplest type imaginable, namely an aggregate of Hertzian vibrators,[1] each with one proper frequency. Each vibrator absorbs energy from any surrounding radiation which is nearly of its own proper frequency, and acts as a resonator, emitting radiant energy.

He first calculated (by classical electrodynamics) the average absorption and emission of a vibrator of frequency[2] ν which is immersed in, and statistically in equilibrium with, a field of radiation, and found that if the average energy-density of the radiation, in the interval of frequency ν to $\nu + d\nu$, is E, then [3]

$$E = \frac{8\pi\nu^2}{c^3} U d\nu, \tag{1}$$

where U is the average energy of the vibrator.

While most of the other workers on radiation were attempting to find the relations between energy, wave-length and temperature, by direct methods, Planck, who was a master of thermodynamics, felt that the concept of entropy must play a fundamental part: and he examined the relation between the energy of a vibrator and its entropy S, showing that if S is known as a function of U, then the law of distribution of energy in the spectrum of pure-temperature radiation can be determined.

We have, from thermodynamics, for a system of constant volume,

$$dS = \frac{dU}{T} \quad \text{or} \quad \frac{dS}{dU} = \frac{1}{T}, \tag{2}$$

while Wien's law of radiation, namely

$$E = a\nu^3 e^{-\beta\nu/T} d\nu$$

[1] This, of course, does not mean (as it has sometimes been wrongly interpreted to mean) that actual matter necessarily has this character.

[2] It will be remembered that ν is the number of oscillations in one second, that is, c multiplied by the wave-number $1/\lambda$.

[3] *Ann. d. Phys.* i (1900), p. 69, equation (34): *Phys. ZS.* ii (1901), p. 530

requires by (1) that we should have

$$U = \gamma \nu e^{-\beta\nu/T}$$

where γ is a constant: so by (2)

$$\frac{dS}{dU} = -\frac{1}{\beta\nu}\log\frac{U}{\gamma\nu}$$

whence

$$\frac{d^2S}{dU^2} = \frac{\text{Constant}}{U}. \qquad (3)$$

Planck had earlier attempted[1] to give a proof of Wien's law of radiation based on this equation (3), which he obtained independently by thermodynamical reasoning: but when confronted by Lummer and Pringsheim's experimental results he realised that Wien's could not be the true law of radiation; and he now proposed to modify (3), which he did by writing

$$\frac{d^2S}{dU^2} = \frac{\alpha}{U(\beta + U)} \qquad (4)$$

where α and β are constants. This is the simplest of all the expressions which give dS/dU as a logarithmic function of U (as suggested by the probability theory of entropy), and which for small values of U agrees with equation (3). Moreover, if Rayleigh's law of radiation had been taken instead of Wien's, we should have obtained

$$\frac{d^2S}{dU^2} = \frac{\text{Constant}}{U^2},$$

which again is a case of (4). From (4) we have, by (2),

$$\frac{1}{T} = \frac{dS}{dU} = \text{Const. } \log\left(\frac{\text{Const.} + U}{U}\right)$$

or

$$U = \frac{\text{Const.}}{e^{\text{Const.}/T} - 1}. \qquad (5)$$

This equation does not give the way in which the frequency ν enters into the formula for U. But as Wien had shown in 1893,[2] E must be of the form $T^5\phi(T\lambda)d\lambda$, or $\nu^3\psi(\nu/T)d\nu$, so by (1), U must be of the form

$$U = \nu\psi\left(\frac{\nu}{T}\right). \qquad (6)$$

[1] *Berlin Sitz.* xxv (1899), p. 440

[2] cf. Vol. I, p. 380

Thus equation (5) must have the form

$$U = \text{Const.}\frac{\nu}{e^{l\nu/T} - 1}$$

and therefore by (1) the average energy-density of the radiation in the frequency-range ν to $\nu + d\nu$ is

$$E = \frac{g\nu^3 d\nu}{e^{l\nu/T} - 1} \qquad (7)$$

where g and l are constants. *This is Planck's formula* which agreed with the experimental determinations of Lummer and Pringsheim, and also of H. Rubens and F. Kurlbaum,[1] and F. Paschen,[2] so well that it soon displaced all other suggested laws of radiation.

It was, however, as yet hardly more than an empirical formula, since equation (4) had no complete theoretical justification. This defect was remedied on 14 December of the same year (1900), when Planck read to the German Physical Society a paper [3] which placed his new law on a sound foundation, and in so doing created a new branch of physics, the quantum theory.

He considered a system consisting of a large number of simple Hertzian vibrators, in a hollow chamber enclosed by reflecting walls: let N of the vibrators have the frequency ν, N′ the frequency ν' and so on. Suppose that an amount A of energy is in the vibrators of frequency ν. Planck assumed that this energy is constituted of equal discrete elements, each of amount ϵ, and that there are altogether P such elements in the N vibrators, so that

$$A = P\epsilon.$$

Thus he assumed that the emission and absorption of radiation by these vibrators takes place not continuously, but by jumps of amount ϵ.

Any distribution of these P elements among the N vibrators may be called a *complexion*. The number of possible complexions is the number of possible ways of distributing P objects among N containers, when we do not take account of which particular objects lie in particular containers, but only of the number contained in each. This number is, by the ordinary theory of permutations and combinations,

$$\frac{(N+P-1)!}{(N-1)!\,P!}.$$

[1] *Berlin Sitz.*, 25 Oct. 1900, p. 929

[2] *Ann. d. Phys.* iv (1901), p. 277

[3] *Verh. deutsch. phys. Ges*, ii (1900), p. 237. This and the paper of 19 October were re-edited and printed in the new form in *Ann. d. Phys.* iv (1901), p. 553.

As N and P are very large numbers, we can use Stirling's approximate value for the factorials, namely,

$$\log \{(z-1)!\} = (z-\tfrac{1}{2}) \log z - z + \tfrac{1}{2} \log (2\pi)$$

so the number of complexions is approximately

$$\left\{\frac{N}{2\pi P(N+P)}\right\}^{\frac{1}{2}} \frac{(N+P)^{N+P}}{N^N P^P}.$$

We assume that all complexions have equal probability, so the probability W of any state of the system of N vibrators is proportional to the number of complexions corresponding to it ; that is, with sufficient approximation for our present purpose, we have

$$\log W = (N+P) \log (N+P) - N \log N - P \log P.$$

Now the entropy in any state of a system depends on the inequality of the distribution of the total energy among the individual members of the system : and Boltzmann had shown by his work on the kinetic theory of gases[1] that the entropy S_N in any state of a system such as these vibrators is closely connected with the probability W of the state. Planck developed this discovery into the equation

$$S_N = k \log W \qquad (8)$$

where the *thermodynamic probability* W is always an integer, and k denotes the gas-constant for one molecule, or Boltzmann constant.[2] Thus

$$S_N = k\{(N+P) \log (N+P) - N \log N - P \log P\}.$$

Now $P = NU/\epsilon$, where U is the average, taken over the N oscillators, of the energy of one of them. Thus, retaining only the most important terms, and ignoring terms which do not involve U, we have

$$S_N = kN\left\{\left(1+\frac{U}{\epsilon}\right) \log \left(1+\frac{U}{\epsilon}\right) - \frac{U}{\epsilon} \log \frac{U}{\epsilon}\right\}$$

[1] cf. L. Boltzmann, *Vorlesungen über Gastheorie*, i (1896), § 6. This is essentially Boltzmann's ' H-theorem.'

[2] cf. Vol. I, p. 382. With Boltzmann the factor k did not occur, since his calculations referred not to individual molecules but to gramme-molecules, and with him the entropy was undetermined as regards an additive constant (i.e. there was an undetermined factor of proportionality in the probability W), whereas with Planck the entropy had a definite absolute value. This was a step of fundamental importance, and, as we shall see, led directly to the hypothesis of ' quanta.' The occurrence of the logarithm in the formula is explained by the circumstance that in compound systems a multiplication of probabilities corresponds to an addition of entropies.

so the entropy of a single oscillator of the set is

$$S=\frac{S_N}{N}=k\left\{\left(1+\frac{U}{\epsilon}\right)\log\left(1+\frac{U}{\epsilon}\right)-\frac{U}{\epsilon}\log\frac{U}{\epsilon}\right\}.$$

Thus from equation (2) above,

$$\frac{1}{T}=\frac{dS}{dU}=\frac{k}{\epsilon}\log\frac{\epsilon+U}{U}$$

or

$$U=\frac{\epsilon}{e^{\epsilon/kT}-1}. \tag{9}$$

But by equation (5) U must be of the form $\nu\psi(\nu/T)$. This condition can be satisfied only if

$$\epsilon=h\nu \tag{10}$$

where h is a constant independent of ν. Thus *the average energy of any simple-harmonic Hertzian vibrator of frequency* ν *must be an integral multiple of* $h\nu$, *and the smallest amount of energy that can be emitted or absorbed by it is* $h\nu$.

From (9) and (10) we have

$$U=\frac{h\nu}{e^{h\nu/kT}-1}$$

and therefore by (1) *the average energy-density of black-body radiation in the interval of frequency between* ν *and* $\nu+d\nu$ *is*

$$E=\frac{8\pi h}{c^3}\frac{\nu^3 d\nu}{e^{h\nu/kT}-1} \quad \text{or} \quad E=\frac{8\pi ch\lambda^{-5}d\lambda}{e^{ch/k\lambda T}-1}, \tag{11}$$

which is *Planck's formula.* This agrees with his earlier result (7), but the constants which were unknown in (7) are now replaced by h and k, which are important constants of nature and appear in many other connections.

When $\nu\rightarrow 0$, the formula gives

$$E=\frac{8\pi\nu^2}{c^3}kTd\nu, \quad \text{or} \quad E=8\pi k\lambda^{-4}Td\lambda$$

which is Rayleigh's law; and when $\nu\rightarrow\infty$ it gives

$$E=\frac{8\pi h}{c^3}\nu^3 e^{-h\nu/kT}d\nu, \quad \text{or} \quad E=8\pi ch\lambda^{-5}\, e^{-hc/k\lambda T}d\lambda$$

which is Wien's law, now expressed in terms of the constants h and k.

To obtain Wien's displacement law,[1] we proceed as follows : Let λ_m denote the wave-length corresponding to the maximum ordinate of a graph in which energy-density of radiation is plotted against wave-length. Then by Planck's formula (11), λ_m is the value of λ given by

$$0=\frac{\partial}{\partial\lambda}\frac{\lambda^{-5}}{e^{ch/k\lambda\mathrm{T}}-1},$$

or

$$0=-5+\frac{ch/k\lambda\mathrm{T}}{1-e^{ch/k\lambda\mathrm{T}}}.$$

Let q be the root of the equation

$$\frac{x}{1-e^{-x}}=5$$

so

$$q=4{\cdot}965114. \; . \; . \; .$$

Then

$$\frac{hc}{k\lambda_m\mathrm{T}}=q,$$

or

$$\lambda_m\mathrm{T}=\frac{(hc/k)}{q}=\frac{1{\cdot}4384}{4{\cdot}965114}\text{cm. degree}=0.28971 \text{ cm. degree},$$

which is Wien's displacement law.

Planck determined the values of the constants h and k by comparing his formula (11) with the measurements of F. Kurlbaum [2] and O. Lummer and E. Pringsheim,[3] the results obtained being

$$h=6{\cdot}55\times10^{-27} \text{ erg. sec.,} \qquad k=1{\cdot}346\times10^{-16} \text{ ergs per degree.}$$

He used this determination of k in order to calculate the number of molecules in a gramme-molecule (Avogadro's number) [4] : from the equation

$$\mathrm{S}=k\log\mathrm{W}$$

we can calculate the entropy of one gramme-molecule of an ideal gas, and from this can derive thermodynamically the relation

$$p=\frac{k\mathrm{NT}}{\mathrm{V}}$$

where p denotes the pressure of the gas, V its volume, and N denotes Avogadro's number : this shows that if R is the absolute gas-constant, then

$$\mathrm{R}=k\mathrm{N}.$$

[1] cf. Vol. I, p. 380
[2] *Ann. d. Phys.* lxv (1898), p. 759
[3] *Verh. d. deutsch phys. Ges.* ii (1900), p. 176
[4] cf. pp. 8, 18

From the known values of R and k, Planck found

$$N = 6{\cdot}175 \times 10^{23};$$

this agreed satisfactorily with the value $6{\cdot}40 \times 10^{23}$ which had been given by O. E. Meyer.[1]

Moreover, the knowledge of N so obtained leads to a new method of finding the charge of an electron. For the charge which is carried in electrolysis by one gramme-ion, that is by N ions, was known, being at that time believed to be 9,658 electromagnetic units. Thus if e is the charge of an electron in electrostatic units, we have

$$Ne = 9{,}658 \times 3 \times 10^{10}$$

which gives

$$e = 4{\cdot}69 \times 10^{-10} \text{ e.s.u.}$$

J. J. Thomson had found $e = 6 \times 10^{-10}$ e.s.u. two years earlier[2]; Planck's value was actually much nearer to the later determinations, which gave approximately $4{\cdot}77 \times 10^{-10}$ e.s.u.

Planck's law made it possible to give a more accurate formulation of the Stefan-Boltzmann law[3] for the total radiation per second from unit surface of a black body at temperature T. For[4] the element of this radiation in the range of wave-lengths λ to $\lambda + d\lambda$ is, by Planck's law

$$\frac{2\pi hc^2\lambda^{-5}d\lambda}{e^{hc/k\lambda T} - 1}$$

so the total radiation for all wave-lengths is

$$2\pi hc^2 \int_0^\infty \frac{\lambda^{-5}d\lambda}{e^{hc/k\lambda T} - 1}$$

or

$$\frac{2\pi h}{c^2}\int_0^\infty \frac{\nu^3 d\nu}{e^{h\nu/kT} - 1}.$$

Now if B_n is the nth Bernoullian number, we have

$$B_n = 4n\int_0^\infty \frac{t^{2n-1}dt}{e^{2\pi t} - 1},$$

[1] *Die Kinetische Theorie der Gase*, 2 Aufl. (1899), p. 337; Planck's actual result, *Ann. d. Phys.* IV (1901), p. 564, is that the number of oxygen molecules in 1 cm^3 at 760 mm. pressure and 15°C., is $2{\cdot}76 . 10^{19}$.

[2] cf. Vol. I, pp. 364–5 [3] cf. Vol. I, p. 374 [4] Vol. I, p. 373

whence, remembering that $B_2 = 1/30$, we have

$$\int_0^\infty \frac{s^3 ds}{e^{2\pi ps} - 1} = \frac{1}{240p^4}.$$

Putting $p = h/2\pi k T$, we see that *the total radiation per second from unit surface of a black body at temperature* T *is*

$$\frac{2\pi^5 k^4}{15c^2h^3} T^4 ;$$

this is the precise expression of the Stefan-Boltzmann formula in terms of the universal constants c, h and k.

A deeper insight into the physical conceptions underlying Planck's law of radiation was furnished by a later proof of it.[1] As is well known, in the kinetic theory of gases, it is shown that the probability that for a particular molecule the x-component of velocity will lie between u and $u + du$, its y-component of velocity between v and $v + dv$ and its z-component between w and $w + dw$, is

$$\left(\frac{m}{2k\pi T}\right)^{\frac{3}{2}} e^{-U/kT} du\ dv\ dw$$

where m is the mass of the molecule, U its kinetic energy, k is Boltzmann's constant, and T the absolute temperature. This result was generalised by Josiah Willard Gibbs [2] (1839–1903) into the following theorem: if we consider a large number of similar dynamical systems (which for simplicity we shall suppose to be linear oscillators), which are in statistical equilibrium with a large reservoir of heat at temperature T, and if q is the co-ordinate in an oscillator (e.g. the elongation of a vibrating electron) and p the momentum (defined as $\partial L/\partial(\partial q/\partial t)$ where L is the kinetic potential), then the probability that for any particular oscillator the co-ordinate lies between q and $q + dq$ and the momentum lies between p and $p + dp$ is [3]

$$\frac{e^{-U/kT} dq\ dp}{\int e^{-U/kT} dq\ dp}$$

where U is the energy of the oscillator, and the integration is to be taken over all possible values of q and p.

The theorem corresponding to this in the quantum theory is that

[1] cf. Lorentz, *Phys. ZS.* xi (1910), p. 1234 ; F. Reiche, *Die Quantentheorie* (Berlin 1921), Note 48

[2] *Elementary Principles in Statistical Mechanics* (New York, 1902)

[3] This is Gibbs's *canonical distribution*

if the energy of an oscillator can take only the discrete set of values U_0, U_1, U_2, U_3, . . ., then the probability that the energy of a particular oscillator is U_s is [1]

$$e^{-U_s/kT} \Big/ \sum_{s=0}^{\infty} e^{-U_s/kT}.$$

Thus if $U_s = sh\nu$ for $s = 0, 1, 2, \ldots$, the probability is

$$e^{-sh\nu/kT}(1 - e^{-h\nu/kT}).$$

The *mean* energy of an oscillator is therefore

$$h\nu(1 - e^{-h\nu/kT}) \sum_{s=0}^{\infty} s e^{-sh\nu/kT},$$

which has the value

$$\frac{h\nu}{e^{h\nu/kT} - 1}.$$

This leads at once, as before, to Planck's formula that the energy-density of black-body radiation in the frequency-interval from ν to $\nu + d\nu$ is

$$E = \frac{8\pi h}{c^3} \frac{\nu^3 d\nu}{e^{h\nu/kT} - 1}.$$

Other derivations of the law, based on many different assumptions, were given by various writers.[2] Some of them will be discussed later.

The next important advance in quantum theory was made by Einstein,[3] in the same volume of the *Annalen der Physik* as his papers

[1] If to an energy-level U_s there belongs a number g_s of permissible states, then the level U_s is said to be *degenerate*, and g_s is called the *weight* of the state. Taking the possibility of degenerate states into account, the above formula should be written

$$\frac{g_s e^{-U_s/kT}}{\sum_{s=0}^{\infty} g_s e^{-U_s/kT}}.$$

[2] Special reference may be made to the following :

J. Larmor, *Proc. R.S.*(A), lxxxiii (1909), p. 82 ; P. Debye, *Ann. d. Phys.* xxxiii (1910), p. 1427, completed by A. Rubinowicz, *Phys. ZS.* xviii (1917), p. 96 ; P. Franck, *Phys. ZS.* xiii (1912), p. 506 ; A. Einstein and O. Stern, *Ann. d. Phys.* xl (1913), p. 551 ; M. Wolfke, *Verh. d. deutsch. phys. Ges.* xv (1913), pp. 1123, 1215 : *Phys. ZS.* xv (1914), pp. 308, 463 ; A. Einstein, *Phys. ZS.* xviii (1917), p. 121 ; C. G. Darwin and R. H. Fowler, *Phil. Mag.* xliv (1922), pp. 450, 823 : *Proc. Camb. Phil. Soc* xxi (1922), p. 262 ; S. N. Bose, *ZS. f. P.* xxvi (1924), p. 178, xxvii (1924), p. 384 ; A. S. Eddington, *Phil. Mag.* l (1925), p. 803

[3] *Ann. d. Phys.* xvii (1905), p. 132 : cf. also *Ann. d. Phys.* xx (1906), p. 199

on the Brownian motion[1] and Relativity.[2] Einstein supposed monochromatic radiation of frequency ν and of small density (within the range of values of ν/T for which Wien's formula of radiation is applicable) to be contained in a hollow chamber of volume v_0 with perfectly-reflecting walls, its total energy being E : and, investigating by use of Wien's formula the dependence of the entropy on the volume, he found for the difference of the entropies when the radiation occupies the volume v_0 and when it occupies a smaller volume v, the equation

$$S - S_0 = \frac{Ek}{h\nu} \log \frac{v}{v_0}.$$

Now by inverting the Boltzmann-Planck relation

$$\text{entropy} = k \times \text{logarithm of probability}$$

he calculated the relative probability from the difference of entropies, and found that the probability that at an arbitrarily-chosen instant of time, the whole of the energy of the radiation should be contained within a part v of the volume v_0, is

$$\left(\frac{v}{v_0}\right)^{\frac{E}{h\nu}}.$$

This formula he studied in the light of a known result in the kinetic theory of gases, namely that if a gas contained in a volume v_0 consists of n molecules, the probability that at an arbitrarily-chosen instant of time, all the n molecules should be collected together within a a part v of the volume, is

$$\left(\frac{v}{v_0}\right)^n.$$

Comparing these formulae, he inferred that the radiation behaves as if it consisted of $E/h\nu$ quanta of energy or *photons*,[3] each of amount $h\nu$. The probability that *all* the photons are found at an arbitrary instant in the part v of the volume v_0 is the product of the probabilities (v/v_0) that a single one of them is in the part v : which shows that they are completely independent of each other.

Now it will be remembered that according to Planck's theory, a vibrator of frequency ν can emit or absorb energy only in multiples of $h\nu$. Planck regarded the quantum property as belonging essentially to the interaction between radiation and matter : free radiation he supposed to consist of electromagnetic waves, in accordance with

[1] cf. p. 9   [2] cf. p. 40

[3] The word *photon* was actually introduced much later, namely, by G. N. Lewis, *Nature*, 18 Dec., 1926 ; but it is so convenient that we shall adopt it now.

Maxwell's theory. Einstein in this paper put forward the hypothesis that parcels of radiant energy of frequency ν and amount $h\nu$ occur not only in emission and absorption, but that they have an independent existence in the aether.

It was shown by P. Ehrenfest[1] of Leiden, by A. Joffé[2] of St Petersburg, by L. Natanson[3] of Cracow and by G. Krutkow[4] of Leiden that Einstein's hypothesis leads not to Planck's law of radiation but to Wien's, at any rate if we assume that each of the light-quanta or photons of frequency ν has energy $h\nu$ and that they are completely independent of each other. In order to obtain Planck's formula it is necessary to assume that the elementary photons of energy $h\nu$ form aggregates, or photo-molecules as we may call them, of energies $2h\nu$, $3h\nu$, . . ., respectively, and that the total energy of radiation is distributed, on the average, in a regular manner between the photons and the different kinds of photo-molecules. This will be discussed more fully later.

Einstein applied his ideas in order to construct a theory of photo-electricity.[5] As we have seen,[6] in 1899 J. J. Thomson and P. Lenard showed independently that the emission from a metal irradiated by ultra-violet light consists of negative electrons: and in 1902 Lenard,[7] continuing his researches, showed that the number of electrons liberated is proportional to the intensity of the incident light, so long as its frequency remains the same: and that the initial velocity of the electrons is altogether independent of the intensity of the light, but depends on its frequency.

Knowledge regarding photo-electricity had reached this stage when in 1905 Einstein's paper appeared. Considering a metal surface illumined by radiation of frequency ν, he asserted that the radiation consists of parcels of energy; when one such parcel or photon falls on the metal, it may be absorbed and liberate a photo-electron: and that the maximum kinetic energy of the photo-electron at emission is $(h\nu - e\phi)$, where $e\phi$ is the energy lost by the electron in escaping from its original location to outside the surface. This of course implies that no photo-electrons will be generated unless the frequency of the light exceeds a certain 'threshold' value $e\phi/h$.

Einstein's equation was verified in 1912 by O. W. Richardson and K. T. Compton[8] and by A. L. Hughes,[9] and with great care in 1916 by R. A. Millikan.[10] For many metals, the threshold frequency is in the ultra-violet: but for the electro-positive metals, such as the alkali metals, it is in the visible spectrum: for sodium, it is in the green.

[1] *Ann. d. Phys.* xxxvi (1911), p. 91
[2] ibid, p. 534
[3] *Phys. ZS.* xii (1911), p. 659
[4] *Phys. ZS.* xv (1914), p. 133
[5] cf. Vol. I, pp. 356–7
[6] cf. Vol. I, p. 365
[7] *Ann. d. Phys.* viii (1902), p. 149; also E. R. Ladenburg (1878–1908), *Ann. d. Phys.* xii (1903), p. 558
[8] *Phil. Mag.* xxiv (1912), p. 575
[9] *Phil. Trans.* ccxii (1912), p. 205
[10] *Phys. Rev.* vii (1916), p. 355. cf. also M. de Broglie, *J. de Phys.* ii (1921), p. 265, and J. Thibaud, *Comptes Rendus*, clxxix (1924), pp. 165, 1053, 1322

The function ϕ is closely connected with the thermionic work-function measured at the same temperature :[1] in fact, the thermionic work-function is equal to h times the least frequency which will eject an electron from the metal :[2] and ϕ is therefore connected with the contact potential-differences between two metallic surfaces :[3] the difference of the functions ϕ for the two metals is equal to the contact difference of potential (reversing the order of the metals) together with the (small) coefficient of the Peltier effect at the junction between them.

Gases and vapours also exhibit the photo-electric effect, if the frequency of the incident radiation is sufficiently great : and the phenomenon can be observed for individual atoms by use of X-rays with the Wilson chamber. This effect is simply ionisation : and the law regarding the threshold frequency becomes the assertion that for ionisation to take place, the energy of the incident photon must be not less than the ionisation energy of the atom or molecule concerned. The electrons chiefly affected photo-electrically are the strongly-bound ones in the K-shell : the electrons in the outer shells, being more feebly-bound, do not absorb radiation to the same degree. The function ϕ in the equation

$$\text{maximum kinetic energy of electron at emission} = h\nu - e\phi$$

is now no longer connected with thermo-electric phenomena or contact differences of potential, but has different values depending on the shell in the atom from which the electron has come.

If a photo-electron is liberated from the K-shell, it may happen that the vacant place is filled by an electron from an outer shell, creating a photon whose energy is equal to the difference of the energies of the electron in the two shells : and this photon may in its turn be absorbed in another shell, giving rise to a second photo-electron, so that two electrons are ejected together. This effect, which was discovered by P. Auger,[4] is called the *compound photo-electric effect.*

The photo-electric effect cannot be explained classically, because the time-lag required by the classical theory, due to the necessity for accumulating sufficient energy from the radiation, is found not to occur.[5]

A hypothesis closely allied to Einstein's light-quantum explanation of the photo-electric effect was put forward in 1908 by J. Stark[6] : namely, that the frequency of the violet edge of the band-spectrum

[1] cf. Vol. I, pp. 426–8
[2] O. W. Richardson and K. T. Compton, *Phil. Mag.* xxiv (1912), p. 595
[3] O. W. Richardson, *Phil. Mag.* xxiii (1912), pp. 263, 594
[4] *Comptes Rendus*, clxxxii (1926), p. 1215
[5] On the time-lag, cf. E. Meyer and W. Gerlach, *Archives des sc. phys. et nat.* xxxvi (1914), p. 253.
[6] *Phys. ZS.* ix (1908), p. 85

of a gas is connected with the ionisation-potential of the gas (measured by the potential-fall necessary to give sufficient kinetic energy to the ionising electron) by the formula

$$I = h\nu.$$

Experiments in agreement with this relation were published in the following year by W. Steubing.[1]

Another hypothesis of the same type, also proposed by Stark [2] in 1908, and elaborated by Einstein [3] in 1912, related to photochemical decomposition: it asserted that when a molecule is dissociated as a result of absorbing radiation of frequency ν, the amount of energy absorbed by the molecule is $h\nu$. There must, therefore, be a lower limit to the frequency of light capable of producing a given chemical reaction, and a relation between the amount of reaction and the amount of light absorbed. The law is applicable only within the range of validity of Wien's law and when the decomposition is purely a thermal effect. Experiments designed to test this hypothesis were made by E. Warburg,[4] with results on the whole favourable.

From Einstein's doctrine that the energy of a photon of frequency ν is $h\nu$, combined with Planck's principle [5] that flux of energy is momentum, it follows at once that in free aether, where the velocity of the photon is c, its momentum [6] must be $h\nu/c$, in the direction of propagation of the light. Long afterwards it was shown experimentally by R. Frisch [7] that when an atom absorbs or emits a photon, the atom experiences a change of momentum of the magnitude and direction attributed to the photon by Einstein: but there had never been any doubt about the matter, since Einstein's value was assumed in the theory of many phenomena, and predictions based on the theory were experimentally verified.

It may be remarked that the above relation between the energy and the momentum of a photon is in agreement with the classical electromagnetic theory of light: for if a beam of light is propagated in free aether in a certain direction, the electric vector **E** and the magnetic vector **H** are equal, and at right angles to each other and to the direction of propagation; and therefore Kelvin's energy-density [8] $1/8\pi\ (E^2 + H^2)$ is $E^2/4\pi$, while J. J. Thomson's momentum-density [9] $1/4\pi c\ [\mathbf{E}.\mathbf{H}]$ is $E^2/4\pi c$: and the latter is equal to the former divided by c.

[1] *Phys. ZS.* x (1909), p. 787

[2] *Phys. ZS.* ix (1908), p. 889: *Ann. d. Phys.* xxxviii (1912), p. 467; cf. E. Warburg, *Verh. d. deutsch. phys. Ges.* ix (1907) p. 753

[3] *Ann. d. Phys.* xxxvii (1912), p. 832: xxxviii (1912), p. 881

[4] *Berlin Sitz.*, 1911, p. 746: 1912, p. 216: 1913, p. 644: 1914, p. 872: 1915, p. 230: 1916, p. 314: 1918, pp. 300, 1228

[5] cf. p. 54

[6] cf. A. Einstein, *Phys. ZS.* x (1909), pp. 185, 817: J. Stark, ibid. pp. 579, 902

[7] *ZS. f. P.* lxxxvi (1933), p. 42

[8] cf. Vol. I, p. 222

[9] cf. Vol. I, p. 317

The formula for the momentum of a photon is also in agreement with the equation

$$\text{momentum} = \text{mass} \times \text{velocity}:$$

for since the energy is $h\nu$, the mass is $h\nu/c^2$; and the velocity is c; so the momentum is $h\nu/c$.

The corpuscular theory of light thus formulated by Einstein leads at once to the relativist formulation of the Doppler effect.[1] For suppose that a star is moving with velocity w relative to an observer P, and that a quantum of light emitted towards the observer P has a frequency ν' as measured by an observer P′ on the star, and a frequency ν as measured by the observer P.

Let the direction-cosines of the line PP′, referred to rectangular axes in P's system of measurement, of which the x-axis is in the direction of the velocity w, be (l, m, n). Then the energy and momentum of the light-quant as observed by P, that is, in a system of reference in which P is at rest, are $(h\nu, -h\nu l/c, -h\nu m/c, -h\nu n/c)$: and the energy of the light-quant as observed by P′, that is, in a system of reference in which P′ is at rest, is $h\nu'$. But the (energy)$/c$ and the three components of momentum form a contravariant four-vector which transforms according to the Lorentz transformation, in which

$$t' = \frac{t - \dfrac{w}{c^2}x}{\left(1 - \dfrac{w^2}{c^2}\right)^{\frac{1}{2}}}$$

and therefore

$$\frac{h\nu'}{c^2} = \frac{\dfrac{h\nu}{c^2} + \dfrac{wh\nu l}{c^3}}{\left(1 - \dfrac{w^2}{c^2}\right)^{\frac{1}{2}}}$$

or

$$\frac{\nu'}{\nu} = \frac{1 + \dfrac{w}{c}l}{\left(1 - \dfrac{w^2}{c^2}\right)^{\frac{1}{2}}}$$

which is the relativist formula for the Doppler effect. Thus *the Doppler effect is simply the Lorentz transformation of the energy-momentum four-vector of the light-quant.*

In the earlier years of the development of quantum-theory, much attention was given to the relation of light-propagation to space,

[1] cf. p. 41

the aim being to find a conception of the mode of propagation which would account both for those experiments which were most naturally explained by the wave-theory, and also for those which seemed to require a corpuscular theory. J. Stark [1] and A. Einstein [2] discussed a fact which seemed very difficult to reconcile with the wave-theory, namely, that when cathode rays fall on a metal plate, and the X-rays there generated fall on a second metal plate, they generate cathode rays whose velocity is of the same order of magnitude as that of the primary cathode rays.

More precisely, let the X-rays be excited by a stream of electrons striking an anticathode (this is, of course, the process inverse to the photo-electric effect). Suppose that the energy of the electrons is what would be obtained by a fall through a potential-difference V, so that the kinetic energy of the electrons is eV. When the electrons are stopped, they give rise to the X-rays, whose frequencies form a continuous [3] spectrum with a limit ν_{max} on the side of high frequencies given by the equation [4]

$$h\nu_{max} = eV.$$

That is, X-rays of frequency ν are not produced unless energy $h\nu$ is available. It is reasonable to suppose that the maximum value of the frequency is obtained when the whole of the energy eV of the electron is converted into energy of radiation (X-radiation of lower frequency is also obtained, because the incident electron may spend part of its energy in causing changes in the atoms of the anticathode). Since the energy of an electron ejected by the X-ray in the photo-electric effect is (save for differences due to other circumstances which need not be considered at the moment) equal to $h\nu_{max}$, we see that it is equal to the energy of the electrons in the cathode rays which had originally excited the X-ray : so that no energy is lost in the changes from electron to X-ray and back to electron again. The X-ray must therefore carry its energy over its whole track in a compact bundle, without any diminution due to spreading : as had been asserted in 1910 by W. H. Bragg (cf. p. 17).

On the other hand, X-rays are certainly of the nature of ordinary light, and can be diffracted : so one would expect them to show the spreading characteristic of waves. The apparent contradiction between the wave-properties of radiation and some of its other properties had been considered by J. J. Thomson in his Silliman lectures of 1903 [5] ; 'Röntgen rays,' he said, 'are able to pass very

[1] *Phys. ZS.* x (1909), pp. 579, 902 : xi (1910), pp. 24, 179

[2] *Phys. ZS.* x (1909), p. 817 ; cf. H. A. Lorentz, *Phys. ZS.* xi (1910), p. 1234

[3] Regarding the discontinuous spectrum of characteristic X-rays, cf. D. L. Webster, *Phys. Rev.* vii (1916), p. 599

[4] cf. W. Duane and F. L. Hunt, *Phys. Rev.* vi (1915), p. 166. The value of *h* was calculated on the basis of this property by F. C. Blake and W. Duane, *Phys. Rev.* ix (1917), p. 568 : x (1917), pp. 93, 624.

[5] J. J. Thomson, *Electricity and Matter* (1904), pp. 63–5, cf. his *Conduction of Electricity through Gases* (1903), p. 258

long distances through gases, and as they pass through the gas they ionise it: the number of molecules so split up is, however, an exceedingly small fraction, less than one-billionth, even for strong rays, of the number of molecules in the gas. Now, if the conditions in the front of the wave are uniform, all the molecules of the gas are exposed to the same conditions: how is it, then, that so small a proportion of them are split up?' His answer was: 'The difficulty in explaining the small ionisation is removed if, instead of supposing the front of the Röntgen ray to be uniform, we suppose that it consists of specks of great intensity separated by considerable intervals where the intensity is very small.'

In this passage Thomson originated the conception of *needle* radiation,[1] i.e. that *in the elementary process of light-emission, the radiations from a source are not distributed equally in all azimuths, but are concentrated in certain directions.* This hypothesis was now adopted by Einstein,[2] who, as we shall see, developed it further in 1916.

When, however, the phenomena of interference were taken into account, the conception of needle radiation, and indeed the whole quantum principle of radiation, met with difficulties which were not resolved for many years. It was shown experimentally [3] that when a classical interference-experiment was performed with light so faint that only a single photon was travelling through the apparatus at any one time, the interference-effects were still produced. This was interpreted at first to mean that a single photon obeys the laws of partial transmission and reflexion at a half-silvered mirror and of subsequent re-combination with the phase-difference required by the wave-theory of light. It is evident, however, that such an explanation would be irreconcilable with the fundamental principle of the quantum theory, according to which interaction between the light and matter at a particular point on the screen can take place only by the absorption or emission of whole quanta of light.

A further objection to the view that coherent beams of light, which are capable of yielding interference-phenomena, could be identical with single photons, appeared when it was found that the volume of a beam of light over which coherence can extend, was much greater than the nineteenth-century physicists had supposed. In 1902 O. Lummer and E. Gehrcke,[4] using green rays from a mercury lamp, obtained interference-phenomena with a phase-difference of 2,600,000 wave-lengths—a distance of the order of one

[1] cf. Thomson's further papers in *Proc. Camb. Phil. Soc.* xiv (1907), p. 417; *Phil. Mag.*(6) xix (1910), p. 301; and N. R. Campbell, *Proc. Camb. Phil. Soc.* xv (1910), p. 310. For an interesting application of the relativity equations to needle radiation, cf. H. Bateman, *Proc. Lond. Math. Soc.*(2) viii (1910), p. 469.

[2] loc. cit.

[3] G. I. Taylor, *Proc. Camb. Phil. Soc.* xv (1909), p. 114; most completely by A. J. Dempster and H. F. Batho, *Phys. Rev.* xxx (1927), p. 644; cf. E. H. Kennard, *J. Franklin Inst.*, ccvii (1929), p. 47

[4] *Verh. deutsch. phys. Ges.* iv (1902), p. 337; cf. M. von Laue, *Ann. d. Phys.*(4) xiii (1904), p. 163, §6

metre. Regarding the lateral extension of a coherent beam, since all the light from a star that enters a telescope-objective takes part in the formation of the image, it is evident that this light must be coherent: and a still greater estimate was obtained in 1920, when interference methods were used at Mount Wilson Observatory to determine the angular diameter of Betelgeuse, and interference was obtained between beams which arrived from the star twenty feet apart. It seemed impossible that these very large coherent beams could be single photons.

The alternative hypothesis seemed to be, that the photons in a coherent beam form a regular aggregate, possessing a quality equivalent to the coherence.[1] One supposition was that the motion of the photons is subject to a system of probability corresponding to the wave-theory explanation of interference, so that a large number of them is directed to the bright places of the interference-pattern, and few or none are directed to the dark places. This explanation was, however, unsatisfactory: for it is not until the two interfering beams of light have actually met that the interference-pattern is determined, and therefore the guiding of the motion of the photons cannot take place during the propagation of the interfering beams, but must happen later, perhaps at the screen itself—a process difficult to imagine. It was therefore suggested that while the photons are being propagated in the beams which are later destined to interfere, they are characterised by a quality corresponding to what in the wave-theory is called *phase*. In order to construct a definite theory based on this idea, however, it would be necessary first to consider whether the photons are to be regarded as points (in which case any particular photon would always retain the same phase, since it travels with the velocity of light, but different photons would have different phases) or whether the photons are to be regarded as extending over finite regions (in which case the phase would presumably vary from one point of a photon to another). In order to account for interference, it would be necessary to suppose that photons, or parts of photons, in opposite phases, neutralise each other. But if the photons are regarded as points, the mutual annulment of two complete photons would be incompatible with the principle of conservation of energy, while if the photons are regarded as extending over finite regions, the annulment of part of one by part of another would seem to be incompatible with the integral character of photons. There is, moreover, a difficulty created by the observation that interference can take place when only a single photon is travelling through the apparatus at any one instant, for this seems to require that the effect of a quantum on an atom persists for some time.

These various attempts—none of them entirely satisfactory—to combine the new and the old conceptions of light, created a doubt

[1] This was first put forward by J. Stark, loc. cit.

as to whether it was possible to construct, within the framework of space and time, a picture or model which would be capable of representing every known phenomenon in optics.[1]

In December 1906 Einstein [2] initiated a new development of quantum theory, by carrying its principles outside the domain of radiation, to which they had hitherto been confined, and applying them to the study of the specific heats of solids. We have seen [3] that according to the classical law of equipartition of energy, in a state of statistical equilibrium at absolute temperature T, with every degree of freedom of a dynamical system there is associated on the average a kinetic energy $\frac{1}{2}kT$, where k is Boltzmann's constant. Now the thermal motions of a crystal are constituted by the elastic vibrations of its atoms about their positions of equilibrium: one of these atoms, since it has three kinetic degrees of freedom and also three potential-energy degrees of freedom, will have a mean kinetic energy $\frac{3}{2}kT$ and a mean potential energy $\frac{3}{2}kT$, or a total mean energy $3kT$. Thus a gramme-atom of the crystal will have a mean energy $3kNT$, where N is Avogadro's number [4]; or 3RT, where R is the gas-constant per gramme-atom. The atomic heat of the crystal (i.e. the amount of heat required to raise the temperature of a gramme-atom by one degree) is therefore 3R. Now

$$R = 8{\cdot}3136 \times 10^7 \text{ ergs} = 1{\cdot}986 \text{ cal.}$$

so the atomic heat (at constant volume) is 5·958 cal. This law had been discovered empirically by P. L. Dulong and A. T. Petit [5] in 1819.

While Dulong and Petit's law is approximately true for a great many elements at ordinary temperatures, exceptions to it had long been known, particularly in the case of elements of low atomic weight, such as C, Bo, Si, for which at ordinary temperatures the atomic heats are much smaller than 5·958 cal.: and shortly before this time it had been shown, particularly by W. Nernst and his pupils, that at very low temperatures *all* bodies have small atomic heats, while at sufficiently high temperatures even the elements of low atomic weight obey the normal Dulong-Petit rule, as was shown e.g. by experiments with graphite at high temperatures.

[1] The state of the coherence problem twenty years after Einstein's paper of 1905 may be gathered from G. P. Thomson, *Proc. R.S.*(A), civ (1923), p. 115; A. Landé, *ZS. f. P.* xxxiii (1925), p. 571; E. C. Stoner, *Proc. Camb. Phil. Soc.* xxii (1925), p. 577; W. Gerlach and A. Landé, *ZS. f. P.* xxxvi (1926), p. 169; E. O. Lawrence and J. W. Beams, *Proc. N.A.S.* xiii (1927), p. 207.

[2] *Ann. d. Phys.* xxii (1907), pp. 180, 800

[3] Vol. I, p. 382. The argument given here is substantially due to Boltzmann, *Wien Sitz.* lxiii (Abth. 2) (1871), p. 712.

[4] cf. pp. 8, 18

[5] *Ann. de chim. et de. phys.* x (1819), p. 395; *Phil. Mag.* liv (1819), p. 267

Einstein now pointed out that if we write Planck's formula in the form :

Energy-density of radiation in the frequency-range ν to $\nu + d\nu$

$$= 8\pi\lambda^{-4} d\lambda \frac{h\nu}{e^{h\nu/kT} - 1},$$

then on comparing this with Rayleigh's derivation of his law of radiation,[1] we see that in order to obtain Planck's formula, a mode of vibration of frequency ν must be counted as possessing the average energy

$$\tfrac{1}{2}kT \frac{x}{e^x - 1},$$

where $x = h\nu/kT$, instead of (as Rayleigh assumed) $\tfrac{1}{2}kT$. If then in the above proof of Dulong and Petit's law we replace $\tfrac{1}{2}kT$ by $\tfrac{1}{2}kT\, x/(e^x - 1)$, we find that a gramme-atom of the crystal will have a mean energy

$$3RT \frac{x}{e^x - 1}$$

(if for simplicity we assume that all the atomic vibrators have the same frequency ν) and therefore the atomic heat is

$$\frac{d}{dT} \cdot \left\{ 3RT \frac{x}{e^x - 1} \right\}, \quad \text{or} \quad 3R \frac{x^2 e^x}{(e^x - 1)^2}, \quad \text{or} \quad 5{\cdot}958 \frac{x^2 e^x}{(e^x - 1)^2}.$$

This is Einstein's formula. As the temperature falls, x increases and $x^2 e^x/(e^x - 1)^2$ decreases, so the decrease of atomic heat with temperature is accounted for. As the absolute zero of temperature is approached, the atomic heats of all solid bodies tend to zero.[2]

The determination of the quantity x, i.e. the determination of ν, was studied by E. Madelung,[3] W. Sutherland,[4] F. A. Lindemann,[5] A. Einstein,[6] W. Nernst,[7] E. Grüneisen,[8] C. E. Blom [9] and H. S. Allen.[10] : relations were found connecting ν approximately with the cubical compressibility of the crystal, with its melting-point and with the ' residual rays ' which are strongly reflected from it.

[1] Vol. I, pp. 383–4

[2] This is a special case of a more general theorem discovered and developed by W. Nernst in 1910 and the following years, namely, that all the properties of solids which depend on the average behaviour of the atoms (including the thermodynamic functions) become independent of the temperature at very low temperatures. Thus, at the absolute zero of temperature, the entropy of every chemically homogeneous body is zero ; cf. Nernst, *Die theoretischen und experimentellen Grundlagen des neuen Wärmesatzes* (Halle, 1918).

[3] *Gött. Nach.* (1909), p. 100 ; *Phys. ZS.* xi (1910), p. 898

[4] *Phil. Mag.* (6) xx (1910), p. 657

[5] *Phys. ZS.* xi (1910), p. 609

[6] *Ann. d. Phys.* xxxiv (1911), p. 170 ; xxxv (1911), p. 679

[7] *Ann. d. Phys.* xxxvi (1911), p. 395

[8] *Ann. d. Phys.* xxxix (1912), p. 257

[9] *Ann. d. Phys.* xlii (1913), p. 1397

[10] *Proc. R.S.* (A), xciv (1917), p. 100 ; *Phil. Mag.* xxxiv (1917), pp. 478, 488

There was, however, one obvious imperfection in all this work, which was pointed out by Einstein in the second of the papers just referred to ; namely, that the vibrations of the atoms in a crystal do not all have the same frequency ν. The mean energy of a gramme-atom of a crystal will therefore not be

$$3RT \frac{x}{e^x - 1}$$

where x has a single definite value, but

$$kT\sum_r \frac{x_r}{e^{x_r} - 1},$$

when the summation is taken over all the frequencies (three for each atom), and k is Boltzmann's constant. The atomic heat, obtained from this by differentiating with respect to T (remembering that $x = h\nu/kT$), is therefore

$$\sum_r \frac{kx_r^2 e^{x_r}}{(e^{x_r} - 1)^2}.$$

In order to determine the frequencies of the natural vibrations of the atoms of a body, and so to be able to evaluate these expressions, P. Debye [1] (*b.* 1884) took, as an approximation to the actual body, an elastic solid, and considered the elastic waves in it. He showed that for a fixed isotropic body of volume V, the number of natural periods or modes of vibration in the frequency-range ν to $\nu + d\nu$ is

$$4\pi V\nu^2 d\nu\left(\frac{2}{c_t^3} + \frac{1}{c_l^3}\right)$$

where c_t is the velocity of transverse waves in the solid and c_l the velocity of longitudinal waves ; and he assumed that the energies of these different sound waves vary in the same way as the energies of the light waves in Planck's formula, so that each of these modes has the energy

$$\frac{h\nu}{e^{h\nu/kT} - 1};$$

thus the energy per unit volume of waves within the frequency-range ν to $\nu + d\nu$ is

$$4\pi h\nu^3\left(\frac{2}{c_t^3} + \frac{1}{c_l^3}\right)\frac{d\nu}{e^{h\nu/kT} - 1}.$$

[1] *Ann. d. Phys.* xxxix (1912), p. 789

The total energy is to be obtained by integrating this with respect to ν. But here a difficulty presents itself: for the number of natural frequencies of a continuous body is infinitely great: ν extends from zero to infinity. Debye (somewhat arbitrarily) dealt with this situation by taking, in the integration with respect to ν, an upper limit ν_m, such that the total number of frequencies less than ν_m is equal to 3N, where N is the number of atoms in the body. Thus ν_m is to be determined from the equation

$$3N = \int_0^{\nu_m} 4\pi V\left(\frac{2}{c_t^3}+\frac{1}{c_l^3}\right)\nu^2 d\nu = \frac{4\pi V}{3}\left(\frac{2}{c_t^3}+\frac{1}{c_l^3}\right)\nu_m^3,$$

and the atomic heat of the body is

$$4\pi k V\left(\frac{2}{c_t^3}+\frac{1}{c_l^3}\right)\int_0^{\nu_m}\frac{x^2 e^x}{(e^x-1)^2}\nu^2 d\nu \quad \text{where} \quad x=\frac{h\nu}{kT}$$

or

$$\frac{9R}{x_m^3}\int_0^{x_m}\frac{x^4 e^x dx}{(e^x-1)^2}, \quad \text{where} \quad x_m=\frac{h\nu_m}{kT}.$$

If we write

$$\frac{h\nu_m}{k}=\Theta,$$

we have $x_m=\Theta/T$, and *the atomic heat is a universal function of the ratio* T/Θ, *that is, the temperature* T *divided by a temperature* Θ *which is characteristic of the body.*

Debye's theory is in good general accord with the experimental results for many elements.[1]

When the temperature T is very great, x_m is very small, and the above formula for the atomic heat becomes

$$\frac{9R}{x_m^3}\int_0^{x_m} x^2 dx \quad \text{or} \quad 3R,$$

which is Dulong and Petit's law, as would be expected.

When on the other hand T is very small, we escape the difficulties which arise from the fact that the body, as contrasted with the continuous elastic solid, has only a finite number of degrees of freedom. The above formula shows that the energy of the body is proportional at low temperatures to the integral

$$\int_0^{\infty}\frac{\nu^3 d\nu}{e^{h\nu/kT}-1} \quad \text{or} \quad \int_0^{\infty}\nu^3 e^{-h\nu/kT}d\nu,$$

that is, it is proportional to T^4; *so the atomic heat of a body at low*

[1] cf. the exhaustive report by E. Schrödinger, *Phys. ZS.* xx (1919), pp. 420, 450, 474, 497, 523

temperatures is proportional to the cube of the absolute temperature. This law has been carefully verified.[1]

In the same year in which Debye's theory of atomic heats was published, Max Born (*b.* 1882) and Theodor v. Kármán (*b.* 1881)[2] attacked the problem from another angle; instead of replacing the body by a continuous elastic medium, as Debye had done, they made a dynamical study of crystals, regarded as Bravais space-lattices of atoms, in order to determine their natural periods. The formulae obtained were of considerable complexity: Debye's method is simpler, though Born and Kármán's is more general and stringent. In the case of low temperatures, Born and Kármán confirmed Deybe's result that the atomic heat is proportional to T^3 when T is small.[3]

The behaviour, as found by experiment, of the molecular heat of gases (i.e. the amount of heat required to raise by one degree the temperature of one gramme-molecule of the gas, at constant volume) can be explained by quantum theory in much the same way as that of solid bodies. According to classical theory,[4] a monatomic gas (such as helium, argon or mercury vapour) has three degrees of freedom (namely, the three required for translatory motion), and to each of them should correspond a mean kinetic energy $\frac{1}{2}kT$, so a gramme-molecule should have an energy $\frac{3}{2}kNT$ when N is Avogadro's number, or $\frac{3}{2}RT$ where R is the gas-constant per gramme-molecule: thus the molecular heat should be $\frac{3}{2}R$ or approximately three calories, a result verified empirically.[5]

For the chief diatomic gases—H, N, O etc.—the molecular heat is 5 calories, which is explained classically by supposing that they have 3 translatory and 2 rotational degrees of freedom. The molecule may be pictured as a rigid dumbbell, having no oscillations along the line joining the atoms, and no rotations about this line as axis. It was, however, found experimentally by A. Eucken[6] that the molecular heat of hydrogen at temperatures below 60° abs. falls to 3 calories, the same value as for monatomic gases. This evidently implies that the part of the molecular heat which is due to the two rotational degrees of freedom of the molecule falls to zero at low temperatures. The quantum theory supplies an obvious explanation

[1] A. Eucken and F. Schwers, *Verh. deutsch phys. Ges.* xv (1913), p. 578; W. Nernst and F. Schwers, *Berlin Ber.* (1914), p. 355; W. H. Keesom and H. Kamerlingh Onnes, *Amsterdam Proc.* xvii (1915), p. 894; xviii (1915), p. 484

[2] *Phys. ZS.* xiii (1912), p. 297; xiv (1913), pp. 15, 65

[3] The diamond was studied specially by Born, *Ann. d. Phys.* xliv (1914), p. 605. cf. Born, *Dynamik der Kristallgitter* (Leipzig, 1915), and many later papers. The detailed study of crystal-theory is beyond the scope of the present work.

[4] cf. Vol. I, p. 383

[5] The quantity which is usually determined directly by experiment (from the velocity of sound in the gas) is the ratio of the molecular heat at constant pressure to the molecular heat at constant volume, or $(2+x)/x$, where x is the molecular heat at constant volume.

[6] *Berlin Sitz.* (1912), p. 141. cf. also K. Scheel and W. Heuse, *Ann. d. Phys.* xl (1913), p. 473, who examined the specific heats of helium, and of nitrogen, oxygen and other diatomic gases, between +20° and −180°; and F. Reiche, *Ann. d. Phys.* lviii (1919), p. 657.

of this behaviour: it is, that no vibration can be excited except by the absorption of quanta of energy that are whole multiples of $h\nu$, where ν is the proper frequency of the vibration: and at low temperatures, the energy communicated by molecular impacts is insufficient to do this, so far as the rotational degrees of freedom are concerned. For the translational degrees of freedom, on the other hand, ν is effectively zero, so no limitation is imposed.

At very high temperatures the molecular heats of the permanent gases rise above 5 calories—to 6 or nearly 7—which evidently signifies that some additional degrees of freedom have come into action, e.g. vibrations along the line joining the two atoms in the molecule. For chlorine and bromine, this phenomenon is observed even at ordinary temperatures, a fact which may be explained by reference to the looser connection of the atoms in the molecules of these elements.

A new prospect opened in 1909, when Einstein [1] discussed the fluctuations in the energy of radiation in an enclosure which is at a given temperature T. From general thermodynamics it can be shown that at any place in the enclosure the mean square of the fluctuations of energy per unit volume in the frequency-range from ν to $\nu + d\nu$, which we may denote by $\overline{\epsilon^2}$, is $kT^2\, dE/dT$, where k is Boltzmann's constant and E is the mean energy per unit volume. Now by Planck's law we have

$$E = \frac{8\pi h\nu^3}{c^3}\frac{d\nu}{e^{h\nu/kT} - 1}$$

whence for $\overline{\epsilon^2}$ we obtain the value

$$\frac{8\pi h^2\nu^4 d\nu}{c^3}\left\{\frac{1}{e^{h\nu/kT} - 1} + \frac{1}{(e^{h\nu/kT} - 1)^2}\right\}$$

or

$$h\nu E + \frac{c^3 E^2}{8\pi\nu^2 d\nu}.$$

If instead of Planck's law of radiation we had taken Wien's law,[2] we should have obtained

$$\overline{\epsilon^2} = h\nu E$$

while if we had taken Rayleigh's law,[3] we should have obtained

$$\overline{\epsilon^2} = \frac{c^3 E^2}{8\pi\nu^2 d\nu}.$$

Thus *the mean-square of the fluctuations according to Planck's law is the sum of the mean-squares of the fluctuations according to Wien's law and Rayleigh's law*, a result which, seen in the light of the principle that fluctuations due to independent causes are additive, suggests that

[1] *Phys. ZS.* x (1909), p. 185 [2] cf. Vol. I, p. 381 [3] cf. Vol. I, p. 384

the causes operative in the case of high frequencies (for which Wien's law holds) are independent of those operative in the case of low frequencies (for which the law is Rayleigh's). Now Rayleigh's law is based on the wave-theory of light, and in fact the value $c^3E^2/(8\pi\nu^2 d\nu)$ for the mean-square fluctuation was shown by Lorentz [1] to be a consequence of the interferences of the wave-trains which, according to the classical picture, are crossing the enclosure in every direction : whereas the value $h\nu E$ for the mean-square fluctuation is what would be obtained if we were to take the formula for the fluctuation of the number of molecules in unit volume of an ideal gas, and suppose that each molecule has energy $h\nu$: that is, the expression is what would be obtained by a corpuscular-quantum theory. Moreover, the ratio of the particle-term to the wave-term in the complete expression for the fluctuation is $e^{h\nu/kT}-1$: so when $h\nu/kT$ is small, i.e. at low frequencies and high temperatures, the wave-term is predominant, and when $h\nu/kT$ is large, i.e. when the energy-density is small, the particle-term is predominant. The formula therefore suggests that light cannot be represented completely either by waves or by particles, although for certain classes of phenomena the wave-representation is practically sufficient, and for other classes of phenomena the particle-representation. The undulatory and corpuscular theories are in some sense both true.

Some illuminating remarks on Einstein's formula were made by Prince Louis Victor de Broglie [2] (*b.* 1892). Planck's formula

$$E=\frac{8\pi h\nu^3}{c^3}\frac{d\nu}{e^{h\nu/kT}-1}$$

may be written

$$E=\frac{8\pi h\nu^3}{c^3}(e^{-h\nu/kT}+e^{-2h\nu/kT}+e^{-3h\nu/kT}+\ldots)d\nu$$

$$=E_1+E_2+E_3+\ldots$$

where

$$E_s=\frac{8\pi h\nu^3}{c^3}e^{-sh\nu/kT}d\nu.$$

Now Einstein's formula is

$$\overline{\epsilon^2}=\frac{8\pi h^2\nu^4 d\nu}{c^3}\left\{\frac{1}{e^{h\nu/kT}-1}+\frac{1}{(e^{h\nu/kT}-1)^2}\right\}$$

$$=\frac{8\pi h^2\nu^4 d\nu}{c^3}\left\{e^{-h\nu/kT}+2e^{-2h\nu/kT}+3e^{-3h\nu/kT}+\ldots.\right\}$$

$$=\sum_{s=1}^{\infty} sh\nu E_s.$$

[1] cf. Lorentz, *Les théories statistiques en thermodynamique* (Leipzig, 1916), p. 114

[2] *Comptes Rendus*, clxxv (1922), p. 811 ; *J. de phys.* iii (1922), p. 422. cf. W. Bothe, *ZS. f. P.* xx (1923), p. 145 ; *Naturwiss.* xi (1923), p. 965

This resembles the first term $h\nu E$ in Einstein's formula, but it is now summed for all values of s. So it is precisely the result we should expect if the energy E_s were made up of light-quanta each of energy $sh\nu$. Thus de Broglie suggested that the term E_1 should be regarded as corresponding to energy existing in the form of quanta of amount $h\nu$, that the second term E_2 should be regarded as corresponding to energy existing in the form of quanta of amount $2h\nu$, and so on. So *Einstein's formula for the fluctuations may be obtained on the basis of a purely corpuscular theory of light, provided the total energy of the radiation is suitably allocated among corpuscles of different energies* $h\nu$, $2h\nu$, $3h\nu$, ...[1]

The theory of the fluctuations of the energy of radiation was developed further in many subsequent papers.[2]

In spite of the many triumphs of the quantum theory, its discoverer Planck was in 1911 still dissatisfied with it, chiefly because it could not be reconciled with Maxwell's electromagnetic theory of light. In that year he proposed[3] a new hypothesis, namely, that although emission of radiation always takes place discontinuously in quanta, absorption on the other hand is a continuous process, which takes place according to the laws of the classical theory. Radiation while in transit might therefore be represented by Maxwell's theory, and the energy of an oscillator at any instant might have any value whatever. When an oscillator has absorbed an amount $h\nu$ of energy, it has a chance of emitting this exact amount: but it does not necessarily take the opportunity, so that emission is a matter of probability. In the new theory, therefore, there were no discontinuities in *space*, although the act of emission involved a discontinuity in *time*.

The system based on these principles is generally called *Planck's Second Theory*. He showed that it can lead to the same formula for black-body radiation as the original theory of 1900; but there is a notable difference, in that the mean-energy of a linear oscillator of frequency ν is now

$$\tfrac{1}{2}h\nu \frac{e^{h\nu/kT}+1}{e^{h\nu/kT}-1},$$

which is greater by $\frac{1}{2}h\nu$ than the value given by the earlier theory: so that *at the absolute zero of temperature, the mean energy of the oscillator is* $\frac{1}{2}h\nu$. This was the first appearance in theoretical physics of the

[1] This corpuscular theory had been proposed in the previous year by M. Wolfke, *Phys. ZS.* xxii (1921), p. 375.

[2] M. von Laue, *Verh. d. deutsch phys. Ges.* xvii (1915), p. 198; W. Bothe, *ZS. f. P.* xx (1923), p. 145; M. Planck, *Berlin Sitz.* xxxiii (1923), p. 355; *Ann. d. Phys.* lxxiii (1924), p. 272; P. Ehrenfest, *ZS. f. P.* xxxiv (1925), p. 362; M. Born, W. Heisenberg, u. P. Jordan, *ZS. f. P.* xxxv (1926), p. 557; S. Jacobsohn, *Phys. Rev.* xxx (1927), pp. 936, 944; J. Solomon, *Ann. de phys.* xvi (1931), p. 411; W. Heisenberg, *Leipzig Ber.* lxxxiii (1931), Math.-Phys. Klasse, p. 3; M. Born and K. Fuchs, *Proc. R.S.* clxxii (1939), p. 465

[3] *Verh. d. deutsch. phys. Ges.* xiii (1911), p. 138

doctrine of *zero-point energy*, which later assumed great importance. In 1913 A. Einstein and O. Stern [1] made it the basis of a new proof of Planck's radiation-formula, and in 1916 W. Nernst [2] suggested that the aether everywhere might be occupied by zero-point energy.

Planck's Second Theory was criticised in 1912 by Poincaré,[3] and in 1914 Planck [4] came to the conclusion that emission by quanta could scarcely be reconciled with classical doctrines, so he now made a new proposal (known as his *Third Theory*), namely, that the emission as well as the absorption of radiation by oscillators is continuous, and is ruled by classical electrodynamics, and that quantum discontinuities take place only in exchanges of energy by collisions between the oscillators and free particles (molecules, ions and electrons). A year later, however, he [5] abandoned the Third Theory, having become convinced by a paper of A. D. Fokker [6] that the calculation of the stationary state of a system of rotating rigid electric dipoles in a given field of radiation, when the calculation was performed according to the rules of classical electrodynamics, led to results that were in direct contradiction with experiment. The Second Theory fell from favour with most physicists about the same time, when the experiments of Franck and Hertz [7] showed the strong analogy between optical absorption and the undoubtedly quantistic phenomena which take place when slow electrons collide with molecules.

Meanwhile, in a Report presented to the Physical Section of the 83rd Congress of German men of science at Karlsruhe on 25 September 1911, Sommerfeld [8] made a suggestion which was the first groping towards a new method. Referring to the name *Quantum of Action* which had been given to the quantity h by Planck, on account of the fact that its dimensions were those of (Energy × Time) or *Action*, he remarked that there should be some connection between h and the integral which appears in Hamilton's Principle, namely,

$$\int (\mathrm{T}\text{-}\mathrm{V})\,dt$$

where T denotes the kinetic and V the potential energy of the mechanical system considered. He proposed to achieve this by making the following hypothesis: *In every purely molecular process, a certain definite amount of Action is absorbed or emitted, namely, the amount*

$$\int_0^\tau \mathrm{L}\,dt = \frac{1}{2\pi}h$$

[1] *Ann. d. Phys.* xl (1913), p. 551

[2] *Verh. d. deutsch. phys. Ges.* xviii (1916), p. 83

[3] *J. de phys.*(5) ii (1912), p. 5, at p. 30

[4] *Berlin Sitz.* 30 July 1914, p. 918

[5] *Berlin Sitz.*, 8 July 1915, p. 512

[6] *Ann. d. Phys.* xliii (1914), p. 810

[7] Described below in Chapter IV

[8] *Verh. deutsch. phys. Ges.* xiii (1911), p. 1074

where L = T − V *is the kinetic potential or Lagrangean function, and where* τ *is the duration of the process*,. A discussion of the photo-electric effect in the light of this principle was given in 1913 by Sommerfeld and Debye.[1] The principle itself, however, was superseded in the later development of the subject.

[1] *Ann. d. Phys.* xli (1913), p. 873

Chapter IV

SPECTROSCOPY IN THE OLDER QUANTUM THEORY

In the nineteenth century, it was generally supposed that the luminous vibrations represented by line spectra were produced in the same way as sounds are produced by the free vibrations of a material body. That is to say, the atom was regarded as an electrical system of some kind, which had a large number of natural periods of oscillation, corresponding to the aggregate of its spectral lines. The first physicist to break with this conception was Arthur William Conway (1875–1950), professor of mathematical physics in University College, Dublin, who in 1907, in a paper of only two and a half pages,[1] enunciated the principles on which the true explanation was to be based : namely, that the spectrum of an atom does not represent the free vibrations of the atom as a whole, but that an atom produces spectral lines one at a time, so that the production of the complete spectrum depends on the presence of a vast number of atoms. In Conway's view, an atom, in order to be able to generate a spectral line, must be in an abnormal or disturbed state : and in this abnormal state, a single electron, situated within the atom, is stimulated to produce vibrations of the frequency corresponding to the spectral line in question. The abnormal state of the atom does not endure permanently, but lasts for a time sufficient to enable the active electron to emit a fairly long train of vibrations.

Conway had not at his disposal in 1907 certain facts about spectra and atoms which were indispensable for the construction of a satisfactory theory of atomic spectra : for until after the publication of Ritz's paper of 1908 [2] physicists did not realise that the frequencies of the lines in the spectrum of an element are the differences, taken in pairs, of certain numbers called ' terms ' ; and it was not until 1911 that Rutherford [3] introduced his model atom, constituted of a central positively-charged nucleus with negative electrons circulating round it. But the revolutionary general principles that Conway introduced were perfectly sound, and showed a remarkable physical insight.

These principles were reaffirmed in 1910 by Penry Vaughan Bevan (1875–1913), in a paper [4] recording experiments on the anomalous dispersion, by potassium vapour, of light in the region of the red lines of the potassium spectrum. He first attempted to explain his results theoretically in accordance with Lorentz's modernised version [5] of the Maxwell-Sellmeier theory, and found

[1] *Sci Proc. R. Dubl. Soc.* xi (March 1907), p. 181
[2] cf. Vol. I, p. 378
[3] cf. p. 22
[4] *Proc. R.S.*(A), lxxxiv (1910), p. 209
[5] cf. Vol. I, p. 401

that it required him to postulate an impossibly great number of electrons per molecule : deriving from this contradiction the correct conclusion, that spectroscopic phenomena are to be explained by the presence of a very great number of atoms, which at any one instant are in different states, and each of which at that instant is concerned not with the whole spectrum, but at most with only one line of it.

The next advances were made by John William Nicholson [1] at that time of Trinity College, Cambridge. He introduced into spectroscopic theory the model atom which had been proposed a few months before by Rutherford, namely a very small nucleus carrying practically all the mass of the atom, surrounded by negative electrons circling round it like planets round the sun. This was so much more precisely defined than earlier model atoms that it might now be possible to calculate exact numerical values for the wave-lengths of lines in atomic spectra. Nicholson's second advance was to recognise the fundamental fact, that the production of atomic spectra is essentially a quantum phenomenon. ' The fundamental physical laws,' he said, ' must lie in the quantum or unit theory of radiation, recently developed by Planck and others, according to which, interchanges of energy between systems of a periodic kind can only take place in certain definite amounts determined by the frequencies of the systems ' [2]; and he discovered the form which the quantum principle should take in its application to the Rutherford atom : ' *the angular momentum of an atom can only rise or fall by discrete amounts*.' [3] Moreover, following Conway and Bevan, he asserted that the different lines of a spectrum are produced by different atoms : ' the lines of a series may not emanate from the same atom, but from atoms whose internal angular momenta have, by radiation or otherwise, run down by various discrete amounts from a standard value. For example, in this view there are various kinds of hydrogen atom, identical in chemical properties and even in weight, but different in their internal motions.' [4] In other words, an atom of a given chemical element may exist in many different *states*, resembling in many ways the energy-levels of Planck's oscillators. And ' the incapacity ' of an atom ' for radiating in a continuous way will secure sharpness of the lines.' [5] Nicholson did not, however, fully appropriate Conway's idea that a single electron (among the many present in the atom) is alone concerned in the production of a spectral line : on the contrary, he studied the vibrations of a number of electrons circulating round a nucleus by methods recalling the work of Maxwell on Saturn's rings, retaining classical ideas for the actual computation of the motion, and identifying the frequency of spectral lines with the frequency of vibration of a dynamical system.

[1] *Mon. Not. R.A.S.* lxxii, pp. 49 (November 1911), 139 (December 1911), 677, 693 (June 1912), 729 (August 1912)
[2] loc. cit. p. 729 [3] loc. cit. p. 679 [4] loc. cit. p. 730 [5] ibid.

The first successful application of quantum principles to spectroscopy was in the domain not of atomic but of molecular spectra. In 1912 Niels Bjerrum[1] applied quantum ideas in order to explain certain characteristics of the absorption spectra that had been observed with hydrochloric and hydrobromic acids in their gaseous forms. For these and similar compounds, two widely-separated regions of absorption had been found in the infra-red, of which one, that of longest wave-length, was assigned by Bjerrum, following Drude,[2] to rotations of the molecules. For the absorption in the short-wave infra-red he gave a new explanation. He assumed that the two atoms constituting a molecule are positively and negatively charged respectively, and that they oscillate relatively to each other along the line joining them, say with frequency ν_0; incident radiation of this frequency is absorbed. Moreover (following a suggestion of Lorentz), he assumed that the line joining the atoms rotates in a plane, and that the rotational energy must be a multiple of $h\nu$, where ν is the number of revolutions per second.[3] Denoting by J the moment of inertia of the rotating system, the rotational energy is $\frac{1}{2}\mathrm{J}\,.\,(2\pi\nu)^2$, and we have

$$\tfrac{1}{2}\mathrm{J}\,.\,(2\pi\nu)^2 = nh\nu \qquad (n=0,\ 1,\ 2,\ 3,\ \ldots):$$

or, denoting this value of ν by ν_n, we have

$$\nu_n = \frac{nh}{2\pi^2\mathrm{J}}.$$

By comparing the linear and rotational motions,[4] oscillations are obtained of frequencies ν_n, $\nu_0+\nu_n$, and $\nu_0-\nu_n$: so the absorption-spectrum should contain in the short-wave infra-red the equidistant frequencies

$$\nu_0, \quad \nu_0 \pm \frac{h}{2\pi^2\mathrm{J}}, \quad \nu_0 \pm \frac{2h}{2\pi^2\mathrm{J}}, \quad \nu_0 \pm \frac{3h}{2\pi^2\mathrm{J}}, \quad \text{etc.}$$

Bjerrum's theory stimulated more careful experimental measurements by W. Burmeister,[5] Eva von Bahr,[6] J. B. Brinsmade and E. C. Kemble,[7] and E. S. Imes.[8] Burmeister, and afterwards Imes, found that the central frequency ν_0 was not observed, which would seem to indicate that rotation is always present.[9]

[1] *Nernst Festschrift*, 1912, p. 90 [2] *Ann. d. Phys.*(4) xiv (1904), p. 677

[3] The quantification of molecular rotations had been suggested by Nernst, *ZS. f. Elektrochem.* xvii (1911), p. 265.

[4] The principle of this composition is due to Rayleigh, *Phil. Mag.*(5) xxiv (1892), p. 410.

[5] *Verh. d. deutsch. phys. Ges.* xv (1913), p. 389

[6] *Verh. d deutsch. phys. Ges.* xv (1913), pp. 710, 731, 1150

[7] *Proc. Nat. Acad. Sci.* iii (1917), p. 420 [8] *Astrophys. J.* l (1919), p. 251

[9] It was later found necessary to modify Bjerrum's theory by associating the lines not with the actual rotations, but with *transitions* from one state of rotation to another.

In the year following Bjerrum's paper, P. Ehrenfest [1] improved the quantum theory of rotation by assuming that if ν is the number of revolutions per second, then the rotational energy must be a multiple of $\frac{1}{2}h\nu$ (not $h\nu$, as Bjerrum had supposed); the factor $\frac{1}{2}$ being inserted because the rotational energy is purely kinetic, and not, as in the case of an oscillator, half potential.[2] If J is the moment of inertia, so that the rotational energy is $\frac{1}{2}J \cdot (2\pi\nu)^2$, we have

$$\tfrac{1}{2}J \cdot (2\pi\nu)^2 = n \cdot \tfrac{1}{2}h\nu \qquad (n=0, 1, 2, 3, \ldots)$$

or

$$\nu = \frac{nh}{4\pi^2 J}.$$

The angular momentum $2\pi\nu J$ therefore has the value $nh/2\pi$. The quantity $h/2\pi$ is now generally denoted [3] by $\hbar$. Thus *the law of quantification of angular momentum is that it must be a whole multiple of* $\hbar$.

The culmination of the efforts to explain atomic spectra came in July 1913, when Niels Bohr (*b.* 1885), a Danish research student of Rutherford's at Manchester, found [4] the true solution of the problem. With unerring instinct Bohr seized upon whatever was right in the ideas of his predecessors, and rejected what was wrong, adjoining to them precisely what was needed in order to make them fruitful, and eventually producing a theory which has been the starting-point of all subsequent work in spectroscopy.

Bohr accepted Conway's principles that (1) atoms produce spectral lines one at a time, and (2) that a single electron is the agent in the process, together with Nicholson's principles that (3) the Rutherford atom provides a satisfactory basis for exact calculations of wave-lengths of spectral lines, (4) the production of atomic spectra is a quantum phenomenon, (5) an atom of a given chemical element may exist in different *states*, characterised by certain discrete values of its angular momentum and also discrete values of its energy. He discovered independently [5] Ehrenfest's principle (6) that in quantum-theory, angular momenta must be whole multiples of $\hbar$. He further adopted a principle which was suggested by Ritz's law of spectral ' terms,' and had been in some degree adumbrated but perhaps not clearly grasped by Nicholson, namely (7) that *two* distinct states of the atom are concerned in the production of a spectral line: and he recognised an exact correspondence between the ' terms ' into which the spectra were analysed

[1] *Verh. d. deutsch. phys. Ges.* xv (1913), p. 451

[2] Ehrenfest's assumption was confirmed by E. C. Kemble, *Phys. Rev.*(2) viii (1916), p. 689.

[3] Read ' crossed *h* '

[4] *Phil. Mag.* xxvi (1913), pp. 1, 476, 875; xxvii (1914), p. 506; xxix (1915), p. 332; xxx (1915), p. 394

[5] Ehrenfest's paper was published on 15 June 1913, and Bohr's in the July number of the *Phil. Mag.*

by Ritz and the states or energy-levels of the atom described by Nicholson. He also assumed (8) that the Planck-Einstein equation $E = h\nu$ connecting energy and radiation-frequency holds for *emission* as well as absorption : and finally he introduced (9) the principle that *we must renounce all attempts to visualise or to explain classically the behaviour of the active electron during a transition of the atom from one stationary state to another.* This last principle, which had not been dreamt of by any of his predecessors, was the decisive new element that was required for the creation of a science of theoretical spectroscopy.

Let us now see how the Balmer series of hydrogen is explained by Bohr's theory. The atom of hydrogen consists of one proton with one negative electron circulating round it, the charges being e and $-e$ respectively. We suppose that by some event such as a collision, the atom has been thrown into an 'excited' state in which the electron describes an orbit more remote from the proton than its normal orbit, and that a spectral line is emitted when the electron falls back into an orbit closer to the proton. Considering any particular orbit, supposed circular for simplicity, let m denote the mass and v the velocity of the electron, and r the radius of the orbit. Then the electrostatic force e^2/r^2 between the proton and electron must be equal to the centripetal force mv^2/r required to hold the electron in its orbit, so

$$mv^2r = e^2.$$

The quantum condition is that the angular momentum of the motion must be a whole multiple of $\hbar$, say

$$mvr = n\hbar.$$

These equations give

$$v = \frac{e^2}{n\hbar}, \qquad r = n^2\frac{\hbar^2}{me^2}.$$

The kinetic energy of the electron, $\frac{1}{2}mv^2$, is $me^4/2n^2\hbar^2$. If the electron makes a transition to an orbit nearer the nucleus for which the angular momentum is $p\hbar$, there is a gain of kinetic energy

$$\frac{me^4}{2\hbar^2}\left(\frac{1}{p^2} - \frac{1}{n^2}\right)$$

but a loss of potential energy

$$e^2\left(\frac{1}{r_p} - \frac{1}{r_n}\right) \quad \text{or} \quad \frac{me^4}{\hbar^2}\left(\frac{1}{p^2} - \frac{1}{n^2}\right),$$

so altogether (remembering that $h=2\pi\hbar$) the loss of energy of the atom is

$$\frac{2\pi^2 me^4}{h^2}\left(\frac{1}{p^2}-\frac{1}{n^2}\right).$$

The equation $E=h\nu$ shows that the frequency of the homogeneous radiation emitted is

$$\nu=\frac{2\pi^2 e^4 m}{h^3}\left(\frac{1}{p^2}-\frac{1}{n^2}\right).$$

This can be identified with the formula for Balmer's series [1]

$$\nu=R\left(\frac{1}{4}-\frac{1}{n^2}\right)$$

provided we take $p=2$ and

$$R=\frac{2\pi^2 e^4 m}{h^3}$$

so we have obtained an expression for Rydberg's constant R in terms of e, m and h.

The value

$$e=4{\cdot}78\times 10^{-10}$$

had been obtained by R. A. Millikan,[2] the value

$$\frac{e}{m}=5{\cdot}31\times 10^{17}$$

by P. Gmelin [3] and A. H. Bucherer.[4] On the basis of Planck's theory, Bohr obtained [5]

$$\frac{e}{h}=7{\cdot}27\times 10^{16}.$$

[1] cf. Vol. I, pp. 376–8. In theoretical researches frequencies are often employed instead of wave-numbers, so e.g. Balmer's formula would be written

$$\nu=R\left(\frac{1}{4}-\frac{1}{m^2}\right)$$

where R stands for c times the R of the wave-number formula.

It may be mentioned that three years previously A. E. Haas, *Wien. Sitz.* cxix (1910), Abth. IIa, p. 119, had conceived the idea that Rydberg's constant should be expressible in terms of the constants e, m, h, which were already known. By an argument which was highly speculative, he obtained the result that this constant has the value $16\pi^2 e^4 m/h^3$, which differs from the correct value only by a simple numerical factor. However, this is not surprising in view of the fact that $e^4 mh^{-3}$ is the only product of powers e, m and h, which has the same dimensions as the Rydberg constant.

[2] *B.A. Rep.* 1912, p. 410

[3] *Ann. d. Phys.* xxviii (1909), p. 1086

[4] *Ann. d. Phys.* xxxvii (1912), p. 597

[5] Calculated from the experiments of E. Warburg, G. Leithaüser, E. Hapka and C. Müller, *Ann. d. Phys.* xl (1913), p. 611

Using these values, he got

$$R = 3{\cdot}26 \times 10^{15}\ (\text{sec})^{-1}$$

in close agreement with observation.[1] This successful prediction had an effect which may be compared with the effect of Maxwell's calculation of the velocity of light from his electromagnetic theory.

Bohr pointed out that while by taking $p=2$ we obtain Balmer's series, by taking $p=3$, $n=4, 5, 6, \ldots$, we obtain a series in the infra-red whose first two members had been discovered by F. Paschen[2] in 1908. At the time of Bohr's paper, the series obtained by taking $p=1$ was not known observationally, but it was discovered in 1914 by Th. Lyman,[3] and the series obtained by taking $p=4$, $n=5, 6, 7, \ldots$, was observed in 1922 by F. Brackett.[4]

Bohr remarked that the frequency of revolution of the electron in the nth state of the hydrogen atom is $v/2\pi r$, or

$$\frac{4\pi^2 me^4}{n^3h^3}.$$

But the frequency of the radiation emitted in the transition from the $(n+1)^{th}$ state to the nth is

$$\frac{2\pi^2 me^4}{h^3}\left\{\frac{1}{n^2}-\frac{1}{(n+1)^2}\right\}$$

which when n is great has approximately the value

$$\frac{4\pi^2 me^4}{n^3h^3}.$$

Thus *for great values of n, the frequency of the radiation emitted in the transition from the nth orbit to the next orbit is equal to the frequency of revolution in the nth orbit.* So we have an asymptotic connection between the classical and quantum theories.

Evidence in favour of Bohr's theory was obtained from certain facts regarding absorption-spectra. It was known that the lines of the Balmer series do not appear in the absorption-spectrum of hydrogen under ordinary terrestial conditions, and this was at once explained by the circumstance that the hydrogen atoms have normally no electrons in the orbits for which $p=2$, and therefore no electrons can be raised from these orbits to higher orbits, as would be necessary for the production of a Balmer absorption line. In the

[1] For a comparison with the observational values of 1950, cf. R. T. Birge, *Phys. Rev.* lxxix (1950), p. 193.

[2] *Ann. d. Phys.* xxvii (1908), p. 565. This is the *Bergmann series* for hydrogen; cf. Vol. I, p. 379.

[3] *Phys. Rev.* iii (1914), p. 504; *Phil. Mag.* xxix (1915), p. 284

[4] *Nature* cix (1922), p. 209

atmospheres of the stars, on the other hand, there are excited hydrogen atoms having electrons in the orbits for which $p=2$, and these are capable of giving the Balmer lines in absorption. Considerations of the same kind can be applied to explain why under ordinary terrestial conditions the principal series of the alkali metals can be obtained in absorption-spectra, but the two subordinate series can not.

In the early days of quantum theory no-one troubled about the fact that most of the equations used were not invariant with respect to the transformations of relativity theory. This defect was evident in the case of Bohr's frequency-equation $\delta E = h\nu$, which connects the loss of energy in the transition with the frequency of the emitted radiation. It was, however, shown in 1924 by P. A. M. Dirac [1] that, *provided the radiation is emitted in a definite direction*, the frequency equation can be expressed in a form which is independent of the frame of reference; it can in fact be written as a vector equation in four-dimensional space-time

$$\delta E^i = h\nu^i$$

where the direction of this vector in space-time is the same as that of the radiation.[2]

Some other problems which had been raised by the observational spectroscopists were solved in Bohr's first papers. In 1896 the American astronomer Edward Charles Pickering (1846–1919) discovered [3] in the spectrum of the star ζ Puppis, together with the Balmer series, a series of lines which had the same convergence-number as the Balmer series. Now it is a property [4] of the 'diffuse' and 'sharp' subordinate series of the alkali metals, that the more refrangible members of the doublets converge to the same limit in the two series; and this fact suggested to Rydberg [5] that the Pickering and Balmer series were actually the 'sharp' and 'diffuse' subordinate series respectively of hydrogen. If this were true, then the wave-lengths of the lines of the principal series would be immediately calculable, the first of them being at λ4687·88: and in fact a line was observed in the spectrum of ζ Puppis at λ4686, very near this position: the higher members of the series could not be expected to be seen, since they were beyond the limit for which our atmosphere is transparent.

The line λ4686 was observed by the English spectroscopist Alfred Fowler [6] (1868–1940) at the Indian eclipse of 22 January 1898, in the spectrum of the sun's chromosphere. In 1912, in an ordinary

[1] *Proc. Camb. Phil. Soc.* xxii (1924), p. 432

[2] On the relation of Bohr's theory to relativity theory, cf. K. Försterling, *ZS. f. P.* iii (1920), p. 404, and E. Schrödinger, *Phys. ZS.* xxiii (1922), p. 301.

[3] *Astroph. J.* iv (1896), p. 369; v (1897), p. 92

[4] cf. Vol. I, p. 377

[5] *Astroph. J.* vi (1897), p. 233; cf. also H. Kayser, ibid. v (1897), pp. 95, 243

[6] *Phil. Trans.* cxcvii (1901), p. 202

discharge tube containing a mixture of hydrogen and helium, he found [1] other lines very near the positions calculated by Rydberg for the supposed 'principal series of hydrogen,' and, moreover, found a new series in the ultra-violet which had the same convergence-limit as this supposed principal series, and which he provisionally named the 'second principal series of hydrogen.' The small discrepancies in wave-length with Rydberg's calculations were, however, unexplained, and also the fact that the presence of helium appeared to be necessary.

Bohr accounted in the most natural way for Pickering's ζ Puppis series, and the two series found by Fowler, by suggesting that they were not due to hydrogen at all, but to ionised helium. The helium nucleus has a charge $2e$, and in ionised helium this is accompanied by one negative electron circulating round it. A calculation similar to that carried out for the hydrogen atom shows that in this case the frequency of the radiation emitted is

$$4R\left\{\frac{1}{p^2}-\frac{1}{n^2}\right\}, \quad \text{where as before } R=\frac{2\pi^2 e^4 m}{h^3}.$$

If in this we take $p=3$ and $n=4, 5, 6, \ldots$, we obtain a series which includes the two series found by Fowler : while if we take $p=4$ and $n=5, 6, 7, \ldots$, we obtain the series observed by Pickering in the spectrum of ζ Puppis. Every alternate line in the series thus calculated would be identical with a line in the Balmer series of hydrogen.

There still, however, remained the difficulty arising from the slight discrepancies in wave-length. This was accounted for by Bohr, who remarked that the geometrical centre of the circular orbits of the electron is, strictly speaking, not the nucleus but the centre of gravity of the nucleus and the electron. This makes it necessary to multiply the value of the Rydberg constant by the ratio of the mass of the nucleus to the combined mass of the nucleus and electron. Remembering that the nucleus of the helium atom has a mass four times as great as the mass of the hydrogen nucleus, the slight difference of the Rydberg constants for hydrogen and for helium can be calculated. As Bohr [2] pointed out, ionised helium must be expected to emit a series of lines closely but not exactly coinciding with lines of the ordinary hydrogen spectrum : the alternate members of the ζ Puppis series cannot be superposed on the Balmer hydrogen lines, but should be slightly displaced with respect to them. Thus near the hydrogen lines H_α(λ6563), H_β(λ4861), H_γ(λ4340·5), H_δ(λ4102), there are helium lines at λ6560·37, λ4859·53, λ4338·86, λ4100·22 respectively, and the observed discrepancies are completely explained.

[1] *Mon. Not. R.A.S.* lxxiii (1912), p. 62 ; *Phil. Trans.* ccxiv (1914), p. 225

[2] *Nature* xcii (1913), p. 231. The attribution of the line λ4686 to helium was confirmed experimentally by E. J. Evans, ibid. p. 5.

A. Fowler[1] noticed that some lines which he had observed in the spectrum of magnesium could be arranged in a series with a Rydberg constant which, like the Rydberg constant for ionised helium, was approximately four times the normal Rydberg constant. The explanation obviously was that they were produced by ionised atoms. The nucleus of magnesium has a positive charge $12e$; and when the metal is ionised, the outermost electron, which describes the orbits concerned, is under the influence of the nucleus together with 10 negative electrons in the inner orbits, that is, the effective central charge is $2e$; and so, as in the case of ionised helium, the Rydberg constant must be multiplied approximately by 4.

In 1923 this principle was carried further by F. Paschen and A. Fowler. Paschen[2] found a series in the spark spectrum of aluminium capable of being represented by a series of the Rydberg type in which the Rydberg constant had nine times its normal value. Since the nucleus of aluminium has a positive charge $13e$, this series obviously belonged to doubly-ionised atoms, for which there would be 10 inner negative electrons, and the effective central charge on the outermost or active electron would be $3e$.

Fowler[3] went further still and discovered in the case of silicon not only series with a Rydberg constant 9 R, due to double ionisation, but also series with a Rydberg constant 16R, due to trebly-ionised atoms of silicon.

Another type of experimental investigation which led to results confirming Bohr's theory must now be mentioned. In the course of an investigation on the ionisation of gases by collisions of electrons with the atoms, James Franck (*b.* 1882) and Gustav Hertz (*b.* 1887) (a nephew of Heinrich Hertz) found[4] that the collision of slow electrons with the atoms of mercury vapour led in some cases to emission of the mercury line $\lambda 2536$. So long as the kinetic energy of the electrons is smaller than $h\nu$, where ν is the frequency of the line, they are reflected elastically by the mercury atoms; but when the kinetic energy is greater than this, light of frequency ν is emitted. Evidently the collision brings about an excited state of the atom, and the radiation is emitted when the atom falls back into its normal state. The subject was pursued in many papers by these and other authors,[5] with the following general results: *Every transition of an electron from one orbit to another, corresponding to a line of the atom's spectrum, can be brought about by the collision of a free electron with the atom, the electron losing an amount $h\nu$ of kinetic energy, where ν is the frequency of the line. Which transitions occur, depends on the state of excitement of the atom. With a normal or unexcited atom, the transitions are those that correspond to the lines of the absorption-spectrum of the unexcited atom.*

[1] *Nature*, xcii (1913), p. 232; *Phil. Trans.* ccxiv (1914), p. 225

[2] *Ann. d. Phys.* lxxi (May 1923), p. 142

[3] *Proc. R.S.* ciii (June 1923), p. 413

[4] *Verh. deutsch. phys. Ges.* xvi (May and June 1914), pp. 457, 517

[5] An extensive bibliography of the experimental work is given at the end of a paper by Franck and Hertz, *Phys. ZS.* xx (1919), p. 132.

Moreover, there is a direct connection between the ionisation of an atom and its spectrum. *If* eV *denotes the energy that must be given to an electron in order that it may ionise the atom, so that* V *is the 'ionisation-potential,' then*

$$eV = h\nu$$

where ν *denotes the frequency of the ultra-violet limit of a series in the spectrum: for the unexcited atom, the ionisation-potential is given by this equation, where* ν *is the limiting frequency of the absorption series of the unexcited atom.* Thus in the case of mercury vapour, the ionisation-potential[1] is 10·27 volts, corresponding to the ultra-violet limit at λ1188 of the series of which λ2536 is the first term.[2]

In one notable case, namely that of helium, the value derived spectroscopically proved more reliable than that obtained in the first place by direct observation of ionisation brought about by electronic bombardment. The latter value was believed until 1922 to be 25·3 volts, whereas Lyman found in that year[3] that the limit of the spectral series concerned was at λ504, which would correspond to an ionisation-potential of 24·5 volts. The discrepancy led Franck[4] to re-examine his experimental data, with the result that he found a source of error which led to reduction of the value 25·3 volts by 0·8 volts, bringing it into coincidence with the spectroscopic value.

Bohr's theory proved adequate also to account for some typical phenomena of fluorescence observed by R. J. Strutt.[5] It was well-known that if sodium vapour is illumined by the D-light emitted by a sodium flame, it emits D-light as a resonance radiation. Now the D-lines constitute the first doublet in the principal series of sodium, the second doublet being in the ultra-violet, at λ3303. Strutt asked the question, whether stimulation of sodium vapour by this second doublet would give rise to D-light? He found that it would. Moreover, he noticed that there is a line in the spectrum of zinc which practically coincides with the less refrangible member of the sodium doublet at λ3303, but that there is no zinc line coinciding with the more refrangible member. So by making use of a zinc spark he was able to stimulate the sodium vapour with light of the wave-length of the less refrangible member only of the sodium doublet. It was found that *both* the D-lines were emitted. The explanation depends on the fact that both lines of both doublets correspond to transitions down to a certain orbit which is the same in all four cases, and which may be called orbit I. By the absorption of the ultra-violet light of the second doublet, an electron is raised from orbit I to a higher orbit, and collisions with other atoms may shake it into somewhat lower orbits, from which it falls into orbit I

[1] F. N. Bishop, *Phys. Rev.* x (1917), p. 244

[2] In making the calculation it is useful to note that according to the equation $eV = h\nu$, one electron volt corresponds to λ12336. This would make λ1188 correspond to 10·4 volts.

[3] *Nature*, cx (1922), p. 278; *Astroph. J.* lx (1924), p. 1

[4] *ZS f. P.* xi (1922), p. 155

[5] *Proc. R.S.*(A), xci (1915), p. 511

with emission of the D-lines. Bohr's theory is thus seen to be of fundamental importance in regard to fluorescence.

In the first year of the century W. Voigt [1] had predicted the existence of an electric analogue to the Zeeman effect—a splitting of spectral lines by an intense electric field. However, discussing the matter by classical physics, he concluded that with a potential-fall of 300 volts per cm. in the field, the effect would be only about the 20,000th part of the separation between the D-lines of sodium, and hence would be unobservable. Notwithstanding this unfavourable opinion, in 1913 Johannes Stark [2] (*b.* 1874), when investigating the light emitted by the particles which constitute the canal rays, examined the influence of an electric field on this light, and observed a measurable effect. By spectroscopic observation in a direction perpendicular to the field, it was found that the hydrogen lines H_β, H_γ were each split into five components, the oscillations of the three inner components (which were of feeble intensity) being parallel to the electric field and the oscillations of the two outer components (which were stronger) being at right angles to the field. The distance between the components was proportional to the electric force. For helium, it was found that the effect of the electric field on the lines of the principal series and the sharp series was very small and hardly distinguishable, but the effect on the lines of the diffuse series [3] was of the same order of magnitude as for the hydrogen lines, though of a different type.

The *Stark-effect*, as it has since been called, was extensively studied from the experimental side in the years immediately following its discovery, chiefly by Stark and his disciples.[4] Some curious properties were noticed, such as that in some cases, where no splitting could be observed with the fields employed, terms were displaced towards the ultra-violet; that terms belonging to the same series were not as a rule affected in the same way, but that there was often a similarity in behaviour between terms which belonged to different but homologous series, and which had the same term-number: and that when the spectroscopic observation was in a direction parallel to the field, only those components of an electrically-split line appeared which, when the observation was at right angles to the field, oscillated linearly perpendicularly to the field; but that these components were now unpolarised.

[1] *Ann. d. Phys.* iv (1901), p. 197

[2] *Berlin Sitz.* 20 November 1913, p. 932; reprinted *Ann. d. Phys.* xliii (1914), p. 965. The effect was discovered independently at the same time by A. Lo Surdo, *Rend. Lincei* xxii (1913), p. 664; xxiii (1914), p. 82.

[3] It may be noted that lines of the diffuse series are broadened when the gas-pressure is increased, while those of the other series are not much affected.

[4] J. Stark, *Verh. deutsch. phys. Ges.* xvi (1914), p. 327; J. Stark and G. Wendt, *Ann. d. Phys.* xliii (1914), p. 983; J. Stark and H. Kirschbaum, ibid. pp. 991, 1017; J. Stark, *Ann. d. Phys.* xlviii (1915), pp. 193, 210; H. Nyquist, *Phys. Rev.* ii (1917), p. 226; J. Stark, O. Hardtke and G. Liebert, *Ann. d. Phys.* lvi (1918), p. 569; J. Stark, ibid. p. 577; G. Liebert, ibid. pp. 589, 610; J. Stark and O. Hardtke, *Ann. d. Phys.* lviii (1919), p. 712; J. Stark, ibid. p. 723

A study of the Stark effect, from the standpoint of Bohr's theory, was undertaken by E. Warburg [1] and by Bohr himself.[2] It was assumed that the field influences the stationary states of the emitting system, and thereby the energy possessed by the system in these states, splitting a term into two or more components. Warburg showed that the effect to be expected, according to quantum theory, of an electric field on the spectral lines of hydrogen, would be of the same order of magnitude as that observed experimentally by Stark ; but the investigations were imperfect as compared with those published two years afterwards by Schwarzschild and Epstein, which will be described later.

The next great advance in theoretical spectroscopy was the removal of a limitation characteristic of the original Bohr theory, namely that it took into consideration only a single set of circular orbits round each atom : it was obviously desirable to extend Bohr's principles by taking into account more than one degree of freedom. This was achieved independently and almost simultaneously in 1915 by William Wilson [3] (*b.* 1875) and Sommerfeld.[4] Their idea recalls that which had inspired Sommerfeld's paper of 1911.[5]

In the circular orbits of the steady states in Bohr's theory of the hydrogen atom, let q denote the angle which the line joining the electron to the proton makes with a fixed line ; then the momentum p corresponding to the co-ordinate q is the angular momentum of the electron round the proton, and for steady states this must be a multiple of $\hbar$. Thus, since $h = 2\pi\hbar$, we have

$$p\int dq = \text{ a multiple of } h$$

where the integration is taken once round the circle ; or, remembering that $\int pdq$ is the definition of the Action, we have

Increase of Action in going once round the orbit = a multiple of h.

Wilson and Sommerfeld generalised this into the statement that under certain circumstances, in a system with several degrees of freedom, if $q_1, q_2, \ldots$ are the co-ordinates, and $p_1, p_2, \ldots$ are the corresponding momenta, then *the steady states of the system are such that* $\int p_1dq_1$, $\int p_2dq_2, \ldots$, *are multiples of* h, when the integrations are extended over periods corresponding to the co-ordinates.[6]

[1] *Verh. d. deutsch. phys. Ges.* xv (December 1913), p. 1259

[2] *Phil. Mag.* xxvii (March 1914), p. 506 ; xxx (1915), p. 404

[3] *Phil. Mag.* xxix (1915), p. 795 ; xxxi (1916), p. 156

[4] *München Sitz.* 1915, pp. 425, 459 ; *Ann. d. Phys.* li (1916), p. 1. At almost the same time Jun Ishiwara, *Tôkyô Sûgaki-But. Kizi*(2) viii, No. 4, p. 106, *Proc. Math. Phys. Soc. Tôkyô*, viii (1915), p. 318 published proposals which in some respects resembled those of Wilson and Sommerfeld ; cf. also the work of Planck on systems with several degrees of freedom, *Verh. d. deutsch. phys. Ges.* xvii (1915), pp. 407, 438 ; *Ann. d. Phys.* l (1916), p. 385.

[5] cf. p. 104

[6] The rule as thus broadly stated is evidently not independent of the choice of co-ordinates, a point on which see Einstein, *Verh. d. deutsch. phys. Ges.* xix (1917), p. 82, and other papers discussed later in the present chapter.

Sommerfeld now considered electrons moving round the nucleus in non-circular orbits, in fact Keplerian ellipses. Denoting by (r,θ) the radius vector and vectorial angle of the electron, by m its mass, and by Ze the positive charge on the nucleus, the Lagrangean function of the motion is

$$L = \tfrac{1}{2}mv^2 - V$$

where

$$v^2 = \left(\frac{dr}{dt}\right)^2 + r^2\left(\frac{d\theta}{dt}\right)^2, \qquad V = -\frac{Ze^2}{r},$$

so the momenta corresponding to θ and r are $mr^2\,d\theta/dt$ and $m\,dr/dt$ respectively. The quantum conditions specifying a steady state were therefore taken by Sommerfeld to be

$$\int mr^2 \frac{d\theta}{dt}\,d\theta = nh \tag{1}$$

and

$$\int m\frac{dr}{dt}\,dr = n'h \tag{2}$$

where n and n' are whole numbers, and the integrations are to be taken once round the orbit.

From (1) and (2) we have

$$(n+n')h = \int mv^2 dt.$$

But from the known properties of Keplerian elliptic motion, we have

$$\int mv^2 dt = 2\pi e(Zma)^{\frac{1}{2}},$$

denoting by a the major semi-axis of the orbit. Therefore

$$(n+n')\,h = 2\pi e(Zma)^{\frac{1}{2}}. \tag{3}$$

Also from (1)

$$nh = mr^2\frac{d\theta}{dt}\int d\theta = 2\pi mr^2\frac{d\theta}{dt}$$

so

$$\frac{n}{n+n'} = \frac{1}{e(Zma)^{\frac{1}{2}}}mr^2\frac{d\theta}{dt}$$

$$= \frac{b}{a} \tag{4}$$

from the known properties of elliptic motion, where b denotes the minor semi-axis. Equations (3) and (4) connect the quantum numbers n and n' with the semi-axes of the orbit.

$$\text{Also, total energy of electron} = \tfrac{1}{2}mv^2 - \frac{Ze^2}{r}$$

$$= -\frac{Ze^2}{2a}$$

$$= -\frac{2\pi^2 e^4 Z^2 m}{h^2(n+n')^2}$$

$$= -\frac{RZ^2 h}{(n+n')^2} \qquad (5)$$

where R is Rydberg's number.

The total number of possible steady orbits has been greatly increased by Sommerfeld's introduction of ellipses in addition to Bohr's circular orbits: but from equation (5) it seems as if the total number of possible values of the energy has *not* been increased, since the energy depends only on the single number $(n+n')$. Thus apparently in the case of the hydrogen atom (for which $Z=1$) we should get exactly the same spectra as before. However, as Sommerfeld pointed out, the theory has so far supposed the orbit to be a Keplerian ellipse, the electron being regarded as a particle of constant mass; whereas, since the velocity in the orbit is great, the relativistic increase of mass with velocity ought to be taken into account.[1] Sommerfeld showed that the orbit is an ellipse with a moving perihelion, the motion of the perihelion being great or small according as (for the quantified ellipses) the quantity $e^2/c\hbar$ is great or small. Actually $e^2/c\hbar$, which is generally denoted by α and called the *fine-structure constant*, has a value[2] which at that time was believed to be about $7 \cdot 10^{-3}$, and has since been found to be 1/137. It represents the ratio of the velocity of the electron in the first Bohr orbit to the velocity of light.

The formula for an energy-level or 'term' of the hydrogen atom now becomes, in the first approximation,

$$T = T_0 + T_1$$

where T_0 is the uncorrected value, so $T_0 = R/n^2$, n being the 'principal quantum number,' and T_1 is a correction term given by

$$T_1 = \frac{R\alpha^2}{n^4}\left(\frac{n}{k} - \frac{3}{4}\right)$$

[1] The necessity for a relativity correction had been pointed out already by Bohr, *Phil. Mag.* xxix (1915), p. 332.

[2] The fact that $e^2/c\hbar$ is a pure number had been pointed out by Jeans, *Brit. Ass. Rep.* 1913, p. 376, who suspected that it might have the value $1/(4\pi)^2$.

where k is a second quantum number. The frequency of a radiation corresponding to the fall of an electron from one energy-level to another will depend on the values of n and k for both levels. It is therefore evident that to each spectral line of the original theory (which depended only on the principal quantum numbers) there will now correspond a group of lines very close together; these predicted lines were found to agree remarkably well with lines observed [1] in what was called the *fine-structure* of the spectrum.

As we shall see later, Sommerfeld's theory is not the complete explanation of the fine-structure, which depends also on a property discovered later, that of the spin of the electron; but it was rightly acclaimed at the time as a great achievement. The quantum number k now introduced in connection with hydrogen was found later to account for the distinction between the principal and subordinate series of e.g. the alkali metals.

Sommerfeld's extension of Bohr's theory led Karl Schwarzschild [2] (1873–1916), director of the Astrophysical Observatory at Potsdam, and Paul Sophus Epstein [3] (*b.* 1883), a former student of Sommerfeld's at Munich, to investigate theoretically the Stark effect. In a stationary state of an atom the active electron has three degrees of freedom, and we may therefore expect that the state will be specified by three quantum numbers. Success in the mathematical treatment of the problem depends on the possibility of finding three pairs of variables q_i, p_i (where q_i is a co-ordinate and p_i the corresponding momentum), such that the Hamilton's partial differential equation belonging to the classical problem [4] can be solved by separation of variables. The integrals $\int p_i dq_i$ corresponding to these separate pairs of variables are then equated to multiples of h. By carrying out this programme, Epstein found in the case of the Balmer series of hydrogen the following simple formula for the displacements of the Stark components from the original position of the spectral line:

$$\Delta\nu = \frac{3h}{8\pi^2 c e \mu} \mathrm{EZ}$$

where

$$\mathrm{Z} = (m_1 + m_2 + m_3)(m_1 - m_2) - (n_1 + n_2 + n_3)(n_1 - n_2).$$

Here E is the applied electric force (in the electrostatic system of units), (m_1, m_2, m_3) are the quantum numbers of the outer orbit from which the electron falls into an inner orbit of quantum numbers (n_1, n_2, n_3), μ is the mass of the electron, and ν is the reciprocal

[1] Especially by the measurements, by F. Paschen, *Ann. d. Phys.* l (1916), p. 901, of the fine structure of Fowler's spectra of ionised helium; cf. E. J. Evans and C. Croxson, *Nature*, xcvii (1916), p. 56

[2] *Berlin Sitz.* April 1916, p. 548

[3] *Phys. ZS.* xvii (1916), p. 148; *Ann. d. Phys.* l (1916), p. 489

[4] cf. Whittaker, *Analytical Dynamics*, § 142

of the wave-length. The sums $(m_1+m_2+m_3)$ and $(n_1+n_2+n_3)$ correspond to the ordinal number of the term in the Balmer series, so $n_1+n_2+n_3=2$, while $m_1+m_2+m_3=3$ for the line H_α, 4 for H_β, 5 for H_γ, etc. In the case of the H_α-line, Z could take the values 0, 1, 2, 3, 4, 5 ; and the wave-numbers calculated by the formula agreed well with Stark's observations, not only as regards the magnitude of the displacements but also as regards polarisation, which depends on whether (m_3-n_3) is odd or even. Similar satisfactory results were found for the other Balmer lines.

The theory of the Stark effect given by Schwarzschild and Epstein is valid so long as the electric field is strong enough to make the Stark separation of the components much greater than the separation due to the fine-structure. The case when the field is so weak that this condition is not satisfied was treated by H. A. Kramers,[1] who found that the fine-structure lines were split into several components, whose displacements were proportional to the *square* of the electric field-strength. Kramers also investigated the Stark effect on series lines in the spectra of elements of higher atomic number.

The methods of quantification employed by Sommerfeld, Schwarzschild and Epstein may be justified by a theory known as the *theory of adiabatic invariants*, which was originally created by the founders of thermodynamics,[2] who studied quantities that are invariant during adiabatic changes. It was now developed by Paul Ehrenfest [3] (1880–1933) of Leiden, whose starting-point was a theorem proved by Wien in the establishment of his displacement law,[4] namely that if radiation is contained in a perfectly-reflecting hollow sphere which is slowly contracting, then the frequency ν_p and the energy ϵ_p of each of the (infinitely-many) principal modes of vibration of the cavity increase together during the compression in such a way that their ratio remains constant. Ehrenfest showed that this property is really the basis of the law that the ratio of energy to frequency ϵ/ν can take only the values 0, h, $2h$, . . . : for a different assumption, such as that ϵ/ν^2 is proportional to 0, h, $2h$, . . ., could be shown to lead to a conflict with the second law of thermodynamics. What impressed Ehrenfest was the circumstance that although Wien's theorem was deduced by purely classical methods, it was valid in quantum theory, and actually indicated a fundamental quantum law. He was thus led to suspect that if in

[1] *ZS. f. Phys.* iii (1920), p. 199

[2] cf. R. Clausius, *Ann. d. Phys.* cxlii (1871), p. 433 ; cxlvi (1872), p. 585 ; cl (1873), p. 106 ; English translations, *Phil. Mag.* xlii (1871), p. 161 ; xliv (1872), p. 365 ; xlvi (1873), pp. 236, 266 ; C. Szily, *Ann. d. Phys.* cxlv (1872), p. 295 = *Phil. Mag.* xliii (1872), p. 339 ; L. Boltzmann, *Vorlesungen über die Principe der Mechanik*, ii Teil (Leipzig, 1904), ch. iv

[3] *Ann. d. Phys.* xxxvi (1911), p. 91 ; li (1916), p. 327 ; *Verh. d. deutsch. phys. Ges.* xv (1913), p. 451 ; *Proc. Amst. Acad*, xvi (1914), p. 591 ; xix (1917), p. 576 ; *Phys. ZS.* xv (1914), p. 657 ; *Phil. Mag.* xxxiii (1917), p. 500

[4] cf. Vol. I, p. 379

a dynamical system some parameter is slowly changed (as was the radius of the sphere in Wien's theorem), and if some quantity J is found to be invariant according to classical physics during this change (as was ϵ/ν in Wien's case), then the system may be quantified by putting J equal to a multiple of h. An example of 'adiabatic change' is furnished by the motion of a simple pendulum when the length of the suspending cord is very slowly altered.

The general problem he formulated as follows. Let there be a system of differential equations (for simplicity taking two independent variables)

$$\frac{dx_1}{dt} = X_1(x_1, x_2, t, m), \qquad \frac{dx_2}{dt} = X_2(x_1, x_2, t, m)$$

where the letter m stands for one or more constant parameters. These equations will possess integrals

$$F_1(x_1, x_2, t, m) = c_1, \qquad F_2(x_1, x_2, t, m) = c_2.$$

Then it is possible that *some functions of the constants of integration* c_1, c_2, *and of the parameters m, exist, which maintain their values unaltered when the parameters m are varied in an arbitrary manner, provided the variation is very slow.* Such functions are called *adiabatic invariants*; and Ehrenfest now showed that *the quantities which had been put equal to whole-number multiples of h in earlier papers on quantum theory were always adiabatic invariants*; and, moreover, that *the rule thus indicated was valid generally.* In other words, *stationary orbits are adiabatically invariant.*

Now it had been shown by J. Willard Gibbs [1] and Paul Hertz,[2] that in the case of a dynamical system with one degree of freedom whose solutions are periodic and whose equations of motion are

$$\frac{dq}{dt} = \frac{\partial H}{\partial p}, \qquad \frac{dp}{dt} = -\frac{\partial H}{\partial q},$$

so that the trajectory is represented in the (q, p) plane by a curve of constant energy

$$H(q, p) = W,$$

then the area $\iint dq\,dp$ enclosed by this curve is an adiabatic invariant; or, $\int p\,dq$ taken round the curve is an adiabatic invariant: that is, *the increment of the Action, in a complete period, is an adiabatic invariant.*

This theorem is true even when the system has more than one degree of freedom, provided the solution is periodic. As an example,[3]

[1] *Principles of statistical mechanics*, 1902, p. 157 [2] *Ann. d. Phys.* xxxiii (1910), p. 537

[3] Other examples of adiabatic invariants are given by W. B. Morton, *Phil. Mag.* viii (1929), p. 186 and by P. L. Bhatnagar and D. S. Kothari, *Indian J. Phys.* xvi (1942), p. 271.

consider the case of a planet describing an elliptic orbit under the Newtonian law of force to the focus; and suppose that the mass m of the planet is slowly increased by moving through a dusty atmosphere, while gradual changes also take place in the strength μ of the centre of force. It is well known that the velocity v, the radius vector r and the mean distance a, are connected by the equation

$$v^2 = \mu\left(\frac{2}{r} - \frac{1}{a}\right).$$

Varying this, we have

$$\frac{2\delta v}{v} = \frac{\delta\mu}{\mu} + \frac{\mu\delta a}{a^2 v^2}$$

or

$$\frac{\delta a}{a^2} = \frac{v^2}{\mu}\left(\frac{2\delta v}{v} - \frac{\delta\mu}{\mu}\right).$$

When the planet picks up a small particle δm previously at rest, by the theorem of conservation of linear momentum we have

$$\delta(mv) = 0, \quad \text{or} \quad \frac{\delta v}{v} = -\frac{\delta m}{m},$$

so the previous equation becomes

$$\frac{\delta a}{a^2} = -\frac{v^2}{\mu}\left(\frac{2\delta m}{m} + \frac{\delta\mu}{\mu}\right).$$

Suppose the changes in m and μ to be brought about so gradually that the increments δm and $\delta\mu$ are spread over a large number of orbital revolutions; we may then, before integrating the last equation, replace v^2 by its time-average. Now since the average value of $1/r$ is $1/a$, we see from the equation

$$v^2 = \mu\left(\frac{2}{r} - \frac{1}{a}\right)$$

that the average value of v^2 is μ/a. When this is inserted, the variational equation becomes

$$\frac{\delta a}{a} + \frac{2\delta m}{m} + \frac{\delta\mu}{\mu} = 0.$$

Integrating, we have

$$m^2 \mu a = \text{Constant}:$$

and since the increment of the Action in a complete period is known to be $2\pi m\mu^{\frac{1}{2}} a^{\frac{1}{2}}$, this equation verifies for Keplerian motion the theorem that the increment of the Action in a complete period is an adiabatic invariant.

In Keplerian motion, the Action

$$\int m\left\{\left(\frac{dr}{dt}\right)^2 + r^2\left(\frac{d\theta}{dt}\right)^2\right\}dt$$

is the sum of two integrals, namely $\int m\,(dr/dt)\,dr$ and $\int mr^2\,(d\theta/dt)\,d\theta$: and since the alterations of μ and m do not affect the angular momentum round the centre of force, we see that the second of these integrals, taken singly, must be an adiabatic invariant. Whence it follows that the first must be also. Thus *a justification is provided for Sommerfeld's choice of integrals to be equated to integral multiples of h.*

We shall now establish a connection (not restricted to Keplerian motion) between the adiabatic invariant representing the increment of Action in a complete period (which we shall denote by J), the total energy of the motion (which we shall denote by W) and the frequency ν of the motion (i.e. the reciprocal of the periodic time).

The solution of the differential equations of motion will consist in representing the original Hamiltonian variables in terms of $Q \equiv \nu t + \epsilon$ (where ϵ is an arbitrary constant) and a variable P which is conjugate to Q and is actually constant: the transformation from (q, p) to (Q, P) being a contact-transformation, so that

$$p\,dq - P\,dQ = d\Omega$$

where $d\Omega$ is the differential of a function which resumes its value after describing the periodic orbit.

Integrating this equation once round the orbit, we have

$$\int p\,dq - P\int dQ = \int d\Omega = 0.$$

But

$$\int dQ = \int \frac{dQ}{dt}\,dt = \nu\int dt = 1.$$

Therefore $P = \int p\,dq$ integrated round the orbit: that is, P is the adiabatic invariant which we have called J. The Hamiltonian equations in terms of the variables P and Q are

$$\frac{dQ}{dt} = \frac{\partial H}{\partial P}, \qquad \frac{dP}{dt} = -\frac{\partial H}{\partial Q} = 0.$$

The former equation is

$$\nu = \frac{dH}{dP}$$

so we have

$$\nu = \frac{dW}{dJ}$$

where W *denotes the total energy.*

The methods employed by Schwarzschild and Epstein in the mathematical treatment of the Stark effect led other workers in quantum theory [1] to study dynamical systems which can be solved by separation of variables, a subject which had been studied extensively in 1891 and the following years by Paul Stäckel of Kiel.[2] In such systems the integral of Action,

$$\int 2T\,dt$$

where T denotes the kinetic energy, can be separated into a sum of integrals each of which depends on one only of the co-ordinates,

$$\sum_k \int \sqrt{\{F_k(q_k)\}}\,dq_k.$$

In general, the motion of each co-ordinate is a *libration*, i.e. it oscillates between two fixed limits, the values of which are determined by the integrated equations of motion.[3] For such systems it was proved by J. M. Burgers [4] that *the single integrals*

$$J_k = \int \sqrt{\{F_k(q_k)\}}\,dq_k,$$

where the integration is taken over a range in which q_k *oscillates once up and down between its limits, are adiabatic invariants* [5]: and therefore the rule that must be followed in order to quantify is to write

$$J_k = n_k h$$

where n_k is a whole number, and is called a *quantum number*.

Denoting by $\bar{T}$ the average value of the kinetic energy, we have

$$2\bar{T} = \frac{1}{A+B}\int_{-A}^{B} 2T\,dt$$

where A and B are large numbers. Hence

$$2\bar{T} = \frac{1}{A+B}\int_{-A}^{B} \sum_k \sqrt{\{F_k(q_k)\}}\,dq_k.$$

Now

$$J_k = \int \sqrt{\{F_k(q_k)\}}\,dq_k$$

[1] P. Debye, *Gött. Nach.* 1916, p. 142; *Phys. ZS.* xvii (1916), pp. 507, 512; A. Sommerfeld, *Phys. ZS.* xvii (1916), p. 491

[2] *Habilitationsschrift*, Halle, 1891; *Comptes Rendus*, cxvi (1893), p. 485; cxxi (1895), p. 489; *Math. Ann.* xlii (1893), p. 545

[3] There is, however, often among the co-ordinates an azimuthal angle which can increase indefinitely but with respect to which the configuration of the system is periodic; an increase of 2π in it takes the place of the libration of the other co-ordinates.

[4] *Phil. Mag.* xxxiii (1917), p. 514; *Ann. d. Phys.* lii (1917), p. 195

[5] Leaving aside some special cases of degeneration

taken over a range of integration described in the periodic time of the oscillation of q_k. So

$$2T = \frac{1}{A+B} \sum_k \frac{A+B}{T_k} J_k$$

or

$$2T = \sum_k \nu_k J_k$$

where ν_k is the frequency associated with the co-ordinate q_k.

Now it was shown in 1887 by Otto Staude[1] of Dorpat (for the case of two degrees of freedom) and by Stäckel[2] in 1901 (for any number of degrees of freedom) that a dynamical system, for which Hamilton's partial differential equation can be integrated by separation of variables, is *multiply periodic*,[3] that is to say, the co-ordinates $(q_1, q_2, \ldots, q_n)$ can be expressed by generalised Fourier series of the type

$$q_r = \sum Q_r \cos\{2\pi(\tau_1\nu_1 + \tau_2\nu_2 + \ldots + \tau_n\nu_n)t + \gamma_r\}$$

where Q_r and γ_r are functions of $\tau_1, \tau_2, \ldots \tau_n$, where the ν_k have the same meanings as before, and where summation is over integral values of the parameters $\tau_1, \tau_2, \ldots \tau_n$.

It can be shown, as in the case of a system with one degree of freedom, that

$$\nu_k = \frac{\partial W}{\partial J_k} \qquad (k = 1, 2, \ldots n)$$

where W denotes the total energy, so that the frequency of the radiation emitted in the transition from a state W_r to a state W_s is given by the equation

$$h\nu_{rs} = W_r - W_s.$$

Now suppose that the quantum numbers belonging to the states W_r and W_s are large, and that the quantum numbers of the state W_r differ from those of the state W_s by $\tau_1, \tau_2, \ldots \tau_n$, respectively, so that

$$\Delta J_1 = \tau_1 h, \quad \Delta J_2 = \tau_2 h, \ldots, \Delta J_n = \tau_n h.$$

Since consecutive orbits with large quantum numbers do not differ greatly from each other, we can represent the increment $W_r - W_s$ of the energy approximately by

$$\Delta W = \frac{\partial W}{\partial J_1}\Delta J_1 + \frac{\partial W}{\partial J_2}\Delta J_2 + \ldots + \frac{\partial W}{\partial J_n}\Delta J_n$$

[1] *Math. Ann.* xxix (1887), p. 468 ; *Sitz. d. Dorpater Naturfor.*, April 1887

[2] *Math. Ann.* liv (1901), p. 86

[3] The German term, introduced by Staude, loc. cit., is *bedingt-periodisch.*

so we have

$$h\nu_{rs} \sim \frac{\partial W}{\partial J_1}\tau_1 h + \frac{\partial W}{\partial J_2}\tau_2 h + \ . \ . \ . \ + \frac{\partial W}{\partial J_n}\tau_n h$$

or

$$\nu_{rs} \sim \nu_1\tau_1 + \nu_2\tau_2 + \ . \ . \ . \ + \nu_n\tau_n :$$

that is to say, *the frequency of the spectroscopic line emitted in the transition* $r \rightarrow s$ *is approximately* $\nu_1\tau_1 + \nu_2\tau_2 + \ . \ . \ . \ + \nu_n\tau_n$, *which is the frequency of one term in the multiple Fourier series representing the classical solution of the problem, namely that for which*

$$\tau_1 = {}^r n_1 - {}^s n_1, \quad \tau_2 = {}^r n_2 - {}^s n_2, \ . \ . \ . \ \tau_n = {}^r n_n - {}^s n_n$$

where $({}^r n_1, {}^r n_2, \ . \ . \ . \ )$ *are the quantum numbers in the state* r, *and* $({}^s n_1, {}^s n_2, \ . \ . \ . \ )$ *are the quantum numbers in the state* s. This was called by Ehrenfest the *correspondence theorem for frequencies.*

In the same year (1916) in which Schwarzschild and Epstein published their explanations of the Stark effect, Sommerfeld [1] and Debye [2] showed that the Zeeman effect also could be brought within the compass of the quantum theory. If we consider the motion of an electron under the influence of a fixed electric charge at the origin, and a magnetic field H parallel to the axis of z, the classical equations of motion may be written in the Hamiltonian form

$$\frac{dq_r}{dt} = \frac{\partial K}{\partial p_r}, \qquad \frac{dp_r}{dt} = -\frac{\partial K}{\partial q_r} \qquad (r = 1, 2, 3)$$

where q_1 is the radius vector from the origin to the electron, q_2 is the angle between q_1 and the intersection of the plane of xy with the plane of instantaneous motion of the electron (which we may call the plane of the orbit), q_3 is the angle between the fixed axis Ox and the intersection of the plane of xy with the plane of the orbit, p_1 is the component of linear momentum along the radius vector, p_2 is the angular momentum of the electron round the origin, and p_3 is the angular momentum of the electron about the axis of z. These variables are separable. During the motion, the plane of the orbit precesses [3] uniformly round the axis of z with angular velocity $e\mathrm{H}/2mc$, so the dynamical equations involving q_3 and p_3 must be

$$\frac{dq_3}{dt} = \frac{e\mathrm{H}}{2mc}, \qquad \frac{dp_3}{dt} = 0$$

[1] *Phys. ZS.* xvii (1916), p. 491

[2] *Phys. ZS.* xvii (1916), p. 507; cf. also A. W. Conway, *Nature,* cxvi (1925), p. 97 and T. van Lohuizen, *Amst. Proc.* xxii (1919), p. 190

[3] This is the *Larmor precession,* described in Vol. I, pp. 415–6

and therefore the term involving q_3 and p_3 in the Hamiltonian function K must be

$$\frac{eHp_3}{2mc}.$$

Proceeding now to the quantification, there will be three quantum conditions of the kind usual in problems for which the variables can be separated, and the third of them will be

$$p_3 = m_j\hbar$$

where m_j is a whole number which will be called the *magnetic quantum number* : so the existence of the magnetic quantum number is an assertion that the component of angular momentum in the direction of the magnetic field can take only values which are whole-number multiples of $\hbar$. Since this component attains its greatest value when it is equal or opposite to the total angular momentum, we see that m_j can take only the values $-j, -j+1, \ldots, j-1, j$, where j is the quantum number which specifies the total angular momentum.

Thus the Hamiltonian function K, which represents the energy of the motion, is increased (as compared with the case when the magnetic field is absent) by

$$\frac{m_j e\hbar H}{2mc}.$$

Supposing that we are dealing with the hydrogen atom, the part of the energy corresponding to the unperturbed motion is (as in Bohr's original theory, neglecting the fine structure)

$$-\frac{2\pi^2 me^4}{n^2h^2}$$

and adding to this the part we have just found, due to the magnetic field, we have for the total energy in the stationary state specified by the quantum numbers n, m_j,

$$-\frac{2\pi^2 me^4}{n^2h^2}+\frac{m_j ehH}{4\pi mc},$$

and therefore the frequency of the spectral line emitted in the transition from the state (n, m_j) to the state (n', m_j') is

$$\nu = \frac{2\pi^2 me^4}{h^3}\left(\frac{1}{n'^2}-\frac{1}{n^2}\right)+\frac{eH}{4\pi cm}\left(m_j - m_j'\right).$$

Thus *when the magnetic field is applied, in place of the single spectral line specified by* (n, n'), *we have a number of lines depending on* m_j *and* m_j'.

These are the Zeeman components. Their number, intensity and state of polarisation are furnished by a principle which was not discovered until 1918, and which will be described presently. It will appear that so far as the number, position and state of polarisation of the components are concerned, the quantum theory gives (for lines such as those we are now considering, namely single lines which are not members of doublets or triplets) exactly the same results as Lorentz's original theory. It will be noted that this became possible because Planck's constant h cancelled out in the magnetic part of the above expression for the frequency.

We may note that since in the classical problem the angle α which the plane of the orbit makes with Oz is given by

$$\cos \alpha = \frac{p_3}{p_2},$$

therefore in the quantified problem, when both the total angular momentum and its component in the direction of Oz are whole-number multiples of $\hbar$, *this angle α can take only certain discrete values.* This is an example of what is called *space quantification* or *direction quantification* ; the plane of the orbit is permitted to be inclined at only certain definite angles to the direction of the field.

The principle of space quantification was strikingly confirmed by an experiment performed in 1921 by O. Stern [1] and W. Gerlach,[2] working in the department of Max Born at Frankfort-on-Main. Let a ray of atoms of silver produced by boiling silver in a furnace and passing the vapour through two fine slits, be travelling in the x-direction, and suppose that the ray encounters a non-uniform magnetic field parallel to the axis of z. In the non-uniform magnetic field H_z, a particle, whose magnetic moment in the z-direction is M, experiences a mechanical force $M\partial H_z/\partial z$ in the z-direction. Space-quantification ensures, however, that atoms orient themselves in the magnetic field in certain ways, in fact that M can take only values which correspond to the directions parallel and antiparallel to H_z, and so are equal and opposite. Hence the original ray of silver atoms is split into two rays in the plane of xz, corresponding to these two opposite values of the z-component of the magnetic moment: in the experiment, these rays strike a plane to which the atoms adhere and so produce an image which is observed.

It may be noted that the Stern-Gerlach effect concerns only a single state of the atom, not (like the Zeeman effect) a transition from one state to another.

The effect of crossed electric and magnetic fields on the radiation from a hydrogen atom was discussed by O. Klein,[3] W. Lenz [4] and N. Bohr.[5]

[1] *ZS. f. P.* vii (1921), p. 249
[2] *ZS. f. P.* viii (1921), p. 110; ix (1922), pp. 349, 352
[3] *ZS. f. P.* xxii (1924), p. 109
[4] *ZS. f. P.* xxiv (1924), p. 197
[5] *Proc. Phys. Soc.* xxxv (1923), p. 275

Before 1918 the Bohr theory had been applied to determine only the *frequencies* of lines in spectra, and had not yielded any results regarding their *intensity*. Some definite questions concerning intensity were, however, suggested by spectroscopic observations ; in particular, certain lines, whose existence might be expected according to the Bohr theory, were found to be absent in the spectrum as observed : which suggested that transitions of the active electron from certain orbits to certain other orbits never took place, so that the corresponding lines could not appear.

The explanation of this phenomenon was given by Adalbert Rubinowicz[1] (*b.* 1889), a Pole then working at Munich, who, considering atoms of the hydrogen type, and defining a stationary orbit by the numbers n, n', introduced by Sommerfeld's quantum conditions

$$\int mr^2 \frac{d\theta}{dt} d\theta = nh, \qquad \int m \frac{dr}{dt} dr = n'h,$$

remarked that the angular momentum of the atom was $n\hbar$, and that in a transition between an orbit of quantum numbers (m, m') and an orbit of quantum numbers (n, n'), this angular momentum would change by $|m-n|\hbar$. But by the principle of conservation of angular momentum, any change in the angular momentum of the atom must be balanced by the angular momentum carried off by the radiation associated with the transition. By an argument based partly on classical electrodynamics and partly on quantum theory (and therefore perhaps not very secure), Rubinowicz found that the angular momentum radiated, when the radiation is circularly polarised, is $\hbar$; and when the radiation is linearly polarised, the angular momentum radiated is zero. Thus we have

$$|m-n| \hbar = 0 \text{ or } 1,$$

and we obtain a *selection-principle*, namely that *the azimuthal quantum number n can only change by* 1, 0 or -1.

In a footnote appended to his paper, Rubinowicz explained that when his paper was ready for press there had appeared the first Part of a memoir by Bohr,[2] in which the same problem was approached from quite a different standpoint, depending on the close relations which exist between the quantum theory and the classical theory for very great quantum numbers. We have seen an example of these relations in the correspondence theorem for frequencies ; Bohr now extended this theorem by assuming that there is a relation between the *intensity* of the spectral line radiated and the amplitude of the corresponding term in the classical multiple-Fourier expansion: in fact, that the *transition-probability* associated with the genesis of

[1] *Phys. ZS.* xix, p. 441 (15 Oct. 1918), and p. 465 (1 November 1918)
[2] *D. Kgl. Danske Vid. Selsk. Skr., Nat. og Math. Afd.*, 8 Raekke, iv, 1 (1918) ; cf. N. Bohr, *Ann. d. Phys.* lxxi (1923), p. 228

the spectral line contains a factor proportional to the square of the corresponding coefficient in the Fourier series. Moreover, he extended this correspondence principle for intensities by assuming its validity not only in the region of high-quantum numbers but over the whole range of quantum numbers: so that *if any term in the classical multiple-Fourier expansion is absent, the spectral line, which corresponds to it according to the correspondence-theorem for frequencies, will also be absent.* This is a *selection-principle* of wide application.

He postulated also that the *polarisation* of the emitted spectral line may be inferred from the nature of the conjugated classical vibration. Thus[1] considering in particular the Zeeman splitting of a spectral line of hydrogen, when the transition is such that the magnetic quantum number is unchanged, the Zeeman component will occupy the same position as the original line, and the radiation will correspond to that emitted in classical electrodynamics by an electron performing linear oscillations parallel to the magnetic field; while in the case when the magnetic quantum number changes by ± 1, (which is the only other possibility permitted by the correspondence-principle), we shall obtain Zeeman components symmetrically situated with respect to the original line, and the radiation will correspond to that emitted by a classical electron describing a circular orbit in a plane at right angles to the magnetic field, in one or the other direction of circulation. The polarisation of the emitted line will therefore in all three cases be the same as that predicted by Lorentz[2] on the classical theory.

An extensive memoir by H. A. Kramers[3] supplied convincing evidence of the validity of Bohr's correspondence-principle for the calculation of the intensities of spectral lines: while W. Kossel and A. Sommerfeld[4] showed that the deductions from the selection principle were confirmed by experiment in the case of many different kinds of atoms.

The correspondence principle was extended to absorption by J. H. van Vleck.[5]

We must now consider developments in the theory of quantum numbers. We have seen that Sommerfeld specified an energy-level or 'term' of an atom by two quantum numbers (leaving aside for the moment the magnetic quantum number). The first of these is the *principal quantum number n*, which had been introduced by Bohr in 1913, and which increases by unity when we pass from a term of a spectral series to the next higher term: and the other is the *azimuthal quantum number k*, which had been introduced by Sommerfeld himself in 1915, and which distinguishes the different series from each other: thus in the sodium spectrum[6] the terms of the 'sharp' series have the frequencies $R/(n+s)^2$, where R is Rydberg's constant

[1] N. Bohr, *Proc. Phys. Soc.* xxxv (1923), p. 275
[2] cf. Vol. I, p. 412
[3] *D. Kgl. Dansk. Vid. Selsk. Skr., Nat. og Math. Afd.*, 8 Raekke, iii, 3 (1919)
[4] *Verh. d. deutsch phys. Ges.* xxi (1919), p. 240
[5] *Phys. Rev.* xxiv (1924), p. 330
[6] cf. Vol. I, p. 378

and n is a positive whole number: those of the 'principal' series have the frequencies $R/(n+p_1)^2$ and $R/(n+p_2)^2$; those of the 'diffuse' series have the frequencies $R/(n+d)^2$, and those of the 'fundamental' series have the frequencies $R/(n+f)^2$: and these series correspond respectively to $k=1, 2, 3, 4$. It has become customary to use in place of k the letter l, where $l=k-1$: the reason being that when there is one active electron, its orbital angular momentum is $l\hbar$. The series of energy-levels of the atom for which $l=0, 1, 2, 3, \ldots$, were denoted by $s, p, d, f, \ldots$, these being the initial letters of the words *sharp*, *principal*, *diffuse*, *fundamental*, etc. l has the selection rule that in a transition it can change only to $l+1$ or $l-1$.

Evidently, however, the two quantum numbers n and l did not suffice for the description of the terms of the alkali spectra, for the principal series was a series of doublets. To meet this situation, Sommerfeld [1] in 1920 introduced a third number j, which he called the *inner quantum number* and which is different for the two terms of a doublet. This number must arise from the quantification of a motion in some third degree of freedom, and it was natural to suppose that besides the orbital angular momentum of the atom, which was accounted for by l, there was yet another independent angular momentum. This was at first conjectured to be the angular momentum of the atom's core, was denoted by $s\hbar$ and was supposed to have (for the alkalis) the value $\frac{1}{2}\hbar$; so that when it was compounded with the angular momentum $l\hbar$, with which space-quantification compels it to be either parallel or anti-parallel, the resultant total angular momentum of the atom, $j\hbar$, could have either of the two values [2] $(l-\frac{1}{2})\hbar$ and $(l+\frac{1}{2})\hbar$. There is a selection-rule that j may pass only to $(j+1)$, j, or $(j-1)$, and moreover transitions in which j remains zero are forbidden.

In the spectra of the alkaline earths there are series of triplets which were accounted for in the same way by supposing that the angular momentum $s\hbar$ can take the values 0 and $\hbar$, giving for the total angular momentum the three possibilities $(l-1)\hbar$, $l\hbar$, $(l+1)\hbar$, and so for j the three possibilities $j=l-1$, $j=l$, $j=l+1$. In other atoms, every value [3] of l was supposed to yield a set of energy-levels or terms corresponding to the values

$$j=l+s, \quad l+s-1, \quad \ldots \quad |l-s|+1, \quad |l-s|.$$

We have said that at first the independent angular momentum $s\hbar$, which is compounded with the orbital angular momentum $l\hbar$ in order to produce the resultant total angular momentum $j\hbar$, was supposed to be the angular momentum of the 'core' of the atom, i.e. possibly the nucleus together with the innermost closed shells

[1] *Ann. d. Phys.* lxiii (1920), p. 221; lxx (1923), p. 32

[2] Except when $l=0$, in which case there is only one value of j, namely $\frac{1}{2}$.

[3] $l=0$ yields only a single term

of electrons. This hypothesis was overthrown by Wolfgang Pauli (*b.* 1900), a Viennese who, after studying with Sommerfeld at Munich and with Bohr at Copenhagen, had become a privat-dozent in the University of Hamburg. Pauli showed [1] that if the angular momentum $s\hbar$ belonged to the atomic core, there would follow a certain dependence of the Zeeman effect on the atomic number, and this effect was not observed : he inferred that the angular momentum $s\hbar$ must be due to a new quantum-theoretic property of the electron, which he called 'a two-valuedness not describable classically.' This remark suggested later in the same year to two pupils of Ehrenfest, G. E. Uhlenbeck and S. Goudsmit of Leiden [2] the adoption of a proposal which had been made in 1921 by Arthur H. Compton,[3] an American who was at that time working with Rutherford at Cambridge, namely that *the electron itself possesses an angular momentum or spin, and a magnetic moment.* Uhlenbeck and Goudsmit proposed as the amount of angular momentum $\frac{1}{2}\hbar$; and they suggested that the values $(l+\frac{1}{2})\hbar$ and $(l-\frac{1}{2})\hbar$ which are possible for the total angular momentum $j\hbar$ in e.g. the alkali spectra, are obtained by compounding the angular momentum $l\hbar$ with the electron-spin, which (since it exists in the magnetic field created by the orbital revolution of the electron) is compelled by space-quantification to take orientations either parallel or anti-parallel to $l\hbar$. Associated with the spin there is a magnetic moment whose value they asserted (for reasons to be discussed presently) to be $e\hbar/2mc$.

The discovery of electron-spin raised a question as to the validity of Sommerfeld's explanation of the fine-structure of the hydrogen lines. For if the electron has a spin with a magnetic moment, then two different orientations of the spin must be allowed, and to these must correspond two different energy-levels for the atom, causing a further resolution of each fine-structure component into a doublet. The measurements of Paschen had shown, however, that Sommerfeld's formula expressed the experimental data for the hydrogen spectrum satisfactorily : and it was found that the two corrections which in a more complete theory should be made to Sommerfeld's analysis, namely (1) replacing the particle-dynamics of Sommerfeld by quantum-mechanics, and (2) taking account of the spin magnetic moment, more or less neutralised each other, producing only a replacement of the quantum number k by $(j+\frac{1}{2})$.

The theory of spectra was much advanced by investigations arising out of the new experimental work on the Zeeman effect.

In the first decade of the twentieth century no great progress was made in the observational field, though in 1907 C. Runge [4]

[1] *ZS. f. P.* xxxi (1925), p. 373

[2] *Naturwiss*, xiii (1925), p. 953 ; *Nature*, cxvii (1926), p. 264. It is said that R. de L. Kronig had the same idea somewhat earlier, but finding it received unsympathetically by a colleague, did not publish it.

[3] *Phil. Mag.* xli (1921), p. 279 ; *J. Frankl. Inst.* cxcii (1921), p. 144

[4] *Phys. ZS.* viii (1907), p. 232

studied a number of spectral lines which show complex types of resolution, and found that the distances (measured as differences of frequency) of the components from the centre of the undisturbed line were connected by simple numerical relations with the frequency-difference which was given by Lorentz's theory of the Zeeman triplet, namely $eH/4\pi cm$, where H is the external magnetic force in electromagnetic units. In 1912, however, F. Paschen and E. Back [1] studied the Zeeman effect in lines which are members of doublets or triplets in series spectra (e.g. of the alkali atoms), and found that so long as the magnetic field is not strong the Zeeman splitting is very complicated. The different lines of a doublet or triplet behave differently, though the separations between the components always increase proportionally to the field-strength. When, however, the field has increased to such a strength that the separations between the Zeeman components are of the same order of magnitude as the separations between the components of the original doublet or triplet, the individual Zeeman components become diffuse and tend to amalgamate: and ultimately, at very great field-strengths, the whole system reduces to three components constituting a normal Zeeman triplet,[2] and having its centre at the centre of the original doublet or triplet. The difference in character between the Zeeman effect in weak and strong external magnetic fields was seen at once to be connected with the fact that the space-quantification of the electron spin is governed by the magnetic field of the orbital motions when the external magnetic field is weak, but is governed by the external field when that is sufficiently strong.

The experimental knowledge regarding the splitting that is found with comparatively weak fields in doublets and triplets, or the *anomalous Zeeman effect* as it was called, was reduced to a mathematical formula by Alfred Landé [3] of Tübingen in 1923. He showed that the frequency of any one of the Zeeman components can, like the frequencies of the lines of the original spectrum, be represented by the difference of two 'terms' or energy-levels. In a magnetic field, each term W of the original spectrum is split into several terms W+Z. If (supposing for simplicity that there is only one active or valence electron) the term W has the quantum numbers n, l, j, s, then for its Zeeman components we have

$$Z = \frac{m_j e g \hbar H}{2mc} \text{ ergs},$$

where e, $\hbar$, m, c have their usual meanings; g, which is called the *Landé splitting factor*, is given by the equation

$$g = 1 + \frac{j(j+1) + s(s+1) - l(l+1)}{2j(j+1)}$$

[1] *Ann. d. Phys.* xxxix (1912), p. 897; xl (1913), p. 960

[2] For a discussion of the phenomena taking place during the passage from weak to strong fields, cf. W. Pauli, *ZS. f. P.* xx (1923), p. 371.

[3] *ZS. f. P.* xv (1923), p. 189

and m_j is the magnetic quantum number,[1] which, as we have already seen, can take the values

$$m_j = j, j-1, \quad . \; . \; ., \; -j$$

so one spectral term splits into $(2j+1)$ equidistant terms. As we have seen, in transitions m_j can change only by 0 or ± 1, and the lines so arising are polarised in the same way as the lines of a Lorentz Zeeman triplet.

If the original term is a singlet level, we have $s=0$ and $l=j$, so $g=1$; and lines resulting from combinations of levels in singlet series have the appearance of Lorentz Zeeman triplets.

Observation of the Zeeman effect is of great assistance in determining the character of any particular spectral line, since the observation furnishes the value of the Landé splitting factor, and this knowledge generally leads to the determination of the quantum number l.

One naturally inquires why the quantum theory of the Zeeman effect given earlier (which in fact is valid only for lines which belong to series of singlets) does not apply in general. The reason must obviously be, that the Larmor procession, whose value was assumed to be $e\text{H}/2mc$ in the earlier proof, has not always this value : and this again can only mean that the ratio of magnetic moment to mechanical angular momentum is somehow different in the case of the anomalous Zeeman effect from what is asserted in Larmor's theory, where the ratio is that corresponding to the revolution of an electron in an orbit large compared with its own size. We conclude therefore that the existence of a g-factor different from unity indicates that the ratio of magnetic moment to angular momentum is not the same for the intrinsic spin of the electron as it is for the orbital motion. Now a state of the atom in which the angular momentum is due solely to electron-spin is specified by $l=0$, $s=\frac{1}{2}$, $j=\frac{1}{2}$: and the g-factor then has the value 2 : so we are led to suspect that for the electron-spin, the ratio of magnetic moment to angular momentum is twice as great as it is in the case of an electron circulating in an orbit : that is, it is e/mc instead of $e/2mc$. *The magnetic moment of the spinning electron is therefore conjectured to be* $e\hbar/2mc$; and it is found that with this assumption the Landé g-factor can be satisfactorily explained in all cases. The cause of the anomalous Zeeman effect was therefore now revealed.

Some striking confirmations of the validity of the correspondence-principle were obtained in a series of papers which followed a discovery made in 1922 by Miguel A. Catalán,[2] a Spanish research student of Alfred Fowler's at the Imperial College in London. When investigating the spectrum of manganese, Catalán noticed that there was a marked tendency for lines of similar character to

[1] cf. A. Sommerfeld and W. Heisenberg, *ZS. f. P.* xxxi (1922), p.131
[2] *Phil. Trans.* ccxxiii (1922), p. 127 (at p. 146)

appear in groups, and that these groups included some of the most intense lines in the spectrum: for instance, a group of nine lines between λ4455 and λ4462. For this kind of regularity he suggested the name *multiplet.* The lines arise from the combination of multiplet energy-levels. Spectral lines, particularly in multiplets, their relative intensities and their Zeeman components, were studied by many writers [1] in 1924–5, with results satisfactorily in accord with the predictions of the correspondence principle.

The rapid development of spectroscopy from 1913 onwards led to a much fuller understanding of the system of electrons which surrounds the nucleus of an atom. Investigators of this subject naturally based their work on the Newlands-Mendeléev periodic table,[2] and were stimulated by the attempts of A. M. Mayer and J. J. Thomson [3] to explain it in terms of stable configurations of electrons. It was obvious from chemical evidence that the two electrons possessed by the helium atom form a very stable configuration, which may be regarded as constituting a complete 'shell' of electrons surrounding the nucleus. The next atom in order of atomic number, lithium, must have this shell together with one loosely-attached electron outside it, and for the succeeding elements further electrons are added to this second shell until it contains eight electrons, when the tenth element neon is formed, thereby completing the shell and arriving again at a very stable configuration. The eleventh element sodium has these two complete shells together with one loosely-attached electron outside them, and so on.

The chemical evidence on atomic structure was marshalled in 1916–19 by two Americans, G. N. Lewis [4] and Irving Langmuir [5] (*b.* 1881). Lewis began by considering the different kinds of bonds that unite atoms into molecules, interpreting one kind of chemical bond as a couple of electrons held in common by two atoms; many facts, such as the tetrahedral carbon atom which is necessary for

[1] F. M. Walters, *J. Opt. Soc. Amer.* viii (1924), p. 245; O. Laporte, *ZS. f. P.* xxiii (1924), p. 135; xxvi (1924), p. 1. These two writers analysed the iron spectrum. H. C. Burger and H. B. Dorgels, *ZS. f. P.* xxiii (1924), p. 258; L. S. Ornstein and H. C. Burger, *ZS. f. P.* xxiv (1924), p. 41; xxviii (1924), p. 135; xxix (1924), p. 241; xxxi (1925), p. 355; W. Heisenberg, *ZS. f. P.* xxxi (1925), p. 617; xxxii (1925), p. 841; S. Goudsmit and R. de L. Kronig, *Proc. Amst. Ac.* xxviii (1925), p. 418; H. Hönl, *ZS. f. P.* xxxi (1925), p. 340; R. de L. Kronig, *ZS. f. P.* xxxi (1925), p. 885; xxxiii (1925), p. 261; A. Sommerfeld and H. Hönl, *Berlin Sitz.* (1925), p. 141; H. N. Russell, *Nature,* cxv (1925), p. 835; *Proc. N.A.S.* xi (1925), pp. 314, 322; H. N. Russell and F. A. Saunders, *Astroph. J.* lxi (1925), p. 38. This paper, arising out of an investigation of groups of lines in the arc spectrum of calcium, was of great importance for the study of complex spectra. It took into account the simultaneous action of two displaced electrons; F. Hund, *ZS. f. P.* xxxiii (1925), p. 345; xxxiv (1925), p. 296.

[2] cf. p. 11

[3] cf. p. 21

[4] *Journ. Amer. Chem. Soc.* xxxviii (1916), p. 762; Lewis, *Valence and the Structure of Atoms and Molecules*, New York, 1923

[5] *Journ. Amer. Chem. Soc.* xli (1919), p. 868. Somewhat similar ideas were published by W. Kossel, *Ann. d. Phys.* (1916), p. 229, who studied the transfer of electrons from electropositive to electronegative atoms, resulting in the formation of ions.

the understanding of chemical processes in organic substances, indicate that the atom must have a structure in three-dimensional space (in contradistinction to a flat ring system) [1]; and Lewis favoured a cubical form. Langmuir, continuing Lewis's work, concluded that the electrons in any atom are arranged in a series of nearly spherical shells. The outermost occupied shell consists of those electrons that do not belong to a closed configuration. These play the principal part in spectroscopic phenomena, and are known as the *active electrons* : they also play the principal part in chemical phenomena, in which connection they are known as *valence electrons.* The properties of the atoms depend much on the ease with which they are able to revert to more stable forms by giving or taking up electrons. The shells that are completed in helium and neon respectively have already been mentioned : argon (atomic number 18) in addition to the innermost shell of 2 and the next shell of 8, has a third shell of 8 : while krypton has four shells of 2, 8, 8 and 18 electrons : and so on. In the light of this model atom, Langmuir explained the chemical properties of the elements, and also their physical properties such as boiling-points, electric conductivity and magnetic behaviour.

It was early realised, however, that the formation of the shells cannot be quite regular : it was suggested in 1920 by R. Ladenburg [2] that in the case of the elements of atomic numbers 21 to 28 inclusive (scandium to nickel) the electrons newly added are not placed in the outermost shell but are used in building up a shell interior to this. This implies that a shell begins to be formed with potassium (atomic number 19) and calcium (atomic number 20) before the third shell is really complete.

It was, however, from the study of X-ray spectra (i.e. the characteristic X-rays which are emitted by solid chemical elements, or compounds of them, when bombarded by a beam of high-energy electrons) [3] that the greatest help was obtained in relating the shell-structure to the chemical elements. It was known that the characteristic X-rays constituted an additive atomic property and therefore the X-ray spectra must belong to the *atoms* of the anti-cathode. Moreover, it was known that they consisted of lines which could be arranged in series, like those of optical spectra : and Moseley's law connecting the progression of X-ray spectra from element to element with the amount of nuclear charge suggested that their origin should be sought in the innermost layers of the atom.

Almost immediately after the publication of Bohr's theory, W. Kossel [4] explained them as being due, like the radiations of optical spectra, to transition-processes : they arise when the atom

[1] On this see also Born, *Verh. d. deutsch. phys. Ges.* xx (1918), p. 230 ; Landé, *Verh. d. deutsch. phys. Ges.* xxi (1919), p. 2 ; E. Madelung, *Phys. ZS.* xix (1918), p. 524

[2] *Naturwiss*, viii (1920), p. 5 ; *ZS. f. Elektrochem*, xxvi (1920), p. 262

[3] cf. p. 16

[4] *Verh. d. deutsch. phys. Ges.* xvi (November 1914), p. 953 ; xviii (1916), p. 339

is restored to its original state after a disturbance which consists primarily in dislodging an electron from one of the innermost shells : the place thus vacated is filled by an electron which falls from a shell at a greater distance from the nucleus, and the place of this is again filled by an electron from a shell still more remote, and so on. The lowest level may be called, in harmony with Barkla's nomenclature of three years earlier,[1] the K-shell, and the fall of an electron into a vacant place in this shell yields an X-ray of the K-series. The next lowest levels are three known as L_I, L_{II} and L_{III}, and so for the others.[2]

Now if Z denotes the atomic number of the atom considered, by taking account of Z in the calculation on page 111, we see that according to Bohr's original theory, the frequency of the radiation emitted when an electron passes from a circular orbit of angular momentum $n\hbar$ to one of angular momentum $p\hbar$ is

$$\frac{2\pi^2 me^4 Z^2}{h^3}\left(\frac{1}{p^2}-\frac{1}{n^2}\right).$$

But if we put $p=1$, this formula represents precisely the frequencies of the K-lines. If $p=2$, it represents the L-lines ; and so on : from which fact we infer that the electrons in the K, L, M, . . . shells move in orbits which have respectively the principal quantum numbers 1, 2, 3, . . .[3] *The principal quantum number increases by unity from each shell to the next.* That the characteristic X-rays are of high frequency as compared with the radiations in optical spectra is explained by the presence of the factor Z^2 in the formula : a factor whose presence accounts at once for Moseley's law that the square root of the frequency of any particular line, such as the K_α-line, is proportional to the atomic number.

In the early days of X-ray spectroscopy, lines of the K-series could not be observed for the lighter elements (atomic numbers below 11), because their wave-lengths were longer than those of X-rays and yet shorter than that of ultra-violet light. The gap between X-rays and the ultra-violet was filled about 1928 by Jean Thibaud [4] of Paris, Erik Bäcklin [5] of Upsala, and A. P. R. Wadlund [6] of Chicago, and it was then found possible to trace the K-lines continuously down to the lightest elements. As might be expected, since the K-lines represent transitions down to the level of principal number 1, they pass into the Lyman series of hydrogen, while the L-lines, which represent transitions down to the principal number 2,

[1] cf. p. 16

[2] The γ-rays emitted by radio-active bodies come from the nucleus of the atom, and depend on nuclear levels, so they constitute a phenomenon altogether different from the characteristic X-rays of the K. L, M, . . . series.

[3] Sommerfeld, *Ann. d. Phys.*(4) li (1916), pp. 1 and 125

[4] *Phys. ZS.* xxix (1928), p. 241 ; *Journ. Opt. Soc. Amer.* xvii (1928), p. 145

[5] *Inaug. Diss. Uppsala Universitets Årsskrift,* 1928

[6] *Proc. Nat. Ac. Sc.* xiv (1928), p. 588

pass into the Balmer series, and the M-lines, which represent transitions down to the principal number 3, pass into the Paschen series of hydrogen.

The Bohr theory, as applied by Kossel, yields at once the laws of absorption of X-rays. Remembering that the first line of the K-series is emitted when an electron falls from the L-shell to the K-shell, it is obvious that this line could not be expected to appear in the absorption-spectrum : for if in absorption an electron were dislodged from the K-shell, there would normally be no vacant place in the L-shell to receive it : indeed the K-absorption only sets in suddenly, when the incident energy is sufficient to separate a K-electron completely from the atom : that is, *the absorption-edge coincides with the series-limit.* This explains why in X-ray spectra there are no absorption-lines,[1] but only absorption-edges : and *the frequency of every line in the X-ray emission-spectrum is the difference of the frequencies of two absorption-edges.*

What was known in 1921 regarding the structure of the atom in relation to the physical and chemical properties of the elements was set forth in an extensive survey by Bohr,[2] whose principles were vindicated in the following year in a somewhat dramatic way. One of the missing elements in the Newlands-Mendeléev table was that of atomic number 72. Now the elements immediately preceding (of atomic numbers 57 to 71 inclusive) belong to the group of the 'rare earths,' and it was expected by many chemists that number 72 would also belong to this group. Indeed in 1911 G. Urbain,[3] by fractionation of the earths of gadolinite, believed that he had discovered a new rare earth to which he gave the name of *Celtium* ; and this was later identified by Urbain and Dauvillier with the missing element 72. Bohr, however, gave a rational interpretation of the occurrence of the rare earths in the periodic system, asserting that they represent a gradual completion of the shell of electrons for which the principal quantum number is 4, while the number of electrons in the shells of principal quantum numbers 5 and 6 remains unchanged. With the rare-earth lutecium (71) the shell $n=4$ attains its full complement of 32 electrons, and it follows that the element 72 cannot be a rare earth, but must have an additional electron in the shells $n=5$ or $n=6$: it must in fact be a homologue of zirconium. In 1922 D. Coster and G. Hevesy of Copenhagen verified this prediction[4] : examining the X-ray spectrum of a Norwegian zirconium mineral, they found lines which, by Moseley's rule, must certainly belong to an element of atomic number 72 : and for this they proposed the name of *Hafnium* (Hafniae = Copenhagen). Its chemical properties showed that it was undoubtedly analogous to titanium (22) and zirconium (40).

[1] For a refinement of this general statement cf. Kossel, *ZS. f. P.* i (1920), p. 119.
[2] *Fysisk Tideskrift*, xix (1921), p. 153 = *Theory of Spectra and Atomic Constitution*, Cambridge, 1922 [3] *Comptes Rendus*, clii (1911), p. 141 ; *Chem. News*, ciii (1911), p. 73
[4] *Nature*, cxi (1923), p. 79

In 1922 and the following years great additions were made to the accurate knowledge of X-ray spectra and their relation to atomic number.[1] Guided by this work, in 1923 N. Bohr and D. Coster [2] introduced new symbols for the layers of electrons in the atoms, based on their spectroscopic behaviour with respect to X-rays. The innermost group was now denoted by $1(1, 1)$K, the next groups by $2(1, 1)L_{I}$, $2(2, 1)L_{II}$, $2(2, 2)L_{III}$, and so on outward. Here the letters with the Roman subscript numbers indicate the previously-recognised shells with their subsidiary levels, while the symbols of the form $n(k, j)$ define the subsidiary levels more closely, n being Bohr's principal quantum number, while k and j are whole numbers which were later to be identified with Sommerfeld's azimuthal quantum number and a function of Sommerfeld's inner quantum number respectively. Then in January 1924 A. Dauvillier [3] showed experimentally (by examining the absorption relative to the level) that the 8 electrons in the L-level must be partitioned into sub-groups of 2, 2 and 4. These results quickly led to the understanding of the orbits and energies of the various groups of electrons which was finally accepted, and which was proposed in 1924–5 by Edmund C. Stoner of Cambridge [4] (whose arguments were based on physical reasoning) and J. D. Main Smith of Birmingham [5] (who approached the matter from the chemical side). According to this system, to each complete shell corresponds a definite value of the principal quantum number n. Within this shell the subsidiary quantum number l can take the values $0, 1, \ldots (n-1)$: and the inner quantum number j of an electron can then take the values $(l+\frac{1}{2})$ or $(l-\frac{1}{2})$, (unless $l=0$, in which case j can take only the value $\frac{1}{2}$). The number of electrons in the sub-group (n, l, j) is simply $(2j+1)$, and therefore the total number of electrons with the quantum numbers n and l is $2(2l+1)$, and the total number of electrons in the n-shell is

$$2\{1+3+5+7+\ldots+(2n-1)\}$$

or

$$2n^2.$$

Helium has altogether 2 electrons, forming a complete K-shell. For neon there are 10 electrons, namely 2 forming a complete K-shell and 8 forming a complete L-shell. For argon there are 18 electrons, of which 2 form a complete K-shell, 8 form a complete L-shell and 8 form an incomplete M-shell : the M-shell is, however, complete

[1] Dirk Coster, *Phil. Mag.* xliii (1922), p. 1070 : xliv (1922), p. 546 ; A. Landé, *ZS. f. P.* xvi (1923), p. 391 ; Manne Siegbahn and A. Žáček, *Ann. d. Phys.* lxxi (1923), p. 187 ; M. Siegbahn and B. B. Ray, *Ark. f. Mat, Ast. och Fys.* xviii (1924), No. 19 ; M. Siegbahn and R. Thoraeus, *Phil. Mag.* xlix (1925), p. 513 ; cf. L. de Broglie and A. Dauvillier, *Phil. Mag.* xlix (1925), p. 752

[2] *ZS. f. P.* xii (1923), p. 342

[3] *Comptes Rendus*, clxxviii (1924), p. 476

[4] *Phil. Mag.* xlviii (1924), p. 719

[5] *J. Chem. Ind.* xliii (1924), p. 323 ; xliv (1925), p. 944 ; *Chemistry and Atomic Structure*, London, 1924 ; cf. A. Sommerfeld, *Ann. d. Phys.* lxxvi (1925), p. 284 ; *Phys. ZS.* xxvi (1925), p. 70

in the next noble gas, krypton, which has 2 K-, 8 L-, 18 M-, and 8 N-electrons. For any complete shell the orbital, spin and total angular momentum are all zero.

We have seen that the number of electrons in the sub-group (n, l, j) is $(2j+1)$: but for a given value of j, the magnetic quantum number m_j can take precisely the $(2j+1)$ values $j, j-1, \ldots, -j$. Thus there is one and only one electron corresponding to each distinct state or energy-level, i.e. each distinct set of the four quantum numbers (n, l, j, m_j). In 1924 Pauli [1] based on this fact a general principle, that *two electrons in a central field can never be in states of binding which have the same four quantum numbers.* This assertion can be extended to systems in which there is not a single central field : e.g. it applies to electrons which are in the field of two nuclei at the same time : the states of these electrons can be described by quantum numbers, and it is still true that no two electrons can have the same state, i.e. be described by the same set of quantum numbers. The statement in this general form is called *Pauli's exclusion principle.* It is valid for protons as well as for electrons, and indeed for all elementary particles whose spin is $\frac{1}{2}\hbar$.

Another discovery of Pauli's, made in the same year,[2] related to a structure much finer than the ordinary multiplet structure, which is observed in some spectra, and which is called the *hyperfine structure* of spectral lines. Pauli showed that this is to be ascribed not to the electron-shells, but to the influence of the atomic nucleus, which may itself have an angular momentum and a magnetic moment : and in the case when the spectrum is that of a mixture of isotopes, the differences in nuclear mass of the isotopes will also cause small differences of position of lines in their spectra, and so contribute to the hyperfine structure.

In 1923 the domain of quantum theory was enlarged, when the diffraction of a parallel beam of radiation by a grating was explained on quantum principles by William Duane.[3] Consider an infinite grating with the spacing d between its rulings. If the grating moves with constant velocity in a direction in its own plane perpendicular to the rulings, it will return to its original aspect when it has moved through a distance d : so we can regard it as a periodic system to which the Wilson-Sommerfeld quantum rule

$$\int p\,dq = nh$$

can be applied, where p denotes momentum in the direction of the surface of the grating perpendicular to its rulings, the spacing d being the domain over which the integration must be extended : and therefore

$$pd = nh,$$

[1] *ZS. f. P.* xxxi (1925), p. 765
[2] *Naturwiss,* xii (1924), p. 741
[3] *Proc. N. A. S.* ix (1923), p. 158. Duane's treatment was considerably improved as regards its justification by A. H. Compton, *ibid.* p. 359.

so *the grating can pick up momentum* p *only in multiples of* h/d: the momentum of radiation is transferred to and from matter in quanta. If a photon of energy $h\nu$, and therefore of momentum $h\nu/c$, falls on the grating in a direction making an angle i with the normal, and is diffracted in a direction making an angle r with the normal, then taking the components of momentum in a direction of the surface of the grating at right angles to its rulings, we have the equation of conservation of momentum

$$(h\nu/c)\ (\sin i - \sin r) = nh/d$$

which in the language of the wave-theory would be

$$n\lambda = d\ (\sin i - \sin r).$$

This is the ordinary equation giving the directions of the diffracted radiation, now obtained from the corpuscular (photon) theory of light.

It may be noted that Duane's equation $pd = nh$ can be applied to a photon, if d be interpreted as wave-length, so we obtain $p\lambda = h$: since $\lambda = c/\nu$, this shows that the momentum of the photon is $h\nu/c$.

Duane's principle is closely related to a principle introduced later in the same year (1923) by L. de Broglie, which will be considered in Chapter VI.

The Duane method was extended to finite gratings, including even the case of only two reflecting points, by P. S. Epstein and P. Ehrenfest in 1924.[1]

[1] *Proc. N. A. S.* x (1924), p. 133; xiii (1927), p. 400

Chapter V

GRAVITATION

We have seen[1] that for many years after its first publication, the Newtonian doctrine of gravitation was not well received. Even in Newton's own University of Cambridge, the textbook of physics in general use during the first quarter of the eighteenth century was still Cartesian: while all the great mathematicians of the Continent—Huygens in Holland, Leibnitz in Germany, Johann Bernoulli in Switzerland, Cassini in France—rejected the Newtonian theory altogether.

This must not be set down entirely to prejudice: many well-informed astronomers believed, apparently with good reason, that the Newtonian law was not reconcilable with the observed motions of the heavenly bodies. They admitted that it explained satisfactorily the first approximation to the planetary orbits, namely that they are ellipses with the sun in one focus: but by the end of the seventeenth century much was known observationally about the departures from elliptic motion, or *inequalities* as they were called, which were presumably due to mutual gravitational interaction: and some of these seemed to resist every attempt to explain them as consequences of the Newtonian law.

The inequalities were of two kinds: first, there were disturbances which righted themselves after a time, so as to have no cumulative effect: these were called *periodic* inequalities. Much more serious were those derangements which proceeded continually in the same sense, always increasing the departure from the original type of motion: these were called *secular* inequalities. The best known of them was what was called the *great inequality of Jupiter and Saturn*, of which an account must now be given.

A comparison of the ancient observations cited by Ptolemy in the *Almagest* with those of the earlier astronomers of Western Europe and their more recent successors, showed that for centuries past the mean motion, or average angular velocity round the sun, of Jupiter, had been continually increasing, while the mean motion of Saturn had been continually decreasing. This indicated some striking consequences in the remote future. Since by Kepler's third law the square of the mean motion is proportional to the inverse cube of the mean distance, the decrease in the mean motion of Saturn implied that the radius of his orbit must be increasing, so that this planet, the most distant of those then known, would be always becoming more remote, and would ultimately, with his attendant ring and

[1] cf. Vol. I, pp. 29–31

satellites, be altogether lost to the solar system. The orbit of Jupiter, on the other hand, must be constantly shrinking, so that he must at some time or other either collide with one of the interior planets, or must be precipitated on the incandescent surface of the sun.

No explanation of the secular inequality of Jupiter and Saturn could be obtained by any simple and straightforward application of Newton's gravitational law, and the French Academy of Sciences offered a prize in 1748, and again in 1752, for a memoir relating to these two planets. On each occasion Euler made considerable advances [1] in the general treatment of planetary perturbations, and received the award : but the result of his investigations was to make the observed secular accelerations of Jupiter and Saturn more mysterious than ever, for they appeared to be quite inconsistent with the tolerably complete theory which he created. Lagrange, who wrote on the problem [2] in 1763, and gave a still more complete discussion, likewise failed to obtain a satisfactory agreement with the observations.

In 1773 the matter was taken up by Laplace.[3] He began by carrying the approximation to a higher order than his predecessors, and was surprised to find that in the final expression for the effect of Jupiter's disturbing action upon the mean motion of Saturn, the terms cancelled each other out. The same result, as he showed, held for the effect of any planet upon the mean motion of any other: thus *the mean motions of the planets cannot have any secular accelerations whatever as a result of their mutual attractions.* Laplace accordingly concluded that the accelerations observed in the case of Jupiter and Saturn could not be genuinely secular : they must really be periodic, though the period might be immensely long.

With this key to the mystery, he completely solved it, in a great memoir of 1784.[4] He realised that an inequality of long period could be produced only by a term of long period in the perturbing function : denoting this term by $p \sin qt$, then in order that it may be of long period, q must be extremely small. By a double integration with respect to the time, such as happens in the course of solving the differential equations, this term would become $(p/q^2) \sin qt$, and (p/q^2) might be quite large even though p were very small. Thus a great inequality of long period might be produced by a term in the perturbing function which was so small that it had been neglected altogether by preceding investigators. This explained why Euler and Lagrange had failed to solve the problem, and nothing remained to be done except to inquire more closely into the identity of the term in the perturbing function. Now five times the mean motion of Saturn is very nearly equal to twice the mean motion of Jupiter: so if n, n' are the mean motions, then $5n-2n'$

[1] *Recueil des pièces qui ont remporté les prix de l'Acad.*, tome vii (1769)
[2] *Mélanges de phil. et de math. de la Soc. Roy. de Turin pour l'année* 1763, p. 179 (1766)
[3] *Mém. des Savans étrang.* vii (1776). Read 10 Feb. 1773
[4] *Mém. de l'Acad.*, 1784, p. 1

is very small: and the term in the perturbing function whose argument is $(5n-2n')t$, which would have an extremely small coefficient, would satisfy all the conditions required. Thus Laplace was able to assert that *the great inequality of Jupiter and Saturn is not a secular inequality, but is an inequality of long period, in fact* 929 *years: and it is due to the fact that the mean motions of the two planets are nearly commensurable.* When the results of his calculations were compared with the observations, the agreement was found to be perfect.[1]

The story of the great inequality of Jupiter and Saturn illustrates a distinctive feature of the situation, namely that the truth of the Newtonian or any other law of gravitation cannot be tested by means of controlled experiments in a laboratory, and its verification must depend on the comparison of astronomical observations, extended over centuries, with mathematical theories of extreme complexity.

After the triumphant conclusion of Laplace's researches on the great inequality of Jupiter and Saturn, there was still outstanding one unsolved problem which formed a serious challenge to the Newtonian theory, namely the secular acceleration of the mean motion of the moon. From a study of ancient eclipses recorded by Ptolemy and the Arab astronomers, Halley[2] had concluded in 1693 that the mean motion of the moon has been becoming continually more rapid ever since the epoch of the earliest recorded observations. The mean distance of our satellite must therefore have been continually decreasing, and it seemed that at some time in the remote future the moon must be precipitated on the earth. The Academy of Sciences of Paris proposed the subject for the prize in 1770, and again in 1772 and 1774, and prizes were awarded to Euler[3] and Lagrange,[4] who made valuable contributions to general dynamical astronomy: on the question proposed, however, they found only the negative results that no secular inequality could be produced by the action of Newtonian gravitation when the heavenly bodies were regarded as spherical, and, moreover, that the observed phenomena could not be explained by taking into account the departures of the figures of the earth and moon from sphericity. Laplace now took up the matter, and showed, first, that the effect was not due to any retardation of the earth's diurnal rotation due to the resistance of the aether: he then investigated the consequences of another supposition, namely that gravitational effects are propagated with a velocity which is finite[5]: but this also led to no satisfactory issue, and at last he found the true solution,[6] which

[1] As Laplace's discovery showed, the existence of 'small divisors' makes it a matter of great difficulty to investigate the convergence of the series that occur in Celestial Mechanics; cf E. T. Whittaker, *Proc. R.S. Ed.* xxxvii (1917), p. 95.

[2] Phil. Trans. xvii (1693), p. 913

[3] *Recueil des pièces qui ont remporté les prix de l'Acad.* ix (1777)

[4] *Recueil des pièces qui ont remporté les prix de l'Acad.* ix (1777), for the competition of 1772; *Mém. des Savans Étrang.* vii, for that of 1774

[5] cf. Vol. I, p. 207

[6] It was presented to the Academy on 19 March 1787; *Mém. de l'Acad.* 1786, p. 235 (published 1788).

may be described as follows. The mean motion of the moon round the earth depends mainly on the moon's gravity to the earth, but is slightly diminished by the action of the sun upon the moon. This solar action, however, depends to a certain extent on the eccentricity of the terrestial orbit, which is slowly diminishing, as a result of the action of the planets on the earth. Consequently the sun's mean action on the moon's mean motion must also be diminishing, and hence the moon's mean motion must be continually increasing, which is precisely the phenomenon that is observed. The acceleration of the moon's mean motion will continue as long as the earth's orbit is approaching a circular form: but as soon as this process ceases, and the orbit again becomes more elliptic, the sun's mean action will increase and the acceleration of the moon's motion will be converted into a retardation. The inequality is therefore not truly secular, but periodic, though the period is immensely long, in fact millions of years. This striking vindication of the Newtonian theory came exactly a century after the publication of the *Principia*.

'The moon, in the present day,' wrote Robert Grant,[1] 'is about two hours later in coming to the meridian than she would have been if she had retained the same mean motion as in the time of the earliest Chaldean observations. It is a wonderful fact in the history of science that those rude notes of the priests of Babylon should escape the ruin of successive empires, and, finally, after the lapse of nearly 3,000 years, should become subservient in establishing a phenomenon of so refined and complicated a character as the inequality we have just been considering.'

In the nineteenth and twentieth centuries, however, many astronomers formed the opinion that Laplace's great memoir had not completely cleared up the situation. The fact to be explained is that the moon has relative to the sun an apparent acceleration of its mean motion of about 22″ per century per century.[2] Laplace's theoretical value was of about this amount, but J. C. Adams [3] found, by including terms of higher order in the calculation, that Laplace's value was much too great, the amount explicable by purely gravitational causes being only 12·2″ per century per century: and his conclusion was substantiated by later workers in lunar theory. In 1905 P. H. Cowell [4] redetermined (from ancient eclipses) the observed secular acceleration, and found that it was almost twice the theoretical value, and that there was also deducible from observation a secular acceleration of the sun (i.e. of the earth's orbital motion). His own tentative explanation [5] depended on the

[1] *History of Physical Astronomy* (London, 1852), p. 63

[2] There is some confusion of language on this subject. The coefficient of t^2 in the expression for the longitude is about 11″, and the true secular acceleration is therefore about 22″; but many writers speak of the acceleration as '11″ in a century.'

[3] *Phil. Trans.* cxliii (1853), p. 397; *Mon. Not. R.A.S.* xl (1880), p. 472

[4] *Mon. Not. R.A.S.* lxv (1905), p. 861

[5] *Mon. Not. R.A.S.* lxvi (1906), p. 352

notion of tidal friction, while J. H. Jeans [1] suggested a modification of the Newtonian law, and E. A. Milne [2] proposed a dependence of the Newtonian constant of gravitation on the age of the universe. These matters were discussed by J. K. Fotheringham [3] in 1920 and in 1939 by H. Spencer Jones,[4] who showed that the secular accelerations of the sun, Mercury, and Venus are proportional to their mean motions, and can be accounted for by the retardation of the earth's axial rotation by tidal friction.[5] This retardation must also produce, in addition to its direct apparent effect, a real secular acceleration of the moon's mean motion, in order that the total angular momentum of the earth-moon system may be conserved : but the amount of this acceleration cannot be predicted theoretically.

We may now refer to other phenomena in the solar system which were not explained with complete satisfaction by Newton's formula. The anomalous motion of the perihelion of the planet Mercury has already been referred to [6] : it might possibly be accounted for if the inverse square law is modified by adding a term involving the velocities of the bodies [7] : or it might, as H. Seeliger [8] showed, be explained by the attraction of the masses forming the zodiacal light. The node of the orbit of Venus was also found by Newcomb [9] to have a secular acceleration which was five times the probable error, and for which no explanation could be offered : and a secular increase in the mean motion of the inner satellite of Mars, discovered in 1945 by B. P. Sharpless,[10] is so far not accounted for.

Certain comets also present problems. Among them the best known is an object which was discovered by Jean-Louis Pons in 1818, but is generally called *Encke's comet* from a long series of memoirs [11] devoted to it by J. F. Encke, who showed in 1819 that it was periodic, with a period of 1,207 days, and later that its motion showed an acceleration which was not explicable by the Newtonian theory. Encke himself proposed to explain this by postulating a resisting medium whose density was inversely proportional to the

[1] *Mon. Not. R.A.S.* lxxxiv (1923), p. 60 (*at* p. 75)

[2] *Proc. R.S.*(A), clvi (1936), p. 62 (*at* p. 81)

[3] *Mon. Not. R.A.S.* lxxx (1920), p. 578

[4] *Mon. Not. R.A.S.* xcix (1939), p. 541

[5] The retardation of the earth's rotation due to tidal friction increases the length of the day by about 1/1000 of a second per century, so each century is 36½ seconds longer than the one preceding. There is, moreover, a variability in the rate of rotation, which is slower in February than in August. This fluctuation, which was discovered by means of clocks formed of vibrating quartz crystals, is probably of meteorological origin.

[6] cf. Vol. I, p. 208

[7] For work more recent than that referred to in Vol. I, cf. Paul Gerber, *ZS. f. Math. u. Phys.* xliii (1898), p. 93 ; *Ann. d. Phys.* lii (1917), p. 415, and the comments on the latter paper by H. Seeliger, *Ann. d. Phys.* liii (1917), p. 31 ; liv (1917), p. 38, and S. Oppenheim, ibid. liii (1917), p. 163

[8] *München Ber.* 1906, p. 595

[9] S. Newcomb, *The elements of the four inner planets* ; Supplement to the American Ephemeris and Nautical Almanac for 1897 ; Washington, 1895

[10] *Ast. J.* li (1945), p. 185

[11] Mostly in the *Berlin Abhandlungen* and the *Astronomische Nachrichten* ; cf. specially *Comptes Rendus*, xlviii (1858), p. 763

square of the distance from the sun : but O. Backlund, who devoted many years to the study of the comet, showed that Encke's assumption is impossible.[1] An alternative hypothesis is that the comet has encounters with a swarm of meteors.

In 1910 P. H. Cowell and A. C. D. Crommelin [2] computed with great care the motion of another periodic comet, that of Halley, between 1759 and 1910, and predicted the time of perihelion for its return in 1910 : the time deduced later from actual observations was about 2·7 days later than this. The discrepancy could not be accounted for by any defect in the calculations, and it would seem therefore that there is some small disturbing cause or causes at work, other than the gravitational attraction expressed by Newton's law.[3]

At the meeting of the Amsterdam Academy of Sciences on 31 March 1900, Lorentz communicated a paper entitled *Considerations on Gravitation*,[4] in which he reviewed the problem as it appeared at that time—a problem which, as the above recital shows, was still far from a completely satisfactory solution. So many phenomena had been successfully accounted for by applications of electromagnetic theory that it seemed natural to seek in the first place an explanation in terms of electric and magnetic actions. As we have seen,[5] the assumptions of Laplace's investigation, which led him to conclude that the velocity of propagation of gravitation must be vastly greater than that of light, do not to twentieth-century minds seem very plausible [6] : and Lorentz felt free to put forward a theory depending on electromagnetic actions propagated with the speed of light. The first possibility he considered was suggested by Le Sage's concept of ultra-mundane corpuscles.[7] Since it had been found that a pressure against a body could be produced as well by trains of electric waves as by moving projectiles, and that the X-rays with their remarkable penetrating power were essentially electric waves, it was natural to replace Le Sage's corpuscles by vibratory motions. Why should there not exist radiations far more penetrating than even the X-rays, which might account for a force which, so far as is known, is independent of all intervening matter ?

Lorentz therefore calculated the interaction between two ions on the assumption that space is traversed in all directions by trains of electric waves of very high frequency. If an ion P is alone in a

[1] Backlund's conclusions are summarised in *Bull. astronomique*, xi (1894), p. 473; cf. also A. Wilkens, *Astr. Nach.* cxcvi (1914), p. 57. On the perturbations of Encke's comet cf. D. Brouwer, *Ast. J.* lii (1947), p. 190.

[2] *Investigation of the motion of Halley's Comet from 1759 to 1910*; Appendix to the 1909 volume of *Greenwich Observations* (1910). *Essay on the return of Halley's comet, Publ. Astr. Ges., Lpz.*, No. 23 (1910)

[3] On the effect of loss of mass by evaporation when a comet is near the sun; cf. F. Whipple, *Astroph. J.* cxi (1950), p. 375

[4] *Proc. Amst. Acad.* ii (1900), p. 559; French translation in *Arch. Néerl.* vii (1902), p. 325

[5] Vol. I, pp. 207–8

[6] The same remark applies to the ideas of R. Lehmann-Filhes, *München. Ber.* xxv (1895), p. 71.

[7] cf. Vol. I, p. 31

field in which the propagation of waves takes place equally in all directions, the mean force on it will vanish. But the situation will be different as soon as a second ion Q has been placed in the neighbourhood of P : for then, in consequence of the vibrations emitted by Q after it has been exposed to the rays, there might be a force on P, of course in the direction of the line QP. It was found, however, that this force could exist only if in some way or other electromagnetic energy were continually disappearing : and after full consideration, Lorentz concluded that the assumptions he had made could not provide a satisfactory explanation of gravitation.

He then considered a second hypothesis, which may be regarded as having been foreshadowed in the one-fluid electrical theory of Watson, Franklin and Aepinus.[1] According to this theory, as developed in 1836 by O. F. Mossotti[2] (1791–1863), electricity is conceived as a continuous fluid, whose atoms repel each other. Material molecules are also supposed to repel each other, but to have with the aether-atoms a mutual attraction, which is somewhat greater than the mutual repulsion of the particles which repel. The composition of these forces accounts for gravitation, except at very small distances, where the same mechanism accounts for cohesion.

Wilhelm Weber (1804–91) of Göttingen and Friedrich Zöllner[3] (1834–82) of Leipzig developed this conception into the idea that all ponderable molecules are associations of positively and negatively charged electrical corpuscles, with the condition that the force of attraction between corpuscles of unlike sign is somewhat greater than the force of repulsion between corpuscles of like sign. If the force between two electric units of like charge at a certain distance is α dynes, and the force between a positive and a negative unit charge at the same distance is γ dynes, then, taking account of the fact that a neutral atom contains as much positive as negative electric charge, it was found that $(\gamma-\alpha)/\alpha$ need only be a quantity of the order 10^{-35} in order to account for gravitation as due to the difference between α and γ.

At the time of Lorentz's paper, no strong physical reason for an assumption of this kind could be given. Many years afterwards Eddington[4] suggested one. He had taken to heart a warning uttered by Mach.[5] 'Even in the simplest case, in which apparently we deal with the mutual action between only two particles, it is impossible to disregard the rest of the universe. Nature does not begin with elements, as we are forced to do. Certainly it is fortunate for us

[1] cf. Vol. I, p. 50

[2] *Sur les forces qui régissent la constitution intérieure des corps, apperçu pour servir à la détermination de la cause et des lois de l'action moléculaire* (Turin, 1836).

[3] *Erklärung der universellen Gravitation*, Leipzig, 1882; pp. 67–82 deal with Weber's contributions; cf. also J. J. Thomson, *Proc. Camb. Phil. Soc.* xv (1910), p. 65

[4] *Fundamental Theory* (Cambridge, 1946), p. 102

[5] E. Mach, *Die Mechanik in ihrer Entwickelung* (5th edn., 1904), p. 249; English translation (London, 1893), p. 235

that we can sometimes turn away from the overwhelming All, and allow ourselves to study isolated facts. But we must not forget ultimately to amend and complete our views by taking into account what had been omitted.' Eddington applied Mach's general principle to the interaction between two electric charges. If they are of opposite sign, all their lines of force run from one to the other, and the two together may be regarded as a self-contained system which is independent of the rest of the universe: but if the two charges are of the same sign, then the lines of force from each of them must terminate on other bodies in the universe, and it is natural to expect that these other bodies will have some influence on the nature of the interaction between the charges. Following up this idea by a calculation, Eddington arrived at the conclusion that when two protons are at a distance r apart, which is of the same order of magnitude as the radius of an atomic nucleus, their mutual energy contains, in addition to the ordinary electrostatic energy corresponding to the inverse-square law, a term of the form

$$Ae^{-\frac{r^2}{k^2}}$$

where A and k are constants: if it could be supposed that this is correct, and is an asymptotic approximation, valid for values of r of nuclear dimensions, to a function whose asymptotic approximation, valid for values of r large compared with nuclear dimensions, is inversely proportional to r, then there would obviously be a possibility of accounting on these lines for gravitation.

From 1904 onwards the Newtonian law of gravitation was examined in the light of the relativity theory of Poincaré and Lorentz. This was done first by Poincaré,[1] who pointed out that if relativity theory were true, gravity must be propagated with the speed of light, and who showed that this supposition was not contradicted by the results of observation, as Laplace had supposed it to be. He suggested modifications of the Newtonian formula, which were afterwards discussed and further developed by H. Minkowski [2] and by W. de Sitter.[3] It was found that relativity theory would require secular motions of the perihelia of the planets, which however would be of appreciable amount only in the case of Mercury, and even in that case not great enough to account for the observed anomalous motion.

In 1907 Planck[4] broke new ground. It had been established by the careful experiments of R. v. Eötvös[5] that *inertial mass* (which determines the acceleration of a body under the action of a given

[1] *Comptes Rendus*, cxl (1905), p. 1504; *Palermo Rend.* xxi (1906), p. 129
[2] *Gött. Nach.* 1908, p. 53
[3] *Mon. Not. R.A.S.* lxxi (1911), p. 388; cf. also F. Wacker, *Inaug. Diss.*, Tübingen, 1909
[4] *Berl. Sitz.* 13 June 1907, p. 542, specially at p. 544
[5] *Math. u. nat. Ber. aus Ungarn*, viii (1891), p. 65

force) and *gravitational mass* (which determines the gravitational forces between the body and other bodies) are always exactly equal : which indicates that *the gravitational properties of a body are essentially of the same nature as its inertial properties.* Now, said Planck, all energy has inertial properties, and therefore *all energy must gravitate.* Six months later Einstein [1] published a memoir in which he introduced [2] what he later called the *Principle of Equivalence*, which may be thus described :

Consider an observer who is enclosed in a chamber without windows, so that he is unable to find out by direct observation whether the chamber is in motion relative to an outside world or not. Suppose the observer finds that any object in the chamber, whatever be its chemical or physical nature, when left unsupported, falls towards one particular side of the chamber with an acceleration f which is constant relative to the chamber. The observer would be justified in putting forward either of two alternative explanations to account for this phenomenon :

(i) he might suppose that the chamber is 'at rest,' and that there is a *field of force*, like the earth's gravitational field, acting on all bodies in the chamber, and causing them if free to fall with acceleration f : or

(ii) he might explain the observed effects by supposing that *the chamber is in motion* : if he postulates that in the outside world there are co-ordinate axes (C) relative to which there is no field of force, and if he moreover supposes the chamber to be in motion relative to these axes (C) with an acceleration equal in magnitude but opposite in direction to f, then it is obvious that free bodies inside the chamber would have an acceleration f relative to the chamber.

The observer has no criterion enabling him to tell which of these two explanations is the true one. If we could say definitely that the chamber is at rest, then explanation (i) would be true, while if we could say definitely that the axes (C) are at rest, then explanation (ii) would be true. But by the Principle of Relativity, we cannot give a preference to one of these sets of axes over the other : we cannot say that one of them is moving and the other at rest : and we must therefore regard the two explanations as equally valid, or, in other words, must assert that a homogeneous field of force is equivalent to an apparent field which is due to the accelerated motion of one set of axes relative to another : *a uniform gravitational field is physically equivalent to a field which is due to a change in the co-ordinate system.*

In this paper Einstein also showed [3] by combining Doppler's principle with the principle of equivalence, that a spectral line generated by an atom situated at a place of very high gravitational potential, e.g. at the sun's surface, has, when observed at a place

[1] *Jahrb. d. Radioakt.* iv (4 Dec. 1907), p. 411 ; cf. Einstein, *Ann. d. Phys.* xxxv (1911), p. 898

[2] At p. 454 [3] At pp. 458–9

of lower potential, e.g. on the earth, a greater wave-length than the corresponding line generated by an identical atom on the earth. This may also be shown very simply as follows. Denoting by Ω the gravitational potential at the sun's surface, the energy lost by a photon of frequency ν in escaping from the sun's gravitational field is $\Omega \times$ the mass of the photon, or $\Omega h\nu/c^2$. Remembering that the energy $h\nu$ is hc/λ, we see that the wave-length of the solar radiation as measured by the terrestial observer is $1+(\Omega/c^2)$ times the wave-length of the same radiation when produced on earth.[1]

In 1911 Einstein followed up this work by an important memoir,[2] in which he argued that since light is a form of (electromagnetic) energy, therefore light must gravitate, that is, *a ray of light passing near a powerfully gravitating body such as the sun, must be curved*: and *the velocity of light must depend on the gravitational field.*

Einstein's paper was the starting-point of a theory published shortly afterwards by Max Abraham.[3] Accepting the principles that the velocity of light c depends on the gravitational potential, and that the law of gravitation might be expressed by a differential equation satisfied by c, he postulated that the negative gradient of c indicates the direction of the gravitational force, and that the energy-density in a statical gravitational field is proportional to $c^{-1}(\text{grad } c)^2$. Einstein himself at almost the same time published[4] a somewhat different theory, in which the equations of motion of a particle in a statical gravitational field, when gravity only is acting, are

$$\frac{d}{dt}\left(\frac{1}{c^2}\frac{dx}{dt}\right) = -\frac{1}{c}\frac{\partial c}{\partial x},$$

and similar equations in y and z.

To the same period belong the theories of G. Nordström[5] and Gustav Mie.[6] Though Mie's theory has not survived as the permanent basis of mathematical physics, it had a marked influence on thought, and some of its ideas appeared later in the researches of other workers. It aimed at being a complete theory of physics,

[1] cf. also J. M. Whittaker, *Proc. Camb. Phil. Soc.* xxiv (1928), p. 414. The red-displacement due to a gravitational field with arbitrary motion of the source and of the observer was calculated by H. Weyl in the fifth edition (1923) of his *Raum Zeit Materie*, Anhang III.

[2] *Ann. d. Phys.*(4) xxxv (21 June 1911), p. 898

[3] *Lincei Atti*, xx (Dec. 1911), p. 678; *Phys. ZS.* xiii (1912), pp. 1, 4, 176, 310, 311, 793; *N. Cimento*(4) iv (Dec. 1912), p. 459

[4] *Ann. d. Phys.* xxxviii (Feb. 1912), pp. 355, 443. A controversy followed, for which see Abraham, *Ann. d. Phys.* xxxviii (1912), p. 1056; xxxix (1912), p. 444; and Einstein, *Ann. d. Phys.* xxxviii (1912), p. 1059; xxxix (1912), p. 704.

[5] *Phys. ZS.* xiii (Nov. 1912), p. 1126; *Ann. d. Phys.*(4) xl (April 1913), p. 856; ibid. xlii (Oct. 1913), p. 533; *Phys. ZS.* xv (1914), p. 375; cf. A. Einstein and A. D. Fokker, *Ann. d. Phys.*(4) xliv (1914), p. 321; M. v. Laue, *Jahrb. d. Rad. u. El.* xiv (1917), p. 263

[6] *Ann. d. Phys.* xxxvii (1912), p. 511; xxxix (1913), p. 1; xl (1913), p. 1; *Phys. ZS.* xv (1914), pp. 115, 169, 263; Festschrift für J. Elster u. H. Geitel (Braunschweig, 1915), pp. 251–68; cf. A. Einstein, *Phys. ZS.* xv (1914), p. 176. There is a good short account of Mie's theory in H. Weyl, *Raum Zeit Materie*, 4th Aufl., (Berlin, 1921), § 26.

based on the principle that electric and magnetic fields and electric charges and currents suffice completely to describe all that happens in the material world, so that matter can be constructed from these elements. Moreover, he originated the notion of a single *world-function* from which, by the aid of the Calculus of Variations, all the laws of physical processes could be derived: as we shall see presently, this conception was developed afterwards by Hilbert.

In Mie's theory all happenings, both in the field and in matter, are described by twenty functions, constituting two six-vectors which describe the field and two four-vectors which describe matter: namely

(i) a six-vector formed of the components of the electric displacement **D** and the magnetic force **H**

(ii) a six-vector formed of the components of the magnetic induction **B** and the electric force **E**

(iii) a four-vector formed of the electric charge and current

(iv) a four-vector formed of the electric scalar and vector potentials.

In the Maxwell-Lorentz theory this last four-vector plays merely a mathematical part: but in Mie's theory its components are physical realities. In Chapter IV of his series of memoirs, Mie discussed quantum theory, and in Chapter V, gravitation.

The next advance owed much to a paper that had been written in 1909 by Harry Bateman[1] (1882–1946). At any place in the earth's gravitational field, take moving rectangular axes (x^1, x^2, x^3) and a measure of time (x^0), such that these axes constitute an inertial system (A), so that the path of a free particle relative to them is (at any rate near the origin) a straight line, and the vanishing of the differential form

$$c^2(dx^0)^2 - (dx^1)^2 - (dx^2)^2 - (dx^3)^2$$

is the condition that a luminous disturbance originating at the point (x^1, x^2, x^3) at the instant x^0, should arrive at the point $(x^1 + dx^1, x^2 + dx^2, x^3 + dx^3)$, at the instant $(x^0 + dx^0)$.

In free aether, where there is no field of force, two different inertial systems either can be derived from each other by simple translation and rotation in ordinary three-dimensional space, or else they have a uniform motion of translation relative to each other (or, of course, a combination of these methods of derivation). But when we move to a distant place in a field of force, e.g. if we move to the antipodes in the earth's gravitational field, although we can here again find axes (say (B)) which are inertial (that is, free particles in their vicinity move relatively to them with uniform velocity in straight lines), a framework (B) does not move with uniform velocity

[1] *Proc. L.M.S.*(2) viii (1910), p. 223; cf. also *Amer. J. Math.* xxxiv (1912), p. 325

relative to a framework (A) (in fact the two frameworks are in accelerated motion relative to each other), so the relation between two inertial frameworks which holds in the relativity theory of Poinacré and Lorentz does not hold when a gravitational field is present. We cannot therefore find co-ordinates (x^0, x^1, x^2, x^3) describing position and time over *the whole* field such that the interval ds at any place in the field is given by the equation

$$(ds)^2 = (dx^0)^2 - \frac{1}{c^2}\left\{(dx^1)^2 + (dx^2)^2 + (dx^3)^2\right\}.$$

Instead of this, we can now find at every place in the field a *local* framework of inertial axes (X^0, X^1, X^2, X^3), such that the interval will be given approximately for points in the neighbourhood of the origin by

$$(ds)^2 = (dX^0)^2 - \frac{1}{c^2}\left\{(dX^1)^2 + (dX^2)^2 + (dX^3)^2\right\}.$$

Let (x^0, x^1, x^2, x^3) now be any co-ordinates specifying position and time over the whole field. Then at each place, the differentials dX^p will be expressible in terms of the x^p and the dx^p by equations of the form

$$dX^p = \sum_{r=0}^{3} a_{pr}\, dx^r.$$

Substituting this in the expression for $(ds)^2$, we have

$$(ds)^2 = \sum_{p,q=0}^{3} g_{pq}\, dx^p dx^q \qquad (1)$$

where

$$g_0 = a_{00}^2 - \frac{1}{c^2}(a_{10}^2 + a_{20}^2 + a_{30}^2) \text{ etc.}$$

The vanishing of this form (1) is now the condition that a luminous disturbance originating at the space-time point (x^0, x^1, x^2, x^3) should arrive at the space-time point $(x^0 + dx^0, x^1 + dx^1, x^2 + dx^2, x^3 + dx^3)$. The form (1) must be invariant for *all* transformations of the co-ordinates (x^0, x^1, x^2, x^3); and its coefficients g_{pq}, which are functions of (x^0, x^1, x^2, x^3), are characteristic of the field.

Bateman realised the connection of his work with the tensor-calculus of Ricci and Levi-Civita[1]: in fact, since (dx^0, dx^1, dx^2, dx^3) is a contravariant vector, it follows from the invariance of the quadratic differential form that the set of the g_{pq} is a symmetric covariant tensor of rank 2.[2]

[1] cf. p. 58

[2] These ideas were applied by Bateman in order to investigate a scheme of fundamental electromagnetic equations which are not altered by very general transformations.

Bateman's ideas were carried over into a more profound treatment of the problem of gravitation in the second half of the year 1913 by Einstein,[1] who in the years 1912–14 worked in partnership (as regarded the mathematics) with a Zürich geometer, Marcel Grossmann. In these papers the theory was put forward, that just as the rectilinear motion of a particle in free aether when there is no field is determined by the equation

$$\delta\left(\int ds\right)=0$$

where δ is the symbol of the Calculus of Variations, and

$$(ds)^2=c^2(dt)^2-(dx)^2-(dy)^2-(dz)^2,$$

so now (making a step analogous to that in Bateman's paper), the motion of a free material particle in a gravitational field is determined by the equation

$$\delta\left(\int ds\right)=0$$

where

$$(ds)^2=\sum_{p,q=0}^{3} g_{pq}\,dx^p\,dx^q,$$

the coefficients g_{pq} being characteristic of the state at the point (x^0, x^1, x^2, x^3) in space-time, and ds being invariant with respect to arbitrary transformations of (x^0, x^1, x^2, x^3). As with Bateman, $ds=0$ is the condition that a luminous disturbance originally at the world-point (x^0, x^1, x^2, x^3) should arrive at the world-point $(x^0+dx^0,\ x^1+dx^1,\ x^2+dx^2,\ x^3+dx^3)$. In geometrical language, *the path of a free material particle in a gravitational field is a geodesic in the four-dimensional curved space whose metric is specified by the equation*[2]

$$(ds)^2=\sum_{p,q=0}^{3} g_{pq}\,dx^p\,dx^q.$$

This was a tremendous innovation, because it implied the abandonment of the time-honoured belief that a gravitational field can be specified by a single scalar potential-function: instead, it proposed

[1] A. Einstein and M. Grossmann, *ZS. f. M. u. P.* lxii (1913), p. 225; lxiii (1914), p. 215; A. Einstein, *Vierteljahr. d. Nat. Ges. Zürich*, lviii (1913), p. 284; *Archives des sc. phys. et nat.*(4) xxxvii (1914), p. 5; *Phys. ZS.* xiv (15 Dec. 1913), p. 1249; *Berlin Sitz.* 1914, p. 1030

[2] A theory that matter consists in ' crinkles ' of space had been published by W. K. Clifford in 1870; cf. *Proc. Camb. Phil. Soc.* ii (1876), p. 157 = Clifford's *Math. Papers*, p. 21.

to specify the gravitational field by the ten functions g_{pq} which could now be spoken of as the *gravitational potentials.*

Einstein justified[1] this new departure by showing that the theory of a single scalar gravitational potential led to inacceptable inferences. He compared, for instance, two systems, in the first of which a moveable hollow box with perfectly-reflecting walls is filled with pure-temperature radiation, while in the second the same radiation is contained inside a fixed vertical pit which is closed at the top and bottom by moveable pistons connected by a rod so as to be always at a fixed distance apart, the pit walls and pistons all being perfectly-reflecting: and he showed that on the single-scalar-potential theory the work necessary to raise the radiation upwards against the force of gravity would in the second system be only one-third of the work required in the first system: a conclusion which was obviously wrong. He admitted, however, that in his own mind the strongest reason for rejecting the single-scalar-potential theory was his conviction that relativity in physics exists not only with respect to the Lorentz group of linear orthogonal transformations but with respect to a much wider group.

The ten coefficients g_{pq} not only specify the force of gravitation, but they determine also the scale of distance in every direction, and the rate of clocks. The metric defined by

$$(ds)^2 = \sum_{p,\, q=0}^{3} g_{pq} dx^p dx^q$$

is not, in general, Euclidean: and since its non-Euclidean qualities determine the gravitational field, we may say that *gravitational theory is reduced to geometry,* in accordance with an idea expressed by FitzGerald[2] in 1894 in the words 'Gravity is probably due to a change of structure of the aether, produced by the presence of matter.' The 'aether' of FitzGerald was called by Einstein simply 'space' or 'space-time': and FitzGerald's somewhat vague term 'structure' became with Einstein the more precise 'curvature.' Thus we obtain the central proposition of the Einsteinian theory: 'Gravity is due to a change in the curvature of space-time, produced by the presence of matter.'

In comparing FitzGerald's statement with Einstein's, it may be remarked that if we consider a gravitational field which is *statical,* i.e. such as would be produced by gravitating masses that are permanently at rest relative to each other, then feeble[3] electromagnetic phenomena taking place in it can be shown to happen exactly in accordance with the ordinary Maxwellian theory of electromagnetic phenomena taking place in a medium whose specific inductive capacity and magnetic permeability are aelotropic and vary from

[1] § 7 of the paper in the *ZS. f. M. u. P.*

[2] FitzGerald's *Works,* p. 313

[3] i.e. so feeble that they do not appreciably change the curvature of the field.

point to point. In particular,[1] if we consider an electric point-charge at rest in the field of a single gravitating point-mass, the electric field is the same as would be obtained, in ordinary electrostatics, by supposing that the specific inductive capacity and magnetic permeability of the medium vary with the distance from the gravitating mass according to the law $(r+1)^3/r^2(r-1)$.

It is possible that when FitzGerald said ' Gravity is probably due to a change of structure of the aether,' he was actually thinking of a change which would show itself in alterations of the dielectric constant and magnetic permeability, and that he had in mind an electrical constitution of matter, on account of which matter would be subject to forces depending on the values of the dielectric constant and magnetic permeability: by analogy with the fact that in a liquid whose dielectric constant varies from point to point, an electrified body moves from places of lower to places of higher dielectric constant.[2]

What differentiates the Einsteinian theory from all previous conceptions is that the older physicists had regarded gravity as merely one among many types of natural force—electric, magnetic, etc.—each of which influenced in its own way the motion of material particles. Space, whose properties were set forth in Euclidean geometry, was, so to speak, the stage on which the forces played their parts. But in the new theory gravity was no longer one of the players, but part of the structure of the stage. A gravitational field consisted essentially in a replacement of the Euclidean properties by a much more complicated kind of geometry: space was no longer homogeneous or isotropic. An analogy may be drawn from the game of bowls. Bowling-greens, in the north of England, are not flat, but rise to a slight elevation in the centre. An observer who failed to notice the central elevation would find that a bowl (supposed without bias) always described a path convex toward the centre of the green, and he might account for this by postulating a centre of repellent force there. A better-informed observer would attribute the phenomenon to a geometrical feature—the slope. The two explanations correspond respectively to the Newtonian and the Einsteinian conceptions of gravity: for Newton it is a force, for Einstein it is a modification of the geometry of space.

When the metric of space-time is specified by an equation

$$(ds)^2 = \sum_{p\ q=0}^{3} g_{pq} dx^p dx^q,$$

an observer moving in any manner will have a *world-line* consisting of the points of space-time which he successively occupies : and at

[1] E. T. Copson, *Proc. R.S.*(A), cxviii (1928), p. 184

[2] This idea was later developed by E. Wiechert, *Ann. d. Phys.* lxiii (1920), p. 301.

any point of his world-line he will have in his immediate neighbourhood an *instantaneous three-dimensional space*, formed by the aggregate of all the elements of length which are orthogonal to his world-line at the point: orthogonality being defined, as already explained,[1] by the statement that two vectors (X) and (Y) are said to be orthogonal if

$$\sum_{p=0}^{3} X_p Y^p = 0,$$

where (X_p) is the covariant form of one vector and (Y^p) is the contravariant form of the other.

Einstein laid down the principle that the equations which describe any physical process must satisfy the condition that their covariance with respect to arbitrary substitutions can be deduced from the invariance of ds. In other words, the laws of nature must be represented by equations which are covariantive for the form $\sum_{p,q} g_{pq} dx^p dx^q$ with respect to all point-transformations of co-ordinates. Laws of nature are assertions of *coincidences* in space-time, and therefore must be expressible by covariant equations.

It might be thought that by following up the consequences of this principle we should obtain important positive results. However, Ricci and Levi-Civita[2] had shown long before that from practically *any* assumed law we can derive another law which does not differ from it in any way that can be tested by observation, but which is covariant. The fact that a formula has the covariant property does not, therefore, tell us anything as to whether it is correct or not. We are, however, perhaps justified in believing that a conjectural law which can be expressed readily and simply in covariant form is more worthy of attention (as being more likely to be true) than one whose covariant form is awkward and complicated.

Not only must the general laws of physics be covariant, it is also necessary that every single assertion which has a physical meaning must be covariant with respect to arbitrary transformations of the co-ordinate system. Thus the assertion that an electron is at rest for an interval of time of duration unity cannot have a physical meaning, since this assertion is not covariant.[3]

In Einstein's general theory, the velocity of light at any place has always the value *c with respect to any inertial frame of reference* for this neighbourhood, and the velocity of any material body is less than *c*. Thus there is no difficulty in the fact that the fixed stars have velocities greater than *c* with respect to axes fixed in the rotating earth: for such axes are not inertial.

[1] See p. 64

[2] *Math. Ann.* liv (1901), p. 125; cf. E. Kretschmann, *Ann. d. Phys.* liii (1917), p. 575, and A. Einstein, *Ann. d. Phys.* lv (1918), p. 241

[3] D. Hilbert, *Math. Ann.* xcii (1924), p. 1

Some physicists called attention to the fact that when light is propagated in a medium where there is anomalous dispersion, the index of refraction may be less than unity, whence it seemed as if the velocity of light in the dispersive medium might be greater than the velocity of light *in vacuo*. The difficulty was removed when it was pointed out by L. Brillouin [1] and A. Sommerfeld [2] that the velocity of light with which the index of refraction is concerned is the *phase* velocity, whereas the velocity of a signal is the *group* velocity, which is never greater than c.

It has sometimes been supposed, by a misunderstanding, that the general Einsteinian theory requires us to regard the Copernican conception of the universe as no more true than the Ptolemaic, and that it is indifferent whether we regard the earth as rotating on her axis or regard the stellar universe as performing a complete revolution about the earth every twenty-four hours. The root of the matter, by which everything is explained, is that the Copernican axes are inertial, while the Ptolemaic are not. The earth rotates *with respect to the local inertial axes.*[3]

In his first paper in the *Zeitschrift für Math. u. Phys.*,[4] Einstein gave the form which Maxwell's equations of the electromagnetic field must take when the metric of space-time is given by a quadratic differential form

$$(ds)^2 = \sum_{p,\,q=0}^{3} g_{pq}\, dx^p\, dx^q.$$

To obtain these, it will be necessary to introduce some other concepts of Ricci's absolute differential calculus, or *tensor-calculus* [5] as Einstein henceforth called it. Suppose that we are given a quadratic differential form in any number of variables

$$(ds)^2 = \sum_{p,\,q=1}^{n} g_{pq}\, dx^p\, dx^q$$

then, following Elwin Bruno Christoffel [6] (1829–1900), we introduce what are called *Christoffel symbols of the first kind*, defined as

$$\begin{bmatrix} p\,q \\ l \end{bmatrix} = \tfrac{1}{2}\left(\frac{\partial g_{pl}}{\partial x^q} + \frac{\partial g_{ql}}{\partial x^p} - \frac{\partial g_{pq}}{\partial x^l}\right) \qquad (p,\, q,\, l = 1, 2, \ldots n),$$

[1] *Comptes Rendus*, clvii (1913), p. 914
[2] *Ann. d. Phys.* xliv (1914), p. 177
[3] cf. G. Giorgi and A. Cabras, *Rend. Lincei*, ix (1929), p. 513
[4] At page 241
[5] The word *tensor* had been used by W. Voigt in 1898, in connection with the elasticity of crystals.
[6] *J. für Math.* lxx (1869), pp. 46, 241

and *Christoffel symbols of the second kind,* defined as

$$\left\{ \begin{matrix} p\,q \\ l \end{matrix} \right\} = \sum_{r=1}^{n} g^{rl} \left[\begin{matrix} p\,q \\ r \end{matrix} \right] \qquad (p, q, l = 1, 2, \ldots n).$$

As Christoffel showed, the Christoffel symbols enable us to form new tensors from known tensors by a process of generalised differentiation. If $(X_1, X_2, \ldots X_n)$ is any covariant vector, then *the quantities* $(X_p)_q$ *defined by*

$$(X_p)_q = \frac{\partial X_p}{\partial x^q} - \sum_{r=1}^{n} \left\{ \begin{matrix} p\,q \\ r \end{matrix} \right\} X_r$$

constitute a covariant tensor of rank 2. This process was called *covariant differentiation* by Ricci. Similarly if (X_{pq}) is any covariant tensor of rank 2, then *the quantities* $(X_{pq})_s$ *defined by the equations*

$$(X_{pq})_s = \frac{\partial X_{pq}}{\partial x^s} - \sum_{r=1}^{n} \left\{ \begin{matrix} p\,s \\ r \end{matrix} \right\} X_{rq} - \sum_{r=1}^{n} \left\{ \begin{matrix} q\,s \\ r \end{matrix} \right\} X_{pr}$$

constitute a covariant tensor of rank 3 *which is called the covariant derivative of* X_{pq}. The covariant differentiation of sums and products is effected by rules similar to those that apply to ordinary differentiation. Moreover if (X^l) is any contravariant vector, then the *quantities* $(X^l)_s$ *defined by*

$$(X^l)_s = \frac{\partial X^l}{\partial x^s} + \sum_{q=1}^{n} \left\{ \begin{matrix} q\,s \\ l \end{matrix} \right\} X^q$$

define a mixed tensor of rank 2 which is called the covariant derivative of (X^l).

We know that Maxwell's equations in Euclidean space consist of the Ampère-Maxwell tetrad

$$\frac{\partial d_x}{\partial x} + \frac{\partial d_y}{\partial y} + \frac{\partial d_z}{\partial z} = 4\pi\rho$$

$$\frac{\partial h_z}{\partial y} - \frac{\partial h_y}{\partial z} = \frac{1}{c}\frac{\partial d_x}{\partial t} + 4\pi\rho v_x$$

$$\frac{\partial h_x}{\partial z} - \frac{\partial h_z}{\partial x} = \frac{1}{c}\frac{\partial d_y}{\partial t} + 4\pi\rho v_y$$

$$\frac{\partial h_y}{\partial x} - \frac{\partial h_x}{\partial y} = \frac{1}{c}\frac{\partial d_z}{\partial t} + 4\pi\rho v_z$$

and the Faraday tetrad

$$\frac{\partial h_x}{\partial x}+\frac{\partial h_y}{\partial y}+\frac{\partial h_z}{\partial z}=0$$

$$\frac{\partial d_z}{\partial y}-\frac{\partial d_y}{\partial z}=-\frac{1}{c}\frac{\partial h_x}{\partial t}$$

$$\frac{\partial d_x}{\partial z}-\frac{\partial d_z}{\partial x}=-\frac{1}{c}\frac{\partial h_y}{\partial t}$$

$$\frac{\partial d_y}{\partial x}-\frac{\partial d_x}{\partial y}=-\frac{1}{c}\frac{\partial h_z}{\partial t}.$$

Now write $x^0=ct$, $x^1=x$, $x^2=y$, $x^3=z$. We have seen in Chapter II that the electric and magnetic vectors together constitute a six-vector

$$d_x=X^{01},\quad d_y=X^{02},\quad d_z=X^{03},\quad h_x=X^{23},\quad h_y=X^{31},\quad h_z=X^{12}.$$

Thus the Ampère-Maxwell tetrad becomes

$$\frac{\partial X^{01}}{\partial x^1}+\frac{\partial X^{02}}{\partial x^2}+\frac{\partial X^{03}}{\partial x^3}=4\pi\rho$$

$$\frac{\partial X^{10}}{\partial x^0}+\frac{\partial X^{12}}{\partial x^2}+\frac{\partial X^{13}}{\partial x^3}=4\pi\rho v_x$$

$$\frac{\partial X^{20}}{\partial x^0}+\frac{\partial X^{21}}{\partial x^1}+\frac{\partial X^{23}}{\partial x^3}=4\pi\rho v_y$$

$$\frac{\partial X^{30}}{\partial x^0}+\frac{\partial X^{31}}{\partial x^1}+\frac{\partial X^{32}}{\partial x^2}=4\pi\rho v_z.$$

Now if (T^p) is a contravariant vector, the quantity $\sum (T^p)_p$ (where the suffix outside the bracket denotes covariant differentiation) is a scalar which is called the *divergence* of (T^p), and denoted by div (T^p): we can easily show that

$$\text{div}\,(T^p)=\frac{1}{\sqrt{g}}\sum_p \frac{\partial}{\partial x^p}\left(\sqrt{g}\;T^p\right).$$

If (T^{pq}) is a contravariant tensor of rank 2, then the quantity

$$\sum_q (T^{pq})_q$$

which is a vector, is called the *vectorial divergence* of (T^{pq}), and is denoted by $\Delta\text{iv}\ (T^{pq})$. We can show that if (T^{pq}) is skew, i.e. is a six-vector, then

$$\Delta\text{iv}\ (T^{pq}) = \frac{1}{\sqrt{g}} \sum_q \frac{\partial}{\partial x^q}\left(\sqrt{g}\ T^{pq}\right).$$

Remembering that when the space is Euclidean, $\sqrt{g}$ is a constant, we see that for a six-vector (T^{pq}) in Euclidean space we have

$$\Delta\text{iv}\ (T^{pq}) = \sum_q \frac{\partial T^{pq}}{\partial x^q}.$$

Comparing this with the above form of the Ampère-Maxwell tetrad, we see that the tensorial (i.e. covariant) form of this tetrad must be

$$\Delta\text{iv}\ (X^{pq}) = 4\pi J^p \tag{A}$$

where J^p denotes the four-vector which represents the electric charge and current, namely $\rho_0\ dx^p/ds$, where ρ_0 is the proper density of the charge, i.e. the charge divided by the volume it occupies, as measured by an observer moving with it.

Next consider the Faraday tetrad of equations, which may clearly be written

$$\frac{\partial X_{23}}{\partial x^1} + \frac{\partial X_{31}}{\partial x^2} + \frac{\partial X_{12}}{\partial x^3} = 0$$

$$\frac{\partial X_{23}}{\partial x^0} + \frac{\partial X_{30}}{\partial x^2} + \frac{\partial X_{02}}{\partial x^3} = 0$$

$$\frac{\partial X_{31}}{\partial x^0} + \frac{\partial X_{10}}{\partial x^3} + \frac{\partial X_{03}}{\partial x^1} = 0$$

$$\frac{\partial X_{12}}{\partial x^0} + \frac{\partial X_{20}}{\partial x^1} + \frac{\partial X_{01}}{\partial x^2} = 0.$$

Now it can be shown that if X^{rs} is a six-vector, and if (p, q, r, s) is an even permutation of the numbers (0, 1, 2, 3), then a six-vector Y_{pq} can be defined by the equations

$$Y_{pq} = \sqrt{(-g)}\ X^{rs}$$

and that we then have

$$X_{pq} = -\sqrt{(-g)}\ Y^{rs}.$$

The two six-vectors X_{pq} and Y_{pq} are said to be *dual* to each other. If Y_{pq} is the six-vector dual to the electromagnetic six-vector X_{pq}, so that in Euclidean space

$$Y^{10}=h_x,\quad Y^{20}=h_y,\quad Y^{30}=h_z,\quad Y^{23}=d_x,\quad Y^{31}=d_y,\quad Y^{12}=d_z,$$

then the Faraday tetrad may evidently be written

$$\frac{\partial Y^{01}}{\partial x^1}+\frac{\partial Y^{02}}{\partial x^2}+\frac{\partial Y^{03}}{\partial x^3}=0$$

$$\frac{\partial Y^{10}}{\partial x^0}+\frac{\partial Y^{12}}{\partial x^2}+\frac{\partial Y^{13}}{\partial x^3}=0$$

$$\frac{\partial Y^{20}}{\partial x^0}+\frac{\partial Y^{21}}{\partial x^1}+\frac{\partial Y^{23}}{\partial x^3}=0$$

$$\frac{\partial Y^{30}}{\partial x^0}+\frac{\partial Y^{31}}{\partial x^1}+\frac{\partial Y^{32}}{\partial x^2}=0$$

or

$$\Delta\text{iv}\,(Y^{pq})=0 \qquad \text{(B)}.$$

Since the equations (A) and (B) are tensor equations, and are therefore covariant with respect to all transformations of the coordinates (x^0, x^1, x^2, x^3), we may assume with Einstein that *they represent the equations of the electromagnetic field in space-time of any metric whatever.*

The six-vector of the electromagnetic field may be expressed in terms of potentials, in the same way as in Euclidean space it is expressed by the equations

$$d_x=-\frac{\partial\phi}{\partial x}-\frac{\partial a_x}{c\partial t}\quad\text{etc.,}\qquad h_x=\frac{\partial a_z}{\partial y}-\frac{\partial a_y}{\partial z}\quad\text{etc.}$$

For if we write $(\phi_0, \phi_1, \phi_2, \phi_3)$ for $(\phi, -a_x, -a_y, -a_z)$, these equations become, as we have seen on page 76,

$$X_{pq}=\frac{\partial\phi_p}{\partial x^q}-\frac{\partial\phi_q}{\partial x^p}\qquad (p, q=0, 1, 2, 3).$$

Writing this

$$X_{pq}=(\phi_p)_q-(\phi_q)_p$$

where the suffixes outside the brackets represent covariant differentiation, we see that *the potential* $(\phi_0, \phi_1, \phi_2, \phi_3)$ *is a covariant vector.*

Electromagnetic theory leads naturally to physical optics, and this again to geometrical optics. Now in the relativity theory of Poincaré and Lorentz, for which the line-element $d\tau$ in the world of space-time is given by

$$(d\tau)^2 = (dt)^2 - \frac{1}{c^2}\left\{(dx)^2 + (dy)^2 + (dz)^2\right\},$$

the geodesics of the world are straight lines, and the null geodesics (i.e. the geodesics for which $d\tau$ vanishes) are the straight lines for which

$$\frac{(dx)^2 + (dy)^2 + (dz)^2}{(dt)^2} = c^2,$$

so the null geodesics are the tracks of rays of light. When Einstein created his new general theory of relativity, in which gravitation was taken into account, he carried over this principle by analogy, and asserted its truth for gravitational fields. The principle was, however, not proved at that time : and indeed there was the obvious difficulty in proving it, that strictly speaking there are no 'rays' of light—that is to say, electromagnetic disturbances which are filiform, or drawn out like a thread—except in the limit when the frequency of the light is infinitely great : in all other cases diffraction causes the 'ray' to spread out.

The matter was investigated in 1920 by M. von Laue,[1] who, starting from the partial differential equations of electromagnetic phenomena in a gravitational field, obtained a particular solution which corresponded to light of infinitely high frequency, and showed that the path of this disturbance satisfied the differential equations of the null geodesics : thus for the first time proving the truth of Einstein's assertion. It was afterwards shown by E. T. Whittaker[2] that the law is really an immediate deduction from the theory of the characteristics of partial differential equations, and that it is not necessary to introduce the notion of frequency at all : in fact, that *in a gravitational field, any electromagnetic disturbance which is filiform must necessarily have the form of a null geodesic of space-time.*

At any point of space-time, the directions which issue from it may be classified into those that have a spatial character and those that have a temporal character : the two classes are separated from each other by a cone whose generators are the paths of rays of light : this is called the *null-cone.*

It can be shown mathematically that if we know the geodesics of space-time (i.e. the paths of free material particles) and also know which of them are null geodesics (i.e. paths of rays of light), then the coefficients g_{pq} of the equation defining the metric

$$(ds)^2 = \sum_{p,q=0}^{3} g_{pq}\,dx^p dx^q$$

[1] *Phys. ZS.* xxi (1920), p. 659 [2] *Proc. Camb. Phil. Soc.* xxiv (1928), p. 32

are completely determinate. In fact, when the null geodesics are given, we can infer the metric save for a factor, say

$$(ds)^2 = \lambda(x^0, x^1, x^2, x^3) \times (\text{a determinate quadratic form in the } dx^p)$$

where λ is unknown: and when we are also given the non-null geodesics, the factor $\lambda(x^0, x^1, x^2, x^3)$ can be determined.

In the papers we have referred to, which are of date earlier than November 1915, Einstein gave, as we have seen, a satisfactory account of the behaviour of mechanical and electrical systems in a field of gravitation which is supposed given: his formulae were derived fundamentally from the principle of equivalence, i.e. the principle that the systems behave just as if there were no gravitational field, but they were referred to a co-ordinate-system with an acceleration equal and opposite to the acceleration of gravity. But he had not as yet succeeded in obtaining an entirely satisfactory set of fundamental equations for the gravitational field itself, i.e. equations which would play the same part in his theory that Poisson's equation [1]

$$\frac{\partial^2 V}{\partial x^2} + \frac{\partial^2 V}{\partial y^2} + \frac{\partial^2 V}{\partial z^2} = -4\pi\rho$$

played in the Newtonian theory. This defect was repaired, and the theory (now known as *General Relativity*) substantially completed, in a series of short papers published in November-December 1915, in the *Berlin Sitzungsberichte*.[2]

Let us first inquire what covariants of the form $\sum_{p,q} g_{pq} dx^p dx^q$ can be formed from the g_{pq}'s and their derivatives alone. It can be shown that these can all be derived from a certain tensor of rank 4, known as the *Riemann tensor*, which must now be introduced.

Let D_s denote the operation which when applied to the Christoffel symbol of the first kind is

$$D_s \begin{bmatrix} rt \\ u \end{bmatrix} = \frac{\partial}{\partial x^s} \begin{bmatrix} rt \\ u \end{bmatrix} - \sum_l \begin{Bmatrix} su \\ l \end{Bmatrix} \begin{bmatrix} rt \\ l \end{bmatrix};$$

then we define the Riemann tensor [3] by the equation

$$K_{pqrs} = D_q \begin{bmatrix} pr \\ s \end{bmatrix} - D_r \begin{bmatrix} pq \\ s \end{bmatrix}.$$

[1] cf. Vol. I, p. 61

[2] *Berlin Sitz.* 1915, pp. 778, 799, 831, 844. An eight-line abstract, dated 25 March 1915, is given at p. 315.

[3] It was discovered by Riemann, in a memoir *Commentata mathematica qua respondere tentatur* . . ., which was sent to the Paris Academy in 1861 and published posthumously, *Werke* (1892), p. 401; it was afterwards used by Christoffel, loc. cit.

From the Riemann tensor we can obtain a tensor of rank 2 which is defined by the equation

$$K_{pq} = \sum_{r,s} g^{rs} K_{pqrs} ;$$

it is called the *Ricci-tensor* or *contracted curvature-tensor* : and from the Ricci tensor we obtain what is called the *scalar curvature* of the space, defined by the equation

$$K = \sum_{p,q} g^{pq} K_{pq}.$$

For a two-dimensional space, e.g. a surface in Euclidean three-dimensional space, $-\frac{1}{2}K$ is the ordinary Gaussian measure of curvature, $1/\rho_1\rho_2$.

In applications to the theory of gravitation, we are dealing with the four-dimensional world of space-time. For such a world, K may be defined geometrically in the following way. At any world-point, take any four directions that are mutually orthogonal with respect to the metric of the world. These four directions, taken in pairs, determine six surface-elements or orientations, in each of which there is an aggregate of geodesics issuing from the point, forming a 'geodesic surface' : then K is minus twice the sum of the Gaussian measures of curvature of these six surfaces at the given world-point. It is independent of the choice of the four orthogonal directions.

Now Mach had introduced long before a principle, that inertia must be reducible to the interaction of bodies : and Einstein generalised this into what he called *Mach's principle*, namely that the field represented by the ten potentials g_{pq} is determined *solely* by the masses of bodies. The word 'mass' is here to be understood in the sense given to it by the theory of relativity, that is, as equivalent to energy : and as energy is expressed covariantively by Minkowski's energy-tensor T_{pq}, it follows that in the fundamental equations of gravitation, corresponding to the equation $\nabla^2 V = -4\pi\rho$ of the Newtonian theory, we may expect the tensor T_{pq} or some linear function of it to take the place of Poisson's ρ. We expect to find on the other side of the equation, corresponding to $\nabla^2 V$, a tensor of the same rank as T_{pq}, that is, the second rank, containing second derivatives of the potentials but no higher derivatives. The only covariant tensors of this character are the Ricci-tensor K_{pq}, with Kg_{pq} and g_{pq}. Einstein first supposed that K_{pq} might be a simple constant multiple of T_{pq} : but this is not satisfactory for reasons that will be more evident later[1] : and he finally proposed the equations

$$K_{pq} = -\kappa(T_{pq} - \tfrac{1}{2}g_{pq}T) \qquad (p, q = 0, 1, 2, 3)$$

[1] The divergence of T_{pq} is zero, and of K_{pq} is not in general zero.

where $T = \sum_{p,q} g^{pq} T_{pq}$, and κ is a constant depending on the Newtonian constant of gravitation. *These are the general field-equations of gravitation.*

Multiplying them by g^{pq}, summing with respect to p and q, and remembering that $\sum_{p,q} g^{pq} g_{pq} = 4$, we have $K = \kappa T$, so the equations may be written

$$K_{pq} - \tfrac{1}{2} g_{pq} K = -\kappa T_{pq} \qquad (p, q = 0, 1, 2, 3).$$

These are ten equations for the ten unknowns g_{pq} : there are four identities between them, as might be expected : for four of the g_{pq}'s can be assigned arbitrarily as functions of the x^p, corresponding to the fact that the equations are invariant under the most general transformation of co-ordinates.

According to Mach's principle as adopted by Einstein, the curvature of space is governed by physical phenomena, and we have to ask whether the metric of space-time may not be determined *wholly* by the masses and energy present in the universe, so that space-time cannot exist at all except in so far as it is due to the existence of matter. The point at issue may be illustrated by the following concrete problem : if all matter were annihilated except one particle which is to be used as a test-body, would this particle have inertia or not? The view of Mach and Einstein is that it would not : and in support of this view it may be urged that, according to the deductions of general relativity, the inertia of a body is increased when it is in the neighbourhood of other large masses : it seems needless, therefore, to postulate other sources of inertia, and simplest to suppose that *all* inertia is due to the presence of other masses. When we confront this hypothesis with the facts of observation, however, it seems that the masses of whose existence we know—the solar system, stars and nebulae—are insufficient to confer on terrestial bodies the inertia that they actually possess : and therefore if Mach's principle were adopted, it would be necessary to postulate the existence of enormous quantities of matter in the universe which have not been detected by astronomical observation, and which are called into being simply in order to account for inertia in other bodies. This is, after all, no better than regarding some part of inertia as intrinsic.

The relation of Einstein's to Newton's laws of motion in the general case was discussed by L. Silberstein,[1] who showed that the differential equations of a geodesic in General Relativity are rigorously identical with the Newtonian equations of motion of a particle,

$$\frac{d^2\xi_p}{dt^2} = \frac{\partial\Omega}{\partial\xi_p} \qquad (p = 1, 2, 3)$$

[1] *Nature*, cxii (1923), p. 788

so long as the frame of reference for the differential equations of the geodesic is a system which is momentarily at rest relative to the particle. For simplicity consider a 'statical' world, specified by a metric

$$(ds)^2 = V^2(dt)^2 - \frac{1}{c^2} \sum_{p,q=1}^{3} a_{pq}\, dx^p dx^q$$

where V and the a_{pq} are functions of (x^1, x^2, x^3) only; and consider an observer who is stationary, i.e. whose co-ordinates (x^1, x^2, x^3) do not vary. Then the components of the gravitational force on the particle can be shown to be

$$g^0 = 0, \qquad g^p = -c^2 \sum_q \frac{a^{pq}}{V} \frac{\partial V}{\partial x^q} \qquad (p = 1, 2, 3),$$

and it can be shown from Einstein's fundamental equation

$$K_{pq} - \tfrac{1}{2} g_{pq} K = -\kappa T_{pq}$$

that

$$\frac{\Delta_2 V}{V} = \tfrac{1}{2}\kappa \left(\frac{T_{00}}{V^2} + \sum_{p,q=1}^{3} a^{pq} T_{pq} \right),$$

where $\Delta_2 V$ denotes the Second Differential Parameter of V in the three-dimensional space.

Now suppose that the field is of the kind considered in Poisson's equation, namely that the space is approximately Euclidean and that there is a volume-density ρ of matter at rest, and no radiation; then the only sensible element of the energy-tensor is the energy-density due to the equivalence of mass and energy, which has the value $c^2\rho$. Moreover we can take $V = 1 + \gamma$, where γ is small; so the above equation becomes

$$\Delta_2 \gamma = \tfrac{1}{2}\kappa c^2 \rho.$$

But writing x, y, z for the co-ordinates, since

$$(ds)^2 = \frac{1}{c^2}\left\{ (dx)^2 + (dy)^2 + (dz)^2 \right\}$$

approximately, we have

$$\Delta_2 \gamma = c^2 \left(\frac{\partial^2 \gamma}{\partial x^2} + \frac{\partial^2 \gamma}{\partial y^2} + \frac{\partial^2 \gamma}{\partial z^2} \right).$$

Thus

$$\frac{\partial^2 \gamma}{\partial x^2} + \frac{\partial^2 \gamma}{\partial y^2} + \frac{\partial^2 \gamma}{\partial z^2} = \tfrac{1}{2}\kappa \rho.$$

Now in the light of the above equations for the gravitational force, we have

$$-c^2\gamma = \text{the gravitational potential } \Omega,$$

so

$$\frac{\partial^2\Omega}{\partial x^2}+\frac{\partial^2\Omega}{\partial y^2}+\frac{\partial^2\Omega}{\partial z^2} = -\tfrac{1}{2}\kappa c^2 \rho.$$

Comparing this with Poisson's equation

$$\frac{\partial^2\Omega}{\partial x^2}+\frac{\partial^2\Omega}{\partial y^2}+\frac{\partial^2\Omega}{\partial z^2} = -4\pi\beta\rho$$

where β is the Newtonian constant of gravitation, we have

$$\kappa = \frac{8\pi\beta}{c^2}$$

an equation connecting the Einsteinian and Newtonian constants of gravitation. Since

$$\beta = 6{\cdot}67 \times 10^{-8} \text{ gr}^{-1} \text{ cm}^3 \text{ sec}^{-2}$$

this gives

$$\kappa = 1{\cdot}87 \times 10^{-27} \text{ cm.gr}^{-1}.$$

Almost simultaneously with Einstein's discovery of General Relativity, David Hilbert[1] (1862–1943) gave a derivation of the whole theory from a unified principle. Defining a point in space-time by generalised co-ordinates (x^0, x^1, x^2, x^3), he adopted Einstein's ten gravitational potentials g_{pq} and a four-vector (ϕ_0, ϕ_1, ϕ_2, ϕ_3) representing the electrodynamic potential : and assumed the following axioms :

Axiom I (Mie's axiom of the world-function). *All physical happenings* (gravitational, electrical, etc.) *in the universe are determined by a scalar function* H (called the *world-function*) *which involves the arguments* g_{pq} *and their first and second derivatives with respect to the x's, and involves also the* ϕ*'s and their first derivatives with respect to the x's : and the laws of physical processes are obtained by annulling the variation of the integral*

$$\iiiint \mathrm{H}\sqrt{g}\, dx^0 dx^1 dx^2 dx^3$$

(where g denotes the determinant of the g_{pq}) *for each of the fourteen potentials* g_{pq}, ϕ_s. The reason for the occurrence of the factor $\sqrt{g}$

[1] *Gött. Nach.*, 1915, p. 395 ; read 20 Nov. 1915. The investigation was carried further by : H. A. Lorentz, *Proc. Amst. Ac.* xix (1916), p. 751 ; J. Tresling, *Proc. Amst. Ac.* xix (1916), p. 892 ; A. Einstein, *Berlin Sitz*, 1916, p. 1111 ; F. Klein, *Gött. Nach.* 1917, p. 469 ; H. Weyl, *Ann. d. Phys.* liv (1917), p. 117 ; A. D. Fokker, *Proc. Amst. Ac.* xix (1917), p. 968 ; A. Palatini, *Palermo Rend.* xliii (1919), p. 203 ; D. Hilbert, *Math. Ann.* xcii (1924), p. 1 ; E. T. Whittaker, *Proc. R.S.*(A), cxiii (1927), p. 496.

is that $\sqrt{g}\, dx^0 dx^1 dx^2 dx^3$, which is called the *invariant hypervolume*, is invariant under all transformations of co-ordinates.

Axiom II (axiom of general invariance). *The world-function* H *is invariant with respect to arbitrary transformations of the* x^r.

These axioms represented a distinct advance on Einstein's methods, in which Hamilton's Principle had played only a very subordinate part.

We suppose that the world-function H is a sum of two terms, of which the first represents gravitation, i.e. whatever is inherent in the intrinsic structure of space-time, while the second term represents all other physical effects. The first term we take to be proportional to K, the scalar curvature of space-time at the point (x^0, x^1, x^2, x^3); this amounts to supposing that the mutual energy of all the gravitating masses in the world can be expressed analytically as an integral taken over the whole of space, namely a numerical multiple of $\iiiint K\sqrt{g}\, dx^0 dx^1 dx^2 dx^3$. With regard to the second term, we shall suppose for simplicity that the only other physical effect to be considered is an electromagnetic field in free aether. Now in Euclidean space there is, in the field, electric energy $(d_x^2 + d_y^2 + d_z^2)/8\pi$ per unit volume, and magnetic energy $(h_x^2 + h_y^2 + h_z^2)/8\pi$ per unit volume; and since these are of opposite type, in the sense in which kinetic and potential energy are of opposite type, we should expect their difference to occur in the world-function. In general co-ordinates this difference is represented by

$$L = -\frac{1}{16\pi}\sum_{b\,p} X^{pq} X_{pq}$$

where X_{pq} is the electromagnetic six-vector. We shall take the second part of the world-function to be a numerical multiple of L. We must therefore have

$$\delta\iiiint (K + 2\kappa L)\sqrt{g}\, dx^0 dx^1 dx^2 dx^3 = 0$$

where κ is a constant: or

$$\iiiint\left\{\delta K \,.\, \sqrt{g} + K \,.\, \delta\sqrt{g} + 2\kappa\delta(L\sqrt{g})\right\} dx^0 dx^1 dx^2 dx^3 = 0.$$

A calculation shows that

$$\iiiint \delta K \,.\, \sqrt{g}\, dx^0 dx^1 dx^2 dx^3 = \iiiint \sum_{p,q} K_{pq}\sqrt{g}\, \delta g^{pq}\, dx^0 dx^1 dx^2 dx^3$$

where K_{pq} is the Ricci tensor: and

$$\delta\sqrt{g} = -\tfrac{1}{2}\sqrt{g}\sum_{p,q} g_{pq}\delta g^{pq}.$$

Let us now find the value of

$$\delta\iiiint \mathrm{L}\sqrt{g}\, dx^0 dx^1 dx^2 dx^3.$$

The part that depends on the variations of the g^{pq} is

$$\iiiint \sum_{p,q} \zeta \frac{\partial(\mathrm{L}\sqrt{g})}{\partial g^{pq}}\, \delta g^{pq}\, dx^0 dx^1 dx^2 dx^3$$

where ζ is 1 or $\frac{1}{2}$ according as p is equal or unequal to q: or

$$\tfrac{1}{2}\sum_{p,q}\iiiint \mathrm{T}_{pq}\, \delta g^{pq}\,.\,\sqrt{g}\, dx^0 dx^1 dx^2 dx^3$$

where

$$\mathrm{T}_{pq} = \frac{2}{\sqrt{g}}\zeta\frac{\partial(\mathrm{L}\sqrt{g})}{\partial g^{pq}} = 2\zeta\frac{\partial \mathrm{L}}{\partial g^{pq}} + \frac{2}{\sqrt{g}}\zeta\frac{\partial(\sqrt{g})}{\partial g^{pq}}\mathrm{L}.$$

Now it may easily be shown that

$$\zeta\frac{\partial g}{\partial g^{pq}} = -g\, g_{pq}$$

so

$$\mathrm{T}_{pq} = 2\zeta\frac{\partial \mathrm{L}}{\partial g^{pq}} - g_{pq}\,\mathrm{L}.$$

Moreover, when we regard L as a function of the g^{pq} and the ϕ_s, remembering that X_{pq} is a function of the ϕ_s only, not involving the g^{pq}, we must write

$$\mathrm{L} = -\frac{1}{16\pi}\sum_{k,l,p,q} g^{pk} g^{ql}\, \mathrm{X}_{pq}\, \mathrm{X}_{kl}$$

so we have

$$\zeta\frac{\partial \mathrm{L}}{\partial g^{pq}} = -\frac{1}{8\pi}\sum_s \mathrm{X}_{qs}\, \mathrm{X}_p{}^s$$

and therefore

$$\mathrm{T}_{pq} = -\frac{1}{4\pi}\sum_s \mathrm{X}_{qs}\, \mathrm{X}_p{}^s + \frac{g_{pq}}{16\pi}\sum_{r,s} \mathrm{X}^{rs}\, \mathrm{X}_{rs}$$

or

$$\mathrm{T}^p_q = -\frac{1}{4\pi}\sum_s \mathrm{X}_{qs}\, \mathrm{X}^{ps} + \frac{\delta^p_q}{16\pi}\sum_{r,s} \mathrm{X}^{rs}\, \mathrm{X}_{rs}$$

which is the expression previously found for Minkowski's electromagnetic energy-tensor.

Thus the variational integral, in so far as the variations of the g^{pq} are concerned, is

$$\sum_{p,\, q=0}^{3} \iiiint \left(K_{pq} - \tfrac{1}{2} g_{pq}\, K + \kappa T_{pq}\right) \delta g^{pq} . \sqrt{g}\; dx^0 dx^1 dx^2 dx^3 = 0$$

and therefore *the variational equations derived from the terms in* g^{pq} *are*

$$K_{pq} - \tfrac{1}{2}\, g_{pq} K = -\kappa T_{pq} \qquad (p,\, q = 0,\, 1,\, 2,\, 3)$$

which are identical with Einstein's gravitational equations, previously given.

Hilbert showed moreover that *the Ricci tensor satisfies the four identical relations expressed by the equation*

$$\Delta\text{iv}\ (K^q_p - \tfrac{1}{2}\, \delta^q_p\ K) = 0.$$

We must now find the part of the variational integral that involves the variations of the electromagnetic potentials. It is (disregarding a constant multiplier)

$$\iiiint \sum_{p,\, q} X^{pq} \sqrt{g}\ \delta X_{pq}\ dx^0 dx^1 dx^2 dx^3.$$

But

$$\delta X_{pq} = \delta \left(\frac{\partial \phi_p}{\partial x^q}\right) - \delta \left(\frac{\partial \phi_q}{\partial x^p}\right)$$

so the part of the variational integral with which we are concerned is

$$\iiiint \sum_{p,\, q} X^{pq} \sqrt{g} \left\{\delta\left(\frac{\partial \phi_p}{\partial x^q}\right) - \delta\left(\frac{\partial \phi_q}{\partial x^p}\right)\right\} dx^0 dx^1 dx^2 dx^3$$

which after integration by parts yields

$$-\iiiint \sum_{p,\, q} \left\{\frac{\partial (X^{pq} \sqrt{g})}{\partial x^q}\, \delta\phi_p - \frac{\partial (X^{pq} \sqrt{g})}{\partial x^p}\, \delta\phi_q\right\} dx^0 dx^1 dx^2 dx^3.$$

Interchanging p and q in the summation of the second term, this becomes

$$-2 \iiiint \sum_{p,\, q} \frac{\partial (X^{pq} \sqrt{g})}{\partial x^q}\ \delta\phi_p\ dx^0 dx^1 dx^2 dx^3,$$

and therefore the variational equations obtained from the variations of the potentials are

$$\sum_q \frac{\partial (X^{pq} \sqrt{g})}{\partial x^q} = 0 \qquad (p = 0,\, 1,\, 2,\, 3)$$

or

$$\Delta\text{iv}\,(X^{pq}) = 0$$

and these are precisely the Ampère-Maxwell tetrad of electromagnetic equations in free aether. The Faraday tetrad of course follows from the expression of the X_{pq} in terms of the potentials. Thus *both Einstein's gravitational equations and the electromagnetic equations can be obtained by the variation of the integral of Hilbert's world-function.*

It was proposed by Cornel Lanczos[1] that the world-function H occurring in the integral $\iiiint H\sqrt{g}\, dx^0 dx^1 dx^2 dx^3$ should be formed in a way different from that adopted by Hilbert, namely that it should be a *quadratic* function of the Ricci tensor K_{pq}, and in fact should be of the form

$$H = \sum_{p,\,q} K_{pq}K^{pq} + CK^2$$

where C is a numerical constant. The advantage of this form is that in the course of the analysis, a four-vector is found to occur naturally, and this vector can be identified with the electromagnetic potential-vector: thus electromagnetic theory can be unified with the theory of General Relativity.

In 1920 a criticism of General Relativity was published in the *Times Educational Supplement*[2] by the mathematician and philosopher Alfred North Whitehead (1861–1947). 'I doubt,' he said, 'the possibility of measurement in space which is heterogeneous as to its properties in different parts. I do not understand how the fixed conditions for measurement are to be obtained.' He followed up this idea by devising an alternative theory which he set forth in a book, *The Principle of Relativity*,[3] in 1922. 'I maintain,' he said, 'the old-fashioned belief in the fundamental character of simultaneity. But I adapt it to the novel outlook by the qualification that the meaning of simultaneity may be different in different individual experiences.' This statement of course admits the relativity theory of Poincaré and Lorentz, but it is not compatible with the General Theory of Relativity, in which an observer's domain of simultaneity is usually confined to a small region of his immediate neighbourhood. Whitehead therefore postulated two fields of natural relations, one of them (namely space and time relations) being isotropic, universally uniform and not conditioned by physical circumstances: the other comprising the physical relations expressed by laws of nature, which are contingent.

A profound study of Whitehead's theory was published in 1952 by J. L. Synge.[4] As he remarked, the theory of Whitehead offers something between the two extremes of Newtonian theory on the

[1] *ZS. f. P.* lxxiii (1931), p. 147; lxxv (1932), p. 63; *Phys. Rev.* xxxix (1932), p. 716
[2] *Times Educ. Suppl.*, 12 Feb. 1920, p. 83
[3] *The Principle of Relativity, with applications to physical science*, Camb. Univ. Press, 1922
[4] *Proc. R.S.*(A), ccxi (1952), p. 303

one hand and the General Theory of Relativity on the other. It conforms to the requirement of Lorentz invariance (thus overcoming the major criticism against the Newtonian theory), but it does not reinstate the concept of force, with the equality of action and reaction, so that its range of applicability remains much lower than that of Newtonian mechanics. However, it does free the theorist from the nearly impossible task of solving a set of non-linear partial differential equations whenever he seeks a gravitational field. It is not a field theory, in the sense commonly understood, but a theory involving action at a distance, propagated with the fundamental velocity *c*.

Whitehead's doctrine, though completely different from Einsteins' in its formulation, may be described very loosely as fitting the Einsteinian laws into a flat space-time ; and no practicable observational test has hitherto been suggested for discriminating between the two theories.[1]

The idea of mapping the curved space of General Relativity on a flat space, and making the latter fundamental, was revived many years after Whitehead by N. Rosen.[2] He and others [3] who developed it claimed that in this way it was possible to explain more directly the conservation of energy, momentum, and angular momentum, and also possibly to account for certain unexplained residuals in repetitions of the Michelson-Morley experiment.[4]

In 1916 K. Schwarzschild made an important advance in the Einsteinian theory of gravitation, by discovering the analytical solution of Einstein's equations for space-time when it is occupied by a single massive particle.[5]

The field being a statical one, we can take the quadratic form which specifies the metric of space-time to be

$$(ds)^2 = V^2(dt)^2 - \frac{1}{c^2}(dl)^2$$

where *dl* is the line-element in the three-dimensional space, S, so

$$(dl)^2 = \sum_{p,q=1}^{3} a_{pq} dx^p dx^q$$

and the functions V, a_{pq}, $(p, q = 1, 2, 3)$, do not involve *t*. There will be a one-to-one correspondence between the points of the space S, which contains the massive particle, and the points of a

[1] cf. G. Temple, *Proc. Phys. Soc.* xxxvi (1924), p. 176 ; W. Band, *Phys. Rev.* lxi (1942), p. 698

[2] *Phys. Rev.* lvii (1940), pp. 147, 150, 154

[3] M. Schoenberg, *Phys. Rev.* lix (1941), p. 616 ; G. D. Birkhoff, *Proc. Nat. Ac. Sci.* xxix (1943), p. 231 ; A. Papapetrou, *Proc. R.I.A.*, lii (A) (1948), p. 11

[4] D. C. Miller, *Proc. Nat. Ac. Sci.* xi (1925), p. 306

[5] K. Schwarzschild, *Berl. Sitz.* 1916, p. 189 ; cf. also : A. Einstein, *Berl. Sitz.* 1915, p. 831 ; D. Hilbert, *Gött. Nach.* 1917, p. 53 ; J. Droste, *Proc. Amst. Ac.* xix (1917), p. 197 ; A. Palatini, *N. Cimento,* xiv (1917), p. 12 ; C. W. Oseen, *Ark. f. Mat. Ast. och. Fys.* xv (1921), No. 9 ; J. L. Synge, *Proc. R.I.A.* liii (1950), p. 83.

space S′ which is completely empty. Now the latter space can be specified by co-ordinates (r, θ, ϕ) in such a way that its line-element is given by

$$(dl')^2 = (dr)^2 + r^2\{(d\theta)^2 + \sin^2\theta(d\phi)^2\},$$

the origin of S′ corresponding to the location of the massive particle in S. The effect of the presence of the particle will be to distort S as compared with S′, but the distortion will be symmetrical with respect to the particle, so the line-element in S will be expressible in the form

$$(dl)^2 = f(r)(dr)^2 + g(r)r^2\{(d\theta)^2 + \sin^2\theta(d\phi)^2\}$$

and if we take a new variable R such that $\sqrt{g(r)}\,.\,r = \mathrm{R}$, this may be written

$$(dl)^2 = \mathrm{A}^2(d\mathrm{R})^2 + \mathrm{R}^2\{(d\theta)^2 + \sin^2\theta(d\phi)^2\}$$

where A is some function of R. By symmetry, V is also a function of R, and the problem is to determine A and V.

Thus, now writing r for R, the quadratic form which defines the metric in the space around the particle is

$$(ds)^2 = \sum_{p,\,q} g_{pq}dx^p dx^q$$

where

$$x^0 = t, \qquad x^1 = r, \qquad x^2 = \theta, \qquad x^3 = \phi$$

$$g_{00} = \mathrm{V}^2, \qquad g_{11} = -\frac{\mathrm{A}^2}{c^2}, \qquad g_{22} = -\frac{r^2}{c^2}, \qquad g_{33} = -\frac{r^2\sin^2\theta}{c^2}$$

and g_{pq} is zero when p is different from q.

Since the energy-tensor is zero everywhere except at the origin, and since $\mathrm{K} = 0$ when $\mathrm{K}_{pq} = 0$, Einstein's gravitational equations reduce to

$$\mathrm{K}_{pq} = 0.$$

Calculating the Ricci tensor, and denoting differentiations by dashes, we find

$$\mathrm{K}_{00} = c^2\left(-\frac{\mathrm{VV}''}{\mathrm{A}^2} + \frac{\mathrm{A'VV'}}{\mathrm{A}^3} - \frac{2\mathrm{VV}'}{\mathrm{A}^2 r}\right)$$

$$\mathrm{K}_{11} = \frac{\mathrm{V}''}{\mathrm{V}} - \frac{\mathrm{A'V'}}{\mathrm{AV}} - \frac{2\mathrm{A}'}{\mathrm{A}r}$$

$$\mathrm{K}_{22} = c^2\left(\frac{1}{\mathrm{A}^2} + \frac{r\mathrm{V}'}{\mathrm{A}^2\mathrm{V}} - \frac{r\mathrm{A}'}{\mathrm{A}^3} - 1\right).$$

Forming the combination $(A^2/c^2V^2)K_{00}+K_{11}$, and equating it to zero, we have

$$\frac{V'}{V}+\frac{A'}{A}=0$$

so AV = a constant ; and since when $r\to\infty$ the space tends to that defined by the quadratic form

$$(ds)^2=(dt)^2-\frac{1}{c^2}\{(dr)^2+r^2(d\theta)^2+r^2\sin^2\theta(d\phi)^2\},$$

that is, $A\to 1$ and $V\to 1$, we must have

$$AV=1.$$

The equation $K_{22}=0$ now becomes

$$\frac{1}{A^2}-\frac{2rA'}{A^3}-1=0.$$

so denoting $(1/A^2)-1$ by u, we have

$$u+r\frac{du}{dr}=0, \qquad \text{or} \qquad ru=\text{Constant},$$

so

$$\frac{1}{A^2}=\left(1-\frac{a}{r}\right).$$

Thus *the metric in the space around the particle is specified by*

$$(ds)^2=\left(1-\frac{a}{r}\right)(dt)^2-\frac{1}{c^2}\left\{\frac{(dr)^2}{1-\frac{a}{r}}+r^2(d\theta)^2+r^2\sin^2\theta(d\phi)^2\right\}.$$

This is Schwarzschild's solution.

If at any instant, in the plane $\theta=\frac{1}{2}\pi$, we consider a circle on which r is constant, we see that the length of an element of its arc is given by the equation

$$(dl)^2=r^2(d\phi)^2$$

so $dl=rd\phi$, and the circumference of the whole circle is $2\pi r$. This determines the physical meaning of the co-ordinate r. But if we consider a radius vector from the origin, the element of length along this radius is given by

$$(dl)^2=\frac{(dr)^2}{1-\frac{a}{r}},$$

so when r, in decreasing, tends to α, then $dl \to \infty$: that is to say, *the region inside the sphere* $r = \alpha$ *is impenetrable.*

The differential equations of motion of a particle (supposed to be so small that it does not disturb the field), under the influence of a single gravitating centre, are the differential equations of the geodesics in space-time with Schwarzschild's matric. These equations can be written down and integrated in terms of elliptic functions.[1]

A particular case is that of *rectilinear orbits along a radius vector from the central mass.* We then have $d\phi = 0$, and can readily derive the integral

$$\left(1 - \frac{\alpha}{r}\right)\frac{dt}{ds} = \text{Constant} = \frac{1}{\mu} \quad \text{say,}$$

and we have

$$(ds)^2 = \left(1 - \frac{\alpha}{r}\right)(dt)^2 - \frac{(dr)^2}{c^2\left(1 - \frac{\alpha}{r}\right)}.$$

Eliminating ds between these equations, we have

$$\frac{1}{c^2\left(1 - \frac{\alpha}{r}\right)^3}\left(\frac{dr}{dt}\right)^2 = \frac{1}{1 - \frac{\alpha}{r}} - \mu^2.$$

Let dl denote the element of length along the radius, and let $d\tau$ denote the element of time, so

$$(dl)^2 = \frac{(dr)^2}{1 - \frac{\alpha}{r}}, \qquad (d\tau)^2 = \left(1 - \frac{\alpha}{r}\right)(dt)^2;$$

then the preceding equation becomes

$$\frac{1}{c^2}\left(\frac{dl}{d\tau}\right)^2 = 1 - \mu^2 + \frac{\alpha\mu^2}{r}$$

which obviously corresponds to the Newtonian equation of the conservation of energy. Differentiating it, we have

$$\frac{d^2 l}{d\tau^2} = -\frac{\alpha\mu^2 c^2}{2r^2}\left(1 - \frac{\alpha}{r}\right)^{\frac{1}{2}}.$$

[1] cf. W. de Sitter, *Proc. Amst. Ac.* xix (1916), p. 367; T. Levi-Civita, *Rend. Lincei*, xxvi (i) (1917), pp. 381, 458; xxvi (ii) (1917), p. 307; xxvii (i) (1918), p. 3; xxvii (ii) (1918), pp. 183, 220, 240, 283, 344; xxviii (i) (1919), pp. 3, 101; A. R. Forsyth, *Proc. R.S.*(A), xcvii (1920), p. 145 (Integration by Jacobian elliptic functions); F. Morley, *Am. J. Math.* xliii (1921), p. 29 (by Weierstrassian elliptic functions); C. de Jans, *Mém. de l'Ac. de Belgique*, vii (1923) (by Weierstrassian functions); K. Ogura, *Jap. J. of Phys.* iii (1924), pp. 75, 85; a very complete study of the trajectories of a small particle in the Schwarzschild field was made by Y. Hagihara, *Jap. J. of Ast. and Geoph.* viii (1931), p. 67.

When r is large compared with α, this becomes approximately

$$\frac{d^2r}{dt^2} = -\frac{\alpha c^2}{2r^2}$$

which is the Newtonian law of attraction.

Comparing it with the Newtonian law of attraction to the sun,

$$\frac{d^2r}{dt^2} = -\frac{\gamma M}{r^2}$$

where γ is the Newtonian constant of attraction and M is the sun's mass, we have

$$\alpha = \frac{2\gamma M}{c^2}$$

which gives

$$\alpha = 2{\cdot}95 \text{ km.}$$

This is the value of the constant α for the sun.

By approximating from the elliptic-function solution for the general case of motion round a gravitating centre, we can study the orbits when the distance of the small particle from the centre of force is large compared with α, in which case we can have orbits differing little from the ellipses described by the planets under the Newtonian attraction of the sun. It is found that the line of apsides of such an orbit is not fixed, but slowly rotates: if l is the semi-latus-rectum, then the advance of the perihelion in one complete revolution is $3\pi\alpha/l$, which for the case of the planet Mercury revolving round the sun amounts to about $0''{\cdot}1$: since Mercury makes about 420 revolutions in a century, the secular advance of the perihelion is $42''$, which agrees well with the observed value of $43''$.[1]

The paths of rays of light in the field of a single gravitating centre are, of course, the null geodesics of the Schwarzschild field. Many of them have remarkable forms. It is found that a ray of light can be propagated perpetually in a circle of radius $\frac{3}{2}\alpha$ about the gravitating centre: and there are trajectories spirally asymptotic to this circle both externally and internally, the other terminus of the trajectory being on the circle $r = \alpha$ in the former case and at infinity in the latter. There are also trajectories of which one

[1] Einstein first showed that the new gravitational theory would explain the anomalous motion of the perihelion of Mercury in the third of his papers in the *Berlin Sitz.* of 1915, p. 831. On the whole subject cf. G. M. Clemence, *Proc. Amer. Phil. Soc.* xciii (1949), p. 532. The secular motion of the earth's perihelion, as revealed by observation, was found by H. R. Morgan, *Ast. J.* li (1945), p. 127, to agree with that calculated by Einstein's theory, which is $3''{\cdot}84$.

terminus is at infinity and the other on the circle $r = \alpha$, trajectories which at both termini meet the circle $r = \alpha$, and quasi-hyperbolic trajectories both of whose extremities are at infinity. For these last-named trajectories it is found that if a is the apsidal distance, the angle between the asymptotes is approximately equal to $2\alpha/a$. A ray coming from a star and passing close to the sun's gravitational field, when observed by a terrestrial observer, will therefore have been deflected through an angle $2\alpha/a$.[1] If we take the radius of the solar corona to be 7×10^5 km., so that $a = 7 \times 10^5$ km. and $\alpha = 3$ km., *then* Einstein found that *the displacement of the star is* $2.3/7.10^5$ *in circular measure, or* 1″·75.

The notion that light possesses gravitating mass, and that therefore a ray of light from a star will be deflected when it passes near the sun, was far from being a new one, for it had been put forward in 1801 by J. Soldner,[2] who calculated that a star viewed near the sun would be displaced by 0″·85. Einstein's prediction was tested by two British expeditions to the solar eclipse of May 1919, who found for the deflection the values 1″·98 ± 0″·12 and 1″·61 ± 0″·30 respectively, so the prediction was regarded as confirmed observationally: and this opinion was strengthened by the first reports regarding the Australian eclipse of 1922 September 21. Three different expeditions found for the shift at the sun's limb the values 1″·72 ± 0″·11, 1″·90 ± 0″·2 and 1″·77 ± 0″·3, all three results differing from Einstein's predicted value by less than their estimated probable errors. However, a re-examination of the 1922 measures gave about 2″·2: at the Sumatra eclipse of 1929 the deflection was found[3] to be 2″·0 to 2″·24: and at the Brazilian eclipse of May 1947 a value of 2″·01 ± 0″·27 was obtained[4]: while it must not be regarded as impossible that the consequences of Einstein's theory may ultimately be reconciled with the results of observation, it must be said that at the present time (1952) there is a discordance.

A second observational test proposed was the anomalous motion of the perihelion of Mercury. This is quantitatively in agreement with Einstein's theory,[5] but as we have seen,[6] there are alternative explanations of it.

A third proposed test was the displacement to the red of spectral lines emitted in a strong gravitational field. This, however, was, as we have seen,[7] explained before General Relativity was discovered, and does not, properly speaking, constitute a test of it in contra-

[1] cf. F. D. Murnaghan, *Phil. Mag.* xliii (1922), p. 580; T. Shimizu, *Jap. J. Phys.* iii (1924), p. 187; R. J. Trumpler, *J.R.A.S. Canada*, xxiii (1929), p. 208

[2] *Berliner Astr. Jahrb.* 1804, p. 161; reprinted *Ann. d. Phys.* lxv (1921), p. 593

[3] E. Freundlich, H. v. Klüber and A. v. Brunn, *ZS. f. Astroph.* iii (1931), p. 171; vi (1933), p. 218; xiv (1937), p. 242; cf. J. Jackson, *Observatory*, liv (1931), p. 292

[4] G. van Biesbroek, *Ast. J.* lv (1950), p. 49; cf. E. Finlay-Freundlich and W. Ledermann, *Mon. Not. R.A.S.* civ (1944), p. 40

[5] cf. G. M. Clemence, *Ast. J.* 1 (1944)

[6] cf. p. 148

[7] p. 152–3; cf. G. Y. Rainich, *Phys. Rev.* xxxi (1928), p. 448

distinction to other theories.[1] In any case, the observed effect so far as solar lines[2] are concerned is complicated by other factors.[3] With regard to stellar lines, the greatest effect might be expected from stars of great density. Now Eddington showed that in a certain class of stars, the 'white dwarfs,' the atoms have lost all their electrons, so that only the nuclei remain: and under the influence of the gravitational field, the nuclei are packed together so tightly that the density is enormous. The companion of Sirius is a star of this class: its radius is not accurately known, but the star might conceivably have a density about 53,000 times that of water, in which case the red-displacement of its lines would be about 30 times that predicted for the sun. The comparison of theory with observation of the red-displacement can, however, hardly be said to furnish a quantitative test of the theory.[4]

We may remark that it is easy to construct models which behave so as to account for the observed facts of astronomy. For example, it has been shown by A. G. Walker[5] that if a particle is supposed to be moving in ordinary Euclidean space in which there is a Newtonian gravitational potential ϕ (so that the contribution to ϕ from a single attracting mass M is $-\gamma M/r$, where γ is the Newtonian constant of attraction) and if the kinetic energy of the particle is

$$T = \tfrac{1}{2}e^{-2\phi/c^2}v^2 \quad \text{where } v^2 = \left(\frac{dx}{dt}\right)^2 + \left(\frac{dy}{dt}\right)^2 + \left(\frac{dz}{dt}\right)^2$$

(which reduces to $T = \tfrac{1}{2}v^2$ when $c \to \infty$), and if the potential energy of the particle is

$$V = \tfrac{1}{2}c^2\left\{1 - e^{-2\phi/c^2}\right\}$$

[1] The derivation of the effect from Einstein's equations may readily be constructed from the following indications. For an atom at rest in a gravitational field (e.g. on the sun) the proper-time is given by

$$d\tau^2 = V^2 dt^2,$$

where, as we have seen, $V = 1 - (1/c^2)\,\Omega$ if Ω is the gravitational potential; while for an atom at rest at a great distance from the gravitational field (e.g. on the earth) we have

$$V = 1 \quad \text{and} \quad d\tau^2 = dt^2.$$

The period of the radiation, as measured at its place of emission, is to its period as measured by the terrestrial observer, in the ratio that $d\tau/dt$ at the place of emission bears to $d\tau/dt$ at the place of the observer, that is, the ratio V to unity; so, as before, we find that the wave-length of the radiation produced in the strong gravitational field and observed by the terrestrial observer is $(1 + \Omega/c^2)$ times the wave-length of the same radiation produced on earth.

[2] C. E. St. John, *Astroph. J.* lxvii (1928), p. 195

[3] cf. Miss M. G. Adam of Oxford, *Mon. Not. R.A.S.* cviii (1948), p. 446

[4] For a discussion as to whether the perihelion motion of Mercury, the deflection of light rays that pass near a gravitating body, and the displacement of spectral lines in the gravitational field, are to be regarded as establishing General Relativity, cf. E. Wiechert, *Phys. ZS.* xvii (1916), p. 442, and *Ann. d. Phys.* lxiii (1920), p. 301.

[5] *Nature*, clxviii (1951), p. 961

(which reduces to ϕ when $c\to\infty$), then the Lagrangean equations of motion of the particle are

$$\frac{d}{dt}\left\{\frac{1}{1-v^2/c^2}\frac{dx}{dt}\right\} = -\frac{1+v^2/c^2}{1-v^2/c^2}\frac{\partial\phi}{\partial x}$$

and two similar equations in y and z (which reduce to $d^2x/dt^2 = -\partial\phi/\partial x$ and two similar equations when $c\to\infty$), and that the trajectories have the following properties:

(i) the velocity of the particle can never exceed c;

(ii) a particle moving initially with velocity c continues to move with this velocity;

(iii) the perihelion of an orbit advances according to the same formula as is derived from General Relativity;

(iv) if the particles moving with speed c are identified with photons, they are deflected towards a massive gravitating body according to the same formulae as is derived from General Relativity.

It is unwise to accept a theory hastily on the ground of agreement between its predictions and the results of observation in a limited number of instances: a remark which perhaps is specially appropriate to the investigations of the present chapter.

An astronomical consequence of General Relativity which has not yet been mentioned was discovered in 1921 by A. D. Fokker,[1] namely that, as a result of the curvature of space produced by the sun's gravitation, the earth's axis has a precession, additional to that deducible from the Newtonian theory, of amount 0″·019 per annum; unlike the ordinary precession, it would be present even if the earth were a perfect sphere. Its existence cannot however be tested observationally, since the shape and internal constitution of the earth are not known with sufficient accuracy to give a reliable theoretical value for the ordinary precession, and the relativistic precession is only about $\frac{1}{2500}$ of this.

Besides the Schwarzschild solution, a number of other particular solutions of the equations of General Relativity were obtained in the years following 1916, notably those corresponding to a particle which has both mass and electric charge,[2] and to fields possessing axial symmetry,[3] especially an infinite rod,[4] and two particles.[5] In 1938 A. Einstein, L. Infeld and B. Hoffman published a method[6] for finding, by successive approximation, the field due to n bodies.

[1] *Proc. Amst. Ac.* xxiii (1921), p. 729; cf. H. A. Kramers, ibid. p. 1052, and G. Thomsen, *Rend. Lincei*, vi (1927), p. 37. The existence of an effect of this kind had been predicted in 1919 by J. A. Schouten, *Proc. Amst. Ac.* xxi (1919), p. 533.

[2] H. Reissner, *Ann. d. Phys.* l (1916), p. 106; H. Weyl, *Ann d. Phys.* liv (1917), p. 117; G. Nordström, *Proc. Amst. Ac.* xx (1918), p. 1236; C. Longo, *N. Cimento*, xv (1918), p. 191; G. B. Jeffery, *Proc. R.S.*(A), xcix (1921), p. 123.

[3] T. Levi-Civita, *Rend. Lincei*(5) xxviii (i) (1919), pp. 4, 101

[4] W. Wilson, *Phil. Mag.*(6) xl (1920), p. 703

[5] H. E. J. Curzon, *Proc. L.M.S.*(2) xxiii (1924–5), pp. xxix and 477

[6] *Ann. of Math.* xxxix (1938), p. 65; cf. H. P. Robertson, ibid., p. 101 and T. Levi-Civita, *Mém. des Sc. Math.*, fasc. cxvi (1950)

Static isotropic solutions of the field equations, and symmetric distributions of matter, have been discussed by M. Wyman.[1]

In 1916 and the following years attention was given [2] to the propagation of disturbances in a gravitational field. If the distribution of matter in space is changed, e.g. by the circular motion of a plate in its own plane, gravitational waves are generated, which are propagated outwards with the speed of light.

If such waves impinge on an electron which is at rest, the principle of equivalence shows that the physical situation is the same as if the electron were moving with a certain acceleration, and therefore *an electron exposed to gravitational waves must radiate.*[3]

In 1917 Einstein [4] pointed out that the field-equations of gravitation, as he had given them in 1915, do not satisfy Mach's Principle, according to which, no space-time could exist except in so far as it is due to the existence of matter (or energy). Einstein's equations of 1915, however, admit the particular solution

$$g_{pq} = \text{Constant}, \qquad \mathrm{T}_{pq} = 0 \qquad (p, q = 0, 1, 2, 3)$$

so that a field is thinkable without any energy to generate it. He therefore proposed now [5] to modify the equations by writing them

$$\mathrm{K}_{pq} - \tfrac{1}{2}g_{pq}\mathrm{K} - \lambda g_{pq} = -\kappa \mathrm{T}_{pq} \qquad (p, q = 0, 1, 2, 3).$$

The effect of the λ-term is to add to the ordinary gravitational attraction between particles a small repulsion from the origin varying directly as the distance: at very great distances this repulsion will no longer be small, but will be sufficient to balance the attraction: and in fact, as Einstein showed, it is possible to have a statical universe, spherical in the spatial co-ordinates, with a uniform distribution of matter in exact equilibrium.[6] This is generally called the *Einstein universe.* The departure from Euclidean metric is measured by the radius of curvature R_0 of the spherical space, and this is connected with the total mass M of the particles constituting the universe by the equation

$$\frac{\gamma \mathrm{M}}{c^2} = \tfrac{1}{4}\pi \mathrm{R}_0$$

[1] *Phys. Rev.* lxvi (1944), p. 267 ; lxxv (1949), p. 1930

[2] A. Einstein, *Berlin Sitz.* 1916, p. 688 ; 1918, p. 154 ; H. Weyl, *Raum Zeit Materie,* 4th edn. (Berlin, 1921) p. 228 (English edn., p. 252) ; A. S. Eddington, *Proc. R.S.*(A), cii (1922), p. 268 ; H. Mineur, *Bull. S.M. Fr.* lvi (1928), p. 50 ; A. Einstein and N. Rosen, *J. Frankl. Inst.* ccxxiii (1937), p. 43 ; M. Brdička, *Proc. R.I.A.* liv (1951), p. 137.

[3] This problem was investigated by W. Alexandrow, *Ann. d. Phys.* lxv (1921), p. 675.

[4] *Berlin Sitz.,* 1917, p. 142 ; *Ann. d. Phys.* lv (1918), p. 241

[5] The new equations can be derived variationally by adding a constant to Hilbert's World-Function.

[6] The suggestion that our universe might be an Einsteinian space-time of constant *spatial* curvature seems to have been made first by Ehrenfest in a conversation with de Sitter about the end of 1916.

where γ denotes the Newtonian constant of gravitation. The total volume of this universe is $2\pi^2 R_0^3$.

It was shown by J. Chazy [1] and by E. Trefftz [2] that when the λ-term is included in the gravitational equations, Schwarzschild's metric for the space-time about a single gravitating centre must be modified to

$$(ds)^2 = \left(1 - \frac{a}{r} - \frac{\lambda}{3} r^2\right)(dt)^2 - \frac{1}{c^2}\left\{\frac{(dr)^2}{1 - \frac{a}{r} - \frac{\lambda}{3} r^2} + r^2(d\theta)^2 + r^2 \sin^2\theta (d\phi)^2\right\}.$$

If in this we put $a = 0$, which amounts to supposing that there is no mass at the origin, so the world is completely empty,[3] we obtain

$$(ds)^2 = \left(1 - \frac{\lambda}{3} r^2\right)(dt)^2 - \frac{1}{c^2}\left\{\frac{(dr)^2}{1 - \frac{\lambda}{3} r^2} + r^2(d\theta)^2 + r^2 \sin^2\theta (d\phi)^2\right\}.$$

Now this metric had been discovered by W. de Sitter [4] in 1917. *It is the metric of a four-dimensional space-time of constant curvature.*

The *de Sitter world*, as it was called, was the subject of many papers, partly on account of its intrinsic geometrical interest to the pure mathematician, and partly because of the possibility that some or all of its features might be similar to those of our actual universe as revealed by astronomical observation.[5]

Let us consider the universe as it would be if all minor irregularities were smoothed out: just as when we say that the earth is a spheroid, we mean that the earth would be a spheroid if all mountains were levelled and all valleys filled up. In the case of the universe the levelling is a more formidable operation, since we have to smooth out the earth, the sun and all the heavenly bodies, and to reduce the world to a complete uniformity. But after all, only a very small fraction of the cosmos is occupied by material bodies: and it is

[1] *Comptes Rendus*, clxxiv (1922), p. 1157

[2] *Math. Ann.* lxxxvi (1922), p. 317; cf. M. von Laue, *Berlin Sitz.*, 1923, p. 27

[3] This, of course, shows the invalidity of the reason Einstein had originally given for introducing the λ-term.

[4] *Proc. Amst. Acad.* xix (31 March 1917), p. 1217; xx (1917), pp. 229, 1309; *Mon. Not. R.A.S.* lxxviii (Nov. 1917), p. 3

[5] cf. F. Klein, *Gött. Nach.* 6 Dec. 1918, = *Ges. Math. Abh.* i, p. 604; K. Lanczos, *Phys. ZS.* xxiii (1922), p. 539; H. Weyl, *Phys. ZS.* xxiv (1923), p. 230; *Phil. Mag.*(6) xlviii (1924), p. 348; *Phil. Mag.*(7) ix (1930), p. 936; P. du Val, *Phil. Mag.*(6) xlvii (1924), p. 930; M. von Laue and N. Sen, *Ann. d. Phys.* lxxiv (1924), p. 252; L. Silberstein, *Phil. Mag.*(6) xlvii (1924), p. 907; H. P. Robertson, *Phil. Mag.*(7) v (1928), p. 835; R. C. Tolman, *Astroph. J.* lxix (1929), p. 245; G. Castelnuovo, *Lincei Rend.* xii (1930), p. 263; M. von Laue, *Berlin Sitz.* 1931, p. 123; E. T. Whittaker, *Proc. R.S.*(A), cxxxiii (1931), p. 93; H.S. Coxeter, *Amer. Math. Monthly*, l (1943), p. 217

interesting to inquire what space-time as a whole is like when we simply ignore them.[1]

The answer must evidently be, that it is a manifold of constant curvature. This means that it is isotropic (i.e. the curvature is the same for all orientations at the same point) and is also homogeneous. As a matter of fact, there is a well-known theorem that any manifold which is isotropic in this sense is necessarily also homogeneous, so that the two properties are connected. A manifold of constant curvature is a projective manifold, i.e. ordinary projective geometry is valid in it when we regard geodesics as straight lines : and it is possible to move about in it any system of points, discrete or continuous, rigidly, i.e. so that the mutual distances are unaltered.

The simplest example of a manifold of constant curvature is the surface of a sphere in ordinary three-dimensional Euclidean space : and the easiest way of constructing a model of the de Sitter world is to take a pseudo-Euclidean manifold of five dimensions in which the line-element is specified by the equation

$$-(ds)^2=(dx)^2+(dy)^2+(dz)^2-(du)^2+(dv)^2,$$

and in this manifold to consider the four-dimensional hyper-pseudosphere [2] whose equation is

$$x^2+y^2+z^2-u^2+v^2=R^2.$$

The pseudospherical world thus defined has a constant Riemannian measure of curvature $-1/R^2$.[3]

The de Sitter world may be regarded from a slightly different mathematical standpoint as having a Cayley-Klein metric, governed by an Absolute whose equation in four-dimensional homogeneous co-ordinates is

$$x^2+y^2+z^2-u^2+v^2=0$$

where u is time. Hyperplanes which do not intersect the Absolute are spatial, so spatial measurements are elliptic, i.e. the three-dimensional world of space has the same kind of geometry as the surface of a sphere, differing from it only in being three-dimensional instead of two-dimensional. In such a geometry there is a natural unit of length, namely the length of the complete straight line,

[1] The curvature of space at any particular place due to the general curvature of the universe is quite small compared to the curvature that may be imposed on it locally by the presence of energy. By a strong magnetic field we can produce a curvature with a radius of less than 100 light-years, and of course in the presence of matter the curvature is stronger still. So the universe is like the earth, on which the local curvature of hills and valleys is far greater than the general curvature of the terrestrial globe.

[2] The prefix *hyper-* indicates that we are dealing with geometry of more than the usual three dimensions, and the prefix *pseudo-* refers to the occurrence of negative signs in the equation.

[3] The world of the Poincaré-Lorentz theory of relativity can be regarded as a four-dimensional hyperplane in the five-dimensional hyperspace.

just as on the surface of a sphere there is a natural unit of length, namely the length of a complete great circle.

We are thus brought to the question of the dimensions of the universe: what is the length of the complete straight line, the circuit of all space? Since 1917 there has seemed to be a possibility that, by the combination of theory with astrophysical observation, this question might be answered.

Different investigations of the de Sitter world, however, reached conclusions which apparently were not concordant. The origin of some of the discordances could be traced to the ambiguities which were involved in the use of the terms ' time,' ' spatial distance ' and ' velocity,' when applied by an observer to an object which is remote from him in curved space-time. The ' interval ' which is defined by $(ds)^2 = \sum_{p,q} g_{pq} dx^p dx^q$ involves space and time blended together: and although any particular observer at any instant perceives in his immediate neighbourhood an ' instantaneous three-dimensional space ' consisting of world-points which he regards as simultaneous, and within which the formulae of the Poincaré-Lorentz relativity theory are valid, yet this space cannot be defined beyond his immediate neighbourhood: for with a general metric defined by a quadratic differential form, it is not in general possible to define simultaneity (with respect to a particular observer) over any *finite* extent of space-time.

The concept of ' spatial distance between two material particles ' is, however, not really dependent on the concept of ' simultaneity.'[1] When the astronomer asserts that ' the distance of the Andromeda nebula is a million light-years,' he is stating a relation between the world-point occupied by ourselves at the present instant and the world-point occupied by the Andromeda nebula at the instant when the light left it which arrives here now, that is, he is asserting a relation between two world-points such that a light-pulse, emitted at one, arrives at the other; or, in geometrical language, between two world-points which lie on the same null geodesic. The spatial distance of two material particles in a general space-time may, then, be thought of as *a relation between two world-points which are on the same null geodesic.* It is obviously right that ' spatial distance ' should exist only between two world-points which are on the same null geodesic, for it is only then that the particles at these points are in direct physical relation with each other. This statement brings out into sharp relief the contrast between ' spatial distance ' and the ' interval ' defined by $(ds)^2 = \sum_{p,q} g_{pq} dx^p dx^q$: for between two points on the same null geodesic the ' interval ' is always zero. Thus *' spatial distance ' exists when, and only when, the ' interval ' is zero.*

In order to define ' spatial distance ' conformably to these ideas, with a general metric for space-time, it is necessary to translate

[1] E. T. Whittaker, *Proc. R.S.*(A), cxxxiii (1931), p. 93

into the language of differential geometry the principle by which astronomers actually calculate the 'distance' of very remote objects such as the spiral nebulae. The principle is this: first, the absolute brightness of the object (the 'star' as we may call it) is determined[1]: then this is compared with the apparent brightness (i.e. the brightness as actually seen by the observer). The *distance* of the star is then defined to be proportional to the square root of the ratio of the absolute brightness to the apparent brightness.

In adopting this principle into differential geometry, we take a 'star' A and an observer B, which are on the same null geodesic, and we consider a thin pencil of null geodesics (rays of light) which issue from A and pass near B. This pencil intersects the observer B's 'instantaneous three-dimensional space,' giving a two-dimensional cross-section: the 'spatial distance AB' is then defined to be proportional to the square root of this cross-section. Distance, as thus defined, is an invariant, i.e. it is independent of the choice of the co-ordinate system. This invariant, however, involves not only the position of the star and the position of the observer, but also the motion of the observer, since his 'instantaneous three-dimensional space' is determinate only when his motion is known. Thus the 'spatial distance' of a star from an observer depends on the motion of the observer: but this is quite as it should be, and indeed had always been recognised in the relativity theory of Poincaré and Lorentz: for in that theory the spatial distance of a star from an observer is $(X^2+Y^2+Z^2)^{\frac{1}{2}}$, where (T, X, Y, Z) are co-ordinates referred to *that particular inertial system with respect to which the observer is at rest*: the necessity for the words in italics shows that the distance depends on the observer's motion.

When the de Sitter world is studied in the light of this definition of distance (the mass of any material particles concerned being supposed to be so small that they do not sensibly affect the geometrical character of the universe), some remarkable results are found. Thus a freely moving star and a freely moving observer cannot remain at a constant spatial distance apart in the de Sitter world. When the observer first sees the star, its spatial distance is equal to the radius of curvature of the universe. After this the star is continuously visible, the distance passing through a minimum value, after which it increases again indefinitely: that is, the star's apparent brightness ultimately decreases to zero, it becomes too faint to be seen. When this happens the star is at a point which is not the terminus of its own world-line, so the star continues to exist after it has ceased to have any relations with this particular observer.

The Einstein world, which as we have seen is a statical solution of the gravitational equations, spherical in the spatial co-ordinates,

[1] For this purpose, astronomers in practice use the known relation between the period and absolute magnitude of Cepheid variables, or (in the method of spectroscopic parallaxes) the known relations between absolute magnitude and spectral behaviour.

was generalised by A. Friedman[1] into a solution in which the curvature depends on the time—what in fact came to be known later as an *expanding universe*. Not much notice was taken of this paper until, five years afterwards, the Abbé Georges Lemaitre,[2] a Belgian priest who had been a research student of Eddington's at Cambridge, proved that the Einstein universe is unstable, and that when its equilibrium is disturbed, the world progresses through a continuous series of intermediate states, towards a limit which is no other than the de Sitter universe.

Eddington, who at the beginning of 1930 was investigating, in conjunction with his research student G. C. McVittie, the stability of the Einstein universe, found in Lemaitre's paper the solution of the problem, and at once[3] saw that it provided an explanation of the observed scattering apart of the spiral nebulae. Since 1930 the theory of the expanding universe has been of central importance in cosmology.

The scheme of general relativity, as put forward by Einstein in 1915, met with some criticism as regards the unsatisfactory position occupied in it by electrical phenomena. While gravitation was completely fused with metric, so that the notion of a mechanical force on ponderable bodies due to gravitational attraction was entirely abolished, the notion of a mechanical force acting on electrified or magnetised bodies placed in an electric or magnetic field still persisted as in the old physics. This seemed, at any rate from the aesthetic point of view, to be an imperfection, and it was felt that sooner or later everything, including electromagnetism, would be re-interpreted and represented in some way as consequences of the pure geometry of space and time. In 1918 Weyl[4] proposed to effect this by rebuilding geometry once more on a new foundation, which we must now examine.

Weyl fixed attention in the first place on the 'light-cone,' or aggregate of directions issuing from a world-point P, in which light-signals can go out from it. The light-cone separates those world-points which can be affected by happenings at P, from those points whose happenings can affect P: it, so to speak, separates past from future. Now the light-cone is represented by the equation $(ds)^2=0$, where ds is the element of interval, and Weyl argued that this equation, rather than the quantity $(ds)^2$ itself, must be taken as the starting-point of the subject: in other words, it is the *ratios* of the ten coefficients g_{pq} in $(ds)^2$, and not the *actual values* of these coefficients,

[1] *ZS. f. P.* x (1922), p. 377. Aberration and parallax in the universes of Einstein, de Sitter and Friedman, were calculated by V. Fréedericksz and A. Schechter, *ZS. f. P.* li (1928), p. 584.

[2] *Ann. de la Soc. sc. de Bruxelles*, xlvii^A (April 1927), p. 49

[3] *Mon Not. R.A.S.* xc (1930), p. 668

[4] *Berl. Sitz.* 1918, Part I, p. 465; *Math. ZS.* ii (1918), p. 384; *Ann. d. Phys.* lix (1919), p. 101; *Phys. ZS.* xxi (1920), p. 649, xxii (1921), p. 473; *Nature*, cvi (1921), p. 781; cf. A. Einstein, *Berl. Sitz.* 1918, p. 478; W. Pauli. *Phys. ZS.* xx (1919), p. 457; *Verh. d. phys. Ges.* xxi (1919), p. 742; A. Einstein, *Berl. Sitz.* 1921, p. 261; L. P. Eisenhart, *Proc. N.A.S.* ix (1923), p. 175

which are to be taken as determined by our most fundamental physical experiences. This leads to the conclusion that comparisons of length at different times and places may yield discordant results according to the route followed in making the comparison.

Following up this principle, Weyl devised a geometry more general than the Riemannian geometry that had been adopted by Einstein: instead of being specified, like the Riemannian geometry, by a single quadratic differential form

$$\sum_{p,q} g_{pq} dx^p dx^q$$

it is specified by a quadratic differential form $\sum_{p,q} g_{pq} dx^p dx^q$ and a linear differential form $\sum_p \phi_p dx^p$ together. The coefficients g_{pq} of the quadratic form can be interpreted, as in Einstein's theory, as the potentials of gravitation, while the four coefficients ϕ_p of the linear form can be interpreted as the four components of the electromagnetic potential-vector. Thus Weyl succeeded in exhibiting both gravitation and electricity as effects of the metric of the world.[1]

The enlargement of geometrical ideas thus achieved was soon followed by a still wider extension due to Eddington.[2] This, and most subsequent constructions in the same field, are based on an analysis of the notion of *parallelism*, which must now be considered.

The question of parallelism in curved spaces was raised in an acute form by the discovery of General Relativity theory: for it now became necessary for the purposes of physics to create a theory of vectors in curved space, and of their variation from point to point of the space. Now in Euclidean space if U and V are two vectors at the same point, P, we can find the vector which is their difference by using the triangle of vectors. But if U is a vector at a point P and V is a vector at a different point Q, and if we want to find the vector which is the difference of U and V, it is necessary (in principle) first to transfer U parallel to itself from the point P to the point Q, and then to find the difference of the two vectors at Q. Thus a process of *parallel transport* is necessary for finding the difference of vectors at different points, and hence for the spatial differentiation of vectors: and this is true whether the space is Euclidean or non-Euclidean.

The spatial differentiation of vectors in curved space has already been discussed[3] analytically under the name of *covariant differentiation.* Evidently this covariant differentiation must really be based on a parallel transport of vectors in the curved space: and in 1917 this particular form of parallel transport was definitely formulated

[1] It does not seem necessary to describe this geometry in detail, since Weyl himself later expressed the opinion (*Amer. J. Math.* lxvi (1944), p. 591) that it does not (at least in its original form) provide a satisfactory unification of electromagnetism and gravitation.

[2] *Proc. R.S.*(A), xcix (1921), p. 104

[3] cf. p. 161

by T. Levi-Civita.[1] It may be regarded as providing a geometrical interpretation for the Christoffel symbols.

After this the idea [2] was developed that a space may be regarded as formed of a great number of small pieces cemented together, so to speak, by a parallel transport, which states the conditions under which a vector in one small piece is to be regarded as parallel to a vector in a neighbouring small piece. Thus every type of differential geometry must have at its basis a definite parallel-transport or ' connection.'

To illustrate these points, let us consider geometry on the surface of the earth, regarded as a sphere. It is obvious that directions U and V at two different points cannot (unless one of the points happens to be the antipodes of the other) be parallel in the ordinary Euclidean sense, i.e. parallel in the three-dimensional space in which the earth is immersed : but a new kind of parallelism can be defined, and that in many different ways. We might, for instance, say that V is derived by parallel transport from U if V and U have the same compass-bearing, so that, e.g. the north-east direction at one place on the earth's surface would be said to be parallel to the north-east direction at any other place.

Having defined parallel-transport on the earth in this way, there are now two different kinds of curve on the earth's surface which may be regarded as analogous to the straight line in the Euclidean plane. If we define a straight line by the property that it is the shortest distance between two points, then its analogue on the earth is a great circle, since this gives the shortest distance between two points on the spherical surface. But if we define a straight line by the property that it preserves the same direction along its whole length, or, more precisely, that its successive elements are derived from each other by parallel transport, then its analogue on the earth is the track of a ship whose compass-bearing is constant throughout her voyage : this is the curve called a *loxodrome*. The existence of the two families of curves, the great circles and the loxodromes, may be assimilated to the fact that if an electrostatic and a gravitational field coexist, the lines of electric force and the lines of gravitational force are two families of curves in space ; and this rough analogy may serve to suggest how different physical phenomena may be represented simultaneously in terms of geometrical conceptions.[3]

Weyl's proposal for a unified theory of gravitation and electromagnetism was followed up in another direction by Th. Kaluza,[4]

[1] *Palermo Rend.* xlii (1917), p. 173

[2] This idea is due essentially to G. Hessenberg, *Math. Ann.* lxxviii (1917), p. 187, though he did not explicitly introduce Levi-Civita's notion of parallel-transport.

[3] Eddington's generalisation of Weyl's theory was the starting-point of important papers by A. Einstein, *Berl. Sitz.* (1923), pp. 32, 76, 137, and by Jan Arnoldus Schouten of Delft (*b.* 1883), *Amst. Proc.* xxvi (1923), p. 850. The latter represented the electromagnetic field as a vortex connected with the *torsion* (in the sense of Cartan) of four-dimensional space ; on this cf. H. Eyraud, *Comptes Rendus*, clxxx (1925), p. 127.

[4] *Berlin Sitz.* (1921), p. 966

in whose theory the ten gravitational potentials g_{pq} of Einstein and the four components ϕ_p of the electromagnetic potential were expressed in terms of the line-element in a space of five dimensions, in such a way that the equations of motion of electrified particles in an electromagnetic field became the equations of geodesics. This *five-dimensional theory of relativity* was afterwards developed by Oskar Klein [1] and others,[2] To ordinary space-time a fifth dimension is adjoined, but the curves in this dimension are very small, of the order of 10^{-30} cm., so that the universe is cylindrical, and indeed filiform, with respect to the fifth dimension, and the variations of the fifth variable are not perceptible to us, its non-appearance in ordinary experiments being due to a kind of averaging over it. A connection with quantum theory was made by assuming that Planck's quantum of Action h is due to periodicity in the fifth dimension: and the atomicity of electricity was presented as a quantum law, the momentum conjugate to the fifth co-ordinate having the two values e and $-e$.

In 1930 Oswald Veblen and Banesh Hoffman of Princeton [3] showed that the Kaluza-Klein theory may be interpreted as a four-dimensional theory based on projective instead of affine geometry.

Weyl's geometry, and especially his type of parallel-transport, suggested to W. Wirtinger [4] philosophical considerations which led him to a very original form of infinitesimal geometry: this, however, perhaps on account of its extreme generality, has not as yet found applications in theoretical physics.

Besides the investigations that have been described, the work of Weyl and Eddington has led, during the last thirty years, to a vast number of other investigations aimed at expressing gravitational and electromagnetic theory together in terms of a system of differential geometry. Some of them, such as those of the Princeton school [5] in America, led by O. Veblen and L. P. Eisenhart, the

[1] *ZS. f. P.* xxxvi (1926), p. 835; xxxvii (1926), p. 895; xlv (1927), p. 285; xlvi (1927), p. 188; *J. de Phys. et le Rad.* viii (1927), p. 242; *Ark. f. Mat. Ast. Fys* xxxiv (1946), No. 1

[2] A. Einstein, *Berlin Sitz.* (1927), pp. 23, 26; L. de Broglie, *J. de Phys. et le Rad.* viii (1927), p. 65; G. Darrieus, *J. de Phys. et le Rad.* viii (1927), p. 444; F. Gonseth et G. Juvet, *Comptes Rendus*, clxxxv (1927), p. 341; H. Mandel, *ZS. f. P.* xxxix (1926), p. 136; xlv (1927), p. 285; xlix (1928), p. 697; liv (1929), p. 564; lx (1930), p. 782; W. Wilson, *Proc. R.S.*(A), cxviii (1928), p. 441; J. W. Fisher, *Proc. R.S.*(A), cxxiii (1929), p. 489; H. T. Flint, *Proc. R.S.*(A), cxxiv (1929), p. 143; J. W. Fisher and H. T. Flint, *Proc. R.S.*(A), cxxvi (1930), p. 644

[3] *Phys. Rev.* xxxvi (1930), p. 810; cf. also J. A. Schouten and D. van Dantzig, *Proc. Amst. Ac.* xxv (1932), pp. 642, 843; *ZS. f. P.* lxxviii (1932), p. 639; J. A. Schouten, *ZS. f. P.* lxxxi (1933), pp. 129, 405; lxxxiv (1933), p. 92; W. Pauli, *Ann. d. Phys.* xviii (1933), p. 305.

[4] *Trans. Camb. Phil. Soc.* xxii (1922), p. 439; *Abh. aus dem math. Sem. der Hamb.* iv (1926), p. 178.

[5] cf. L. P. Eisenhart and O. Veblen, *Proc. N.A.S.* viii (1922), p. 19; O. Veblen, *Proc. N.A.S.* viii (1922), p. 192; ix (1923), p. 3; L. P. Eisenhart, *Proc. N.A.S.* viii (1922), p. 207; ix (1923), p. 4; xi (1925), p. 246; *Annals of Math.* xxiv (1923), p. 367; O. Veblen and J. M. Thomas, *Proc. N.A.S.* xi (1925), p. 204; T. Y. Thomas, *Proc. N.A.S.* xi (1925), p. 199; xiv (1928), p. 728; *Math. ZS.* xxv (1926), p. 723, and many papers in the succeeding years, in the same journals, by these authors and their associates

Dutch school at Delft, founded by J. A. Schouten,[1] and the French school whose most distinguished representative was E. Cartan [2] (1869–1951), have made discoveries which have great potentialities, but the significance of which at present appears to lie in pure mathematics rather than in physics, and which are therefore not described in detail here. The work of E. Bortolotti [3] of Cagliari should also be referred to. The outlook of the Germans, and of some of the Americans, has been, broadly speaking, more physical.[4]

It must be said, however, that, elegant though the mathematical developments have undoubtedly been, their relevance to fundamental physical theory must for the present be regarded as hypothetical.

This chapter has been concerned, for the most part, with General Relativity, which is essentially a geometrisation of physics. It may be closed with some account of a movement in the opposite direction, seeking to abolish the privileged position of geometry in physics, and indeed inquiring how far it may be possible to construct a physics independent of geometry. Since the notion of metric is a complicated one, which requires measurements with clocks and scales, generally with *rigid* bodies, which themselves are systems of great complexity, it seems undesirable to take metric as fundamental, particularly for phenomena which are simpler and actually independent of it.

The movement was initiated by Friedrich Kottler of Vienna, who in 1922 published two papers *Newton'sches Gesetz und Metrik* [5] and *Maxwell'sche Gleichungen und Metrik.*[6] Kottler first remarked on

[1] cf. J. A. Schouten, *Proc. Amst. Ac.* xxvii (1924), p. 407 ; xxix (1926), p. 334 ; *Palermo Rend.* l (1926), p. 142 ; J. A. Schouten and D. van Dantzig, *Proc. Amst. Ac.* xxxiv (1932), p. 1398 ; xxxv (1932), p. 642 ; *ZS. f. P.* lxxviii (1932), p. 639 ; lxxxi (1933), pp. 129, 405

[2] cf. E. Cartan, *Ann. Éc. Norm.* xl (1923), p. 325 ; xli (1923), p. 1 ; *Bull. Soc. M. France,* lii (1924), p. 205

[3] *Atti Inst. Veneto,* lxxxvi (1926–7), p. 459 ; *Rend. Lincei,* ix (1929), p. 530

[4] cf. R. Weitzenbock, *Wien. Ber.* cxxix 2a (1920), pp. 683, 697 ; cxxx 2a (1921), p. 15 ; H. Weyl, *Gött. Nach.* (1921), p. 99 ; F. Jüttner, *Math. Ann.* lxxxvii (1922), p. 270 ; E. Reichenbächer, *ZS. f. P.* xiii (1923), p. 221 ; A. Einstein, *Berlin Sitz.* (1925), p. 414 ; D. J. Struik and O. Wiener, *J. Math. Phys.* vii (1927), p. 1 ; E. Wigner, *ZS. f. P.* liii (1929), p. 592 ; A. Einstein, *Berlin Sitz.* (1928), pp. 217, 224 ; (1929), pp. 2, 156 ; N. Wiener and M. S. Vallarta, *Proc. N.A.S.* xv (1929), pp. 353, 802 ; M. S. Vallarta, *Proc. N.A.S.* xv (1929), p. 784 ; H. Weyl, *Bull. Am. M.S.* xxxv (1929), p. 716 ; N. Rosen and M. S. Vallarta, *Phys. Rev.* xxxvi (1930), p. 110 ; A. Einstein, *Berlin Sitz.* (1930), p. 18 ; *Math. Ann.* cii (1930), p. 685 ; A. Einstein and W. Mayer, *Berlin Sitz.* (1930), p. 110 ; (1931), pp. 257, 541 ; (1932), p. 130 ; W. Pauli, *Ann. d. Phys.*(5) xviii (1933), p. 305 ; E. Schrödinger, *Proc. R.I.A.* xlix (1943), pp. 43, 135 ; G. D. Birkhoff, *Proc. N.A.S.* xxix (1943), p. 231 ; xxx (1944), p. 324 ; A. Barajas, G. D. Birkhoff, C. Graef and M. S. Vallarta, *Phys. Rev.*(2) lxvi (1944), p. 138 ; H. Weyl, *Am. J. Math.* lxvi (1944), p. 591 ; E. Schrödinger, *Nature,* cliii (1944), p. 572 ; *Proc. R.I.A.* xlix (A) (1944), pp. 225, 237, 275 ; A. Einstein and V. Bargmann, *Annals of M.* xlv (1944), p. 1 ; A Einstein, *Annals of M.* xlv (1944), p. 15 ; E. Schrödinger, *Proc. R.I.A.* l (1945), pp. 143, 223 ; li (1946), p. 41 ; li (1947), pp. 147, 163 ; li (1948), p. 205 ; lii (1948), p. 1 ; A. Einstein, Appendix II in the third edition of his *The Meaning of Relativity* (Princeton, 1949) ; on the further development of this, cf. A. Papapetrou and E. Schrödinger, *Nature,* clxvii (1951), p. 40, and W. B. Bonnor, *Proc. R.S.*(A), ccix (1951), p. 353 ; ccx (1952), p. 427

[5] *Wien. Sitz.,* Abt. IIa, cxxxi (1922), p. 1

[6] ibid., p. 119

the independence of metric which characterises the analysis of skew tensors (i.e. tensors for which an interchange of any two indices produces a reversal of sign) : thus the divergence of a six-vector (X^{pq}) (which is a skew tensor of rank 2) is

$$\frac{1}{\sqrt{g}} \sum_q \frac{\partial}{\partial x^q} \left(\sqrt{g}\, X^{pq} \right)$$

and this is clearly unchanged if the metric

$$(ds)^2 = \sum_{p,q} g_{pq}\, dx^p\, dx^q$$

is replaced by

$$(ds)^2 = \lambda^2 \sum_{p,q} g_{pq}\, dx^p\, dx^q$$

where λ is any constant. While he recognised that it is impossible to abolish metric from physics entirely, he aimed at expressing the laws of nature as far as possible in terms of skew tensors, and ascertaining in each case the precise point where they cease to be adequate and the introduction of a metric becomes inevitable. In the first paper he considered the Newtonian theory of gravitation : Poisson's equation for the gravitational potential is

$$\frac{\partial^2 \phi}{\partial x_1^2} + \frac{\partial^2 \phi}{\partial x_2^2} + \frac{\partial^2 \phi}{\partial x_3^2} = -4\pi\rho$$

or, if the metric is given by

$$(ds)^2 = \sum_{p,q} a_{pq} dx^p dx^q \qquad (p, q = 1, 2, 3),$$

it is

$$\frac{1}{\sqrt{a}} \sum_p \frac{\partial}{\partial x^p} \left(\sqrt{a} \sum_q a^{pq} \frac{\partial \phi}{\partial x^q} \right) = -4\pi\rho.$$

But $\sum_q a^{pq} \frac{\partial \phi}{\partial x^q}$ is the contravariant vector corresponding to the covariant vector $\partial\phi/\partial x^p$: call it (L^p). Then Poisson's equation is

$$\sum_p \frac{\partial}{\partial x^p} \left(\sqrt{a} \,.\, L^p \right) = -4\pi\rho\sqrt{a}.$$

In order to transform this into an equation depending on a skew tensor, Kottler made use of a theorem of tensor-calculus which may be stated thus : in space of three dimensions, where the metric is given by

$$(ds)^2 = \sum_{p,q} a_{pq} dx^p dx^q \qquad (p, q = 1, 2, 3),$$

let δ_{pqr} have the value 1 or -1 according as we obtain (pqr) from (123) by an even or an odd substitution: and when p, q, r are not all different, let δ_{pqr} be defined to be zero. *Then* $\sqrt{a} \,.\, \delta_{pqr}$ *is a skew tensor of rank* 3.[1] We shall denote it by G_{pqr}.

Now write

$$F_{pq} = \sum_r G_{pqr} L^r$$

so F_{pq} is a covariant skew tensor of rank 2: its three numerically distinct components are

$$F_{12} = \sqrt{a} \,.\, L^3, \qquad F_{31} = \sqrt{a} \,.\, L^2, \qquad F_{23} = \sqrt{a} \,.\, L^1.$$

Thus Poisson's equation may be written

$$\frac{\partial F_{23}}{\partial x^1} + \frac{\partial F_{31}}{\partial x^2} + \frac{\partial F_{12}}{\partial x^3} = \mu_{123}$$

where $\mu_{123} = -4\pi\rho G_{123}$.

Now if S be any simple surface enclosing a volume V, then by Green's theorem (which is independent of any system of metric) we have

$$\iiint_V \left(\frac{\partial F_{23}}{\partial x^1} + \frac{\partial F_{31}}{\partial x^2} + \frac{\partial F_{12}}{\partial x^3}\right) dx^1 dx^2 dx^3$$

$$= \iint_S \left\{ F_{23} \frac{\partial(x_2, x_3)}{\partial(u, v)} + F_{31} \frac{\partial(x_3, x_1)}{\partial(u, v)} + F_{12} \frac{\partial(x_1, x_2)}{\partial(u, v)} \right\} du dv$$

where (u, v) are any parameters fixing position on the surface. Thus we obtain the equation

$$\iint_S \left\{ F_{23} \frac{\partial(x_2, x_3)}{\partial(u, v)} + F_{31} \frac{\partial(x_3, x_1)}{\partial(u, v)} + F_{12} \frac{\partial(x_1, x_2)}{\partial(u, v)} \right\} du\, dv = \iiint_V \mu_{123} dx^1 dx^2 dx^3.$$

Kottler interpreted the left-hand side of this equation as the total *flux*, through the surface S, of the skew tensor F_{pq}: and he regarded the equation as representing the relation between the occurrences inside S and the field (expressed by a skew tensor) acting on S, *in a form independent of metric*. It is of course essentially a form of Gauss's theorem.[2] Kottler regarded this equation as the starting-point of

[1] It is generally called the *skew fundamental tensor of Ricci and Levi-Civita*.

[2] On Gauss's theorem and the concept of mass in General Relativity, cf: E. T. Whittaker, *Proc. R.S.*(A), cxlix (1935), p. 384; H. S. Ruse, *Proc. Edin. Math. Soc.*(2) iv (1935), p. 144; J. T. Combridge, *Phil. Mag.*(7) xx (1935), p. 971; G. Temple, *Proc. R.S.*(A), cliv (1936), p. 354; J. L. Synge, *Proc. R.S.*(A), clvii (1936), p. 434; *Proc. Edin. Math. Soc.*(2) v (1937), p. 93; A. Lichnerowicz, *Comptes Rendus*, ccv (1937), p. 25

a fundamental non-metrical formulation of gravitational theory: metric is introduced, at a later stage, only when the concept of *work*, as distinguished from that of *flux*, is introduced: that is, when the force acting on a particle comes to be considered.

His second paper, on Maxwell's equations, presented an easier problem, since the electric and magnetic forces together constitute a six-vector, which is the kind of tensor required for a non-metrical presentation of the subject. Moreover there were available the results of papers written in 1910 by E. Cunningham[1] and H. Bateman,[2] who, starting from the proposition that the fundamental equations of electrodynamics are invariant with respect to Lorentz transformations, i.e. to rotations in the four-dimensional world of space-time, had remarked that these electrodynamic equations are invariant with respect to a much wider group, namely all the transformations for which the equation

$$(dx^2 + (dy)^2 + (dz)^2 - c^2(dt)^2 = 0$$

is invariant. By writing $\sqrt{-1} \, . \, cdt = du$, this last equation becomes

$$(dx)^2 + (dy)^2 + (dz)^2 + (du)^2 = 0.$$

The transformations which leave this latter equation invariant are what are called the *conformal* group of transformations in space of four dimensions: they had been studied long before by Sophus Lie, who had shown that they are composed of reflections, translations, rotations, magnifications and inversions with respect to the hyperspheres of the four-dimensional space. Types of motion of an electromagnetic system may be derived from one another by such transformations: they are in general more complicated than those considered in the relativity theory of Poincaré and Lorentz: for in the latter case a fixed three-dimensional configuration is transformed into one, every point of which has the same velocity of translation: but in the conformal case, under the simplest operation of the group, a fixed system becomes one in which the whole is expanding or contracting in a certain way.[3]

Cunningham and Bateman's work was now developed by Kottler,[4] H. Weyl,[5] J. A. Schouten and J. Haantjes,[6] and particularly by D. van Dantzig,[7] into a *complete theory of electromagnetism, independent*

[1] *Proc. L.M.S.*(2) viii, p. 77

[2] ibid., p. 223; cf. also Bateman's book, *Electrical and Optical Wave-motions* (Camb., 1915); H. Bateman, *Phys. Rev.* xii (1918), p. 459; *Proc. L.M.S.*(2) xxi (1920), p. 256; E. Bessel-Hagen, *Math. Ann.* lxxxiv (1921), p. 258; F. D. Murnaghan, *Phys. Rev.* xvii (1921), p. 73; G. Kowalewski, *J. für Math.* clvii (1927), p. 193

[3] It is characteristic of these transformations that a sphere which is expanding with the velocity of light transforms into a sphere expanding with the same velocity.

[4] loc. cit. [5] *Raum, Zeit, Materie*, 4th Aufl., p. 260 [6] *Physica* I (1934), p. 869

[7] *Proc. Camb. Phil. Soc.* xxx (1934), p. 421; *Proc. Amst. Ac.* xxxvii (1934), pp. 521, 526, 644, 825; xxxix (1936), pp. 126, 785; *Cong. Int. des Math.* Oslo (1936), II, p. 225; cf. also J. A. Schouten, *Tensor Calculus for Physicists*, Oxford, 1951

of metrical geometry, and in fact needing the ideas neither of metric nor of parallelism. It is characteristic of theories such as this that *differential* relations are generally replaced by integral relations: thus for the statement that under certain circumstances 'the divergence of the flux of energy vanishes' is substituted the statement that 'the integral of the flux over a closed surface vanishes,' which is[1] a mathematical form of the physical statement that 'the algebraic sum of the energies of all the particles crossing through a closed surface vanishes.'

[1] cf. D. van Dantzig, *Erkenntnis*, vii (1938), p. 142

Chapter VI

RADIATION AND ATOMS IN THE OLDER QUANTUM THEORY

In 1916 Einstein[1] published a new and extremely simple proof of Planck's law of radiation, and at the same time obtained some important results regarding the emission and absorption of light by molecules. The train of thought followed was more or less similar to that adopted by Wien in the derivation[2] of his law of radiation, but Einstein now adapted it to the new situation created by Bohr's theory of spectra.

Consider a molecule of a definite kind, disregarding its orientation and translational motion: according to quantum theory, it can take only a discrete set of states $Z_1, Z_2, \ldots Z_n, \ldots$ whose internal energies may be denoted by $\epsilon_1, \epsilon_2, \ldots \epsilon_n, \ldots$ If molecules of this kind belong to a gas at temperature T, then the relative frequency W_n of the state Z_n is given by the formula of Gibbs's canonical distribution as modified for discrete states,[3] namely,[4]

$$W_n = e^{-\frac{\epsilon_n}{kT}}.$$

Now suppose that the probability of a single molecule in the state Z_m passing in time dt spontaneously, i.e., without excitation by external agencies (as in the emission of γ rays by radio-active bodies) to the state of lower energy Z_n with emission of radiant energy $\epsilon_m - \epsilon_n$ is

$$A_m^n \, dt. \qquad \text{(A)}$$

Suppose also that the probability of a molecule under the influence of radiation of frequency ν and energy-density ρ passing in time dt from the state of lower energy Z_n to the state of higher energy Z_m, by absorbing the radiant energy $\epsilon_m - \epsilon_n$, is

$$B_n^m \rho \, dt \qquad \text{(B)}$$

and suppose that the probability of a molecule under the influence of this radiation-field passing in time dt from the state of higher

[1] *Mitt. d. phys. Ges. Zürich,* No. 18 (1916) ; *Phys. ZS.* xviii (1917), p. 121 ; cf. A. S. Eddington, *Phil. Mag.* l (1925), p. 803

[2] cf. Vol. I, p. 382

[3] cf. *supra,* p. 87

[4] For simplicity we omit consideration of the statistical ' weight ' of the state.

energy Z_m to the state of lower energy Z_n, with emission of the radiant energy $\epsilon_m - \epsilon_n$, is

$$B^n_m \rho \, dt. \qquad (B')$$

This is called *stimulated emission* ; its existence was recognised here for the first time.

Now the exchanges of energy between radiation and molecules must not disturb the canonical distribution of states as given above. So in unit time, on the average, as many elementary processes of type (B) must take place as of types (A) and (B′) together. We must therefore have

$$e^{-\frac{\epsilon_n}{kT}} B^m_n \rho = e^{-\frac{\epsilon_m}{kT}} \left(B^n_m \rho + A^n_m\right).$$

We assume that ρ increases to infinity with T, so this equation gives

$$B^m_n = B^n_m, \qquad (1)$$

and the preceding equation may therefore be written

$$\rho = \frac{(A^n_m / B^n_m)}{e^{(\epsilon_m - \epsilon_n)/kT} - 1}.$$

This is evidently Planck's law of radiation : in order that it may pass asymptotically into Rayleigh's law for long wave-lengths, and Wien's law for short wave-lengths, we must have

$$\epsilon_m - \epsilon_n = h\nu$$

and

$$A^n_m = \frac{8\pi h \nu^3}{c^3} B^n_m. \qquad (2)$$

The two equations (1) *and* (2), first given in this paper of Einstein's, *are fundamental in the theory of the exchanges of energy between matter and radiation*, and have been extensively used in the later development of quantum theory.[1] It may be remarked that as there is spontaneous emission, but not spontaneous absorption, there is asymmetry as between past and future : but so far as transitions stimulated by radiation are concerned, there is a symmetrical probability, $B^n_m = B^m_n$.[2]

[1] If the weights of the energy-levels are g_m, g_n, the relation (1) must be replaced by $g_n B^m_n = g_m B^n_m$. Relation (2) is unaffected.

[2] Einstein's formulae were extended to the case of non-sharp energy-levels by R. Becker, *ZS. f. P.* xxvii (1924), p. 173, and to the laws of interaction between radiation and free electrons by A. Einstein and P. Ehrenfest, *ZS. f. P.* xix (1923), p. 301.

One of the chief problems of quantum theory is to compute coefficients, such as these Einstein coefficients, from data regarding atoms and molecules. The relation (2) has been verified experimentally by a comparison of the strengths of absorption and emission lines. The B's have been found from measures of the intensities of the components of multiplets in spectra by L. S. Ornstein and H. C. Burger.[1]

Another important result established in this paper related to exchanges of momentum between molecules and radiation. Einstein showed that *when a molecule, in making a transition from the state* Z_n *to* Z_m, *receives the energy* $\epsilon_m - \epsilon_n$, *it receives also momentum of amount* $(\epsilon_m - \epsilon_n)/c$ *in a definite direction;* and, moreover, that *when a molecule, in the transition from* Z_m *to the state of lower energy* Z_n, *emits radiant energy of amount* $(\epsilon_m - \epsilon_n)$, *it acquires momentum of amount* $(\epsilon_m - \epsilon_n)/c$ *in the opposite direction.* Thus the processes of emission and absorption are *directed* processes: there seems to be no emission or absorption of spherical waves.

Einstein's theory of the coefficients of emission and absorption enabled W. Bothe[2] to give an instructive fresh calculation of the numbers of quanta $h\nu$ in black-body radiation which are associated as 'photo-molecules' in pairs $2h\nu$, trios $3h\nu$, etc. He considered a cavity filled with black-body radiation at temperature T, in which there were a number of gas-molecules, each of which was either in the state of energy Z_1 or the state of energy Z_2, where $Z_2 - Z_1 = h\nu$, their average relative numbers being given by the law of canonical distribution at temperature T. He assumed that when a single quant $h\nu$ of the radiation causes a stimulated emission, the emitted quant moves with the same velocity and in the same direction as the quant that causes it, so they become a quant-pair $2h\nu$; if the exciting quant itself already belongs to a pair, then there arises a trio $3h\nu$, and so on. The absorption of a quant from a quant-pair leaves a single quant, and a spontaneous emission produces a single quant. Writing down the conditions that the average numbers of single quants, of quant-pairs, etc. do not change in time, and using Einstein's relations between the coefficients of spontaneous emission, stimulated emission and absorption, we obtain a set of equations from which it follows that in unit volume and in the frequency-range $d\nu$ the average number of single quants which are united into s-fold quant-molecules $sh\nu$ is

$$\frac{8\pi\nu^2}{c^3} e^{-\frac{sh\nu}{kT}} d\nu,$$

in agreement with the result previously obtained.[3] The average total energy per unit volume in the range $d\nu$ is therefore

$$\frac{8\pi h\nu^3}{c^3}\left(\sum_{s=1}^{\infty} e^{-\frac{sh\nu}{kT}}\right)d\nu$$

[1] *ZS. f. P.* xxiv (1924), p. 41 [2] *ZS. f. P.* xx (1923), p. 145 [3] cf. p. 103 *supra*

or

$$\frac{8\pi h\nu^3}{c^3}\frac{d\nu}{e^{h\nu/kT}-1}$$

in agreement with Planck's law of radiation.

Einstein's theory of the coefficients of emission and absorption also enabled theoretical physicists within a few years to create a satisfactory quantum theory of the scattering, refraction and dispersion of light.[1] In 1921 Rudolf Ladenburg[2] (*b.* 1882), a former pupil of Röntgen's at Munich, who afterwards settled in America at Princeton, introduced quantum concepts into the theory.

It was necessary first for him to find a quantum-theoretic interpretation for the number that in classical theory represented the number of electrons bound to atoms by forces of restitution, for it was these electrons which were responsible for light-scattering, refraction, and dispersion. Let us call them *dispersion-electrons*. This he achieved by calculating the emitted and absorbed energy of a set of molecules in temperature-equilibrium with radiation, on the basis of classical theory on the one hand and of quantum theory on the other.

Suppose then that there are $\mathfrak{N}$ dispersion-electrons per cm.^{3}, capable of oscillating freely with frequency ν_1. Now for the harmonic oscillator of frequency ν_1 if the displacement x at the instant t is $x_0 \cos 2\pi\nu_1 t$, the mean value of the energy $\frac{1}{2}m\,\{(dx/dt)^2 + 4\pi^2\nu_1{}^2x^2\}$ is $\overline{\mathrm{U}} = 2\pi^2 m\nu_1{}^2 x_0{}^2$, and therefore[3] the average energy radiated per second is

$$\frac{16\pi^4e^2}{3c^3}\nu_1{}^4x_0{}^2 \quad \text{or} \quad \frac{8\pi^2e^2\nu_1{}^2}{3mc^3}\overline{\mathrm{U}},$$

and therefore the energy radiated in one second by the $\mathfrak{N}$ dispersion-electrons is

$$\mathrm{J}_{el} = \frac{8\pi^2e^2\nu_1{}^2}{3mc^3}\mathfrak{N}\overline{\mathrm{U}}.$$

If the molecules are in equilibrium with radiation at temperature T, and if we regard the electrons as spatial oscillators with three degrees

[1] The classical theory of the scattering of light by small particles had been given by Lord Rayleigh in 1871 (*Phil. Mag.*(4) xli, pp. 107, 274, 447) on the basis of the elastic-solid theory of light, and in 1881 (*Phil. Mag.*(5) xii, p. 81) and 1899 (*Nature*, lx, p. 64; *Phil. Mag.*(5) xlvii, p. 375) on the basis of Maxwell's electromagnetic theory; cf. also J. J. Thomson, *Conduction of Electricity through Gases*, 2nd edn. (1906), p. 321.

[2] *ZS. f. P.* iv (1921), p. 451; cf. also R. Ladenburg and F. Reiche, *Naturwiss*, xi (1923), p. 584; R. Ladenburg, *ZS. f. P.* xlviii (1928), p. 15

[3] cf. Vol. I, p. 326

of freedom,[1] then [2] we have between $\overline{U}$ and the radiation-density ρ the relation

$$\overline{U}=\frac{3c^3}{8\pi\nu_1^2}\rho.$$

Thus

$$J_{el}=\frac{\pi e^2}{m}\mathfrak{N}\rho.$$

The energy absorbed in equilibrium is of course equal to the energy radiated.

Now in quantum-theory the emission from a molecule is due to transitions from a state of higher energy (2) to a state of lower energy (1). The number of spontaneous transitions per second is, in the notation we have already used,

$$N_2A_2^1$$

where N_2 is the number of molecules in the state (2). The number of transitions from the lower to the higher state per second is

$$N_1B_1^2\rho,$$

where N_1 is the number of molecules in the state (1) : and therefore the absorbed energy is

$$J_Q=h\nu_1N_1B_1^2\rho.$$

By Einstein's relations we have [3]

$$B_1^2=\frac{c^3}{8\pi h\nu_1^3}A_2^1.$$

Thus we have

$$J_Q=N_1\frac{c^3}{8\pi\nu_1^2}A_2^1\rho.$$

Equating J_Q to J_{el}, we have

$$\mathfrak{N}=N_1\frac{mc^3}{8\pi^2e^2\nu_1^2}A_2^1.$$

[1] For a three-dimensional oscillator in an isotropic radiation-field we obtain three times the value for a linear oscillator in the same field ; on this point, cf. F. Reiche, *Ann. d. Phys.* lviii (1919), p. 693.

[2] cf. Planck, p. 79 *supra*

[3] Supposing for simplicity that the statistical weights of the energy-levels (1) and (2) are equal.

This formula expresses the constant $\mathfrak{N}$ (which may be derived experimentally from measurements of emission, absorption, anomalous dispersion and magnetic rotation, and which in classical theory is interpreted as the number of dispersion-electrons) *in terms of the quantum-theoretic quantities* N_1 (the number of molecules in the lower of the two states) *and* A_2^1 (the probability of the spontaneous transitions which give rise to radiation of frequency ν_1). Thus from measurements of e.g. anomalous dispersion at different lines of a series in a spectrum, we can make inferences regarding the probability of the various transitions.

Ladenburg's result enables us to replace the dispersion-formula found in Vol. I [1] by

$$\mu^2 = 1 + \frac{c^3}{8\pi^3}\sum_i \frac{N_i A_k^i}{\nu_i^2(\nu_i^2 - \nu^2)}$$

where (i) denotes the lower level and (k) the upper level of energy in the transition corresponding to the frequency ν_i.

Now consider the scattering of light by an atom. We suppose the wave-length of the light to be much greater than the dimensions of the atom, so that at any instant the field of force is practically uniform over these dimensions. We suppose also that the atom contains an electron of charge $-e$ and mass m, which is bound to the atom so that when the electron is displaced a distance ξ parallel to the x-axis from its equilibrium position, a force of restitution $4\pi^2 m\nu_1^2\xi$ is called into play. Then when the atom is irradiated by a light-wave whose electric field is

$$Ee^{2\pi i\nu t}$$

it can easily be shown that according to classical theory it radiates secondary wavelets such as would be produced by an electric doublet of moment

$$\frac{e^2 E e^{2\pi i\nu t}}{4\pi^2 m(\nu_1^2 - \nu^2)}.$$

These secondary spherical wavelets, which are coherent with the incident wave, constitute the scattered radiation.

In quantum theory we are concerned with scattering not by a single bound electron but by an atom, and therefore in order to transfer this expression to quantum theory we must first multiply it by $\mathfrak{N}/N_1$, the number of classical dispersion-electrons per atom (the atom being supposed to be in the state (1) when scattering).

[1] At p. 401 : there is a change of notation, the N, k, n of Vol. I being here represented by $\mathfrak{N}_i$, $2\pi\nu_i\sqrt{m}$, and $2\pi\nu$ respectively : and we suppose now that in the classical case there are several kinds of dispersion-electrons with different natural periods. The formula had been confirmed by the experiments of R. W. Wood, *Phil. Mag.*(6) viii (1904), p. 293 and P. V. Bevan, *Proc. R.S.* lxxxiv (1910), p. 209 ; lxxxv (1911), p. 58, on the dispersion of light in the vapours of the alkali metals.

Thus we obtain for the amplitude of the doublet-moment

$$\frac{\mathfrak{N}}{N_1}\frac{Ee^2}{4\pi^2 m(\nu_1{}^2-\nu^2)}.$$

Applying Ladenburg's result, this becomes

$$\frac{c^3 A_2^1 E}{32\pi^4\nu_1{}^2(\nu_1{}^2-\nu^2)}.$$

We must sum over all higher states (2) : and thus we have the result that *an atom in state* (1), *when irradiated by a light-wave whose electric field is* $Ee^{2\pi i\nu t}$, *emits secondary wavelets like an electric oscillator of frequency* ν, *whose electric moment has the amplitude*

$$P=\frac{c^3E}{32\pi^4}\sum_i\frac{A_i^1}{\nu_i{}^2(\nu_i{}^2-\nu^2)},$$

where A_i^1 *denotes the probability of the atom performing spontaneously in unit time the transition to the state* (1) *of energy* E_1 *from a higher state* (i) *of energy* E_i, *and where*

$$\nu_i=\frac{E_i-E_1}{h}.$$

This result was further modified by Henrik Antony Kramers[1] (1894–1952), who, taking the above Ladenburg formula as correct when the atom was in the normal state, took into consideration the case when the atom is excited, and proposed to deal with it by taking the summation not only over the stationary states i which have higher energy-levels than the state (1), but also over the states j which have lower energy-levels than the state (1), so that the formula becomes

$$P=\frac{c^3E}{32\pi^4}\left\{\sum_i\frac{A_i^1}{\nu_i{}^2(\nu_i{}^2-\nu^2)}-\sum\frac{A_1^j}{\nu_j{}^2(\nu_j{}^2-\nu^2)}\right\}$$

where[2]

$$\nu_j=\frac{E_1-E_j}{h}.$$

This formula of course relates to a single atom, and a factor must be adjoined representing the number of atoms in this state.

Now as we have seen, in the classical theory of scattering, the

[1] *Nature*, cxiii (1924), p. 673 ; cxiv (1924), p. 310

[2] The formula as given by Kramers contained an additional factor 3 ; this was a consequence of his assumption that the free oscillations are parallel to the incident field, whereas the above formula assumes all orientations of the atom to be equally probable.

atom behaves like an electric doublet, the amplitude of whose moment is

$$\frac{e^2 E}{4\pi^2 m(\nu_1{}^2 - \nu^2)}.$$

Comparing the above Ladenburg-Kramers formula with this, we see that according to quantum theory the atom behaves with respect to the incident radiation as if it contained a number of bound electric charges constituting harmonic oscillators as in the classical theory, one of these oscillators corresponding to each possible transition between the state of the atom and another stationary state. Thus we can describe the behaviour of an atom in dispersion by means of a doubly-infinite (i.e. depending on two quantum numbers m and n) set of virtual harmonic oscillators, the displacement in the oscillator (m, n) being represented by

$$q(m, n) = Q(m, n)e^{2\pi i\nu(m, n)t}$$

where $\nu(m, n)$ denotes the frequency of this oscillator. The aggregate of these virtual harmonic oscillators was called by A. Landé [1] the *virtual orchestra.* The virtual orchestra is thus a classical substitution-formalism for the radiation, and so indirectly becomes a representative of the quantum radiator itself.

In place of the classical e^2/m we have the value $c^3 A_i^1/8\pi^2\nu_i^2$ for one of the 'absorption oscillators,' i.e. those corresponding to transitions between the state (1) and higher states, but we have the value $-c^3 A_1^j/8\pi^2\nu_j^2$ for one of the 'emission oscillators,' i.e. those corresponding to transitions between the state (1) and lower states: so that there is a kind of 'negative dispersion' arising from the emission oscillators, which may be regarded as analogous to the 'negative absorption' represented by Einstein's coefficient B_2^1. Another way of putting the matter is to say that a quantum-oscillator which is in the higher state, when irradiated by a light-wave which is not markedly absorbed by it, emits a spherical wave, which differs from that emitted in the lower state only by being displaced 180° in phase.

Almost immediately after the appearance in May 1924 of Kramers's paper, Max Born [2] gave a general method, to which he gave the name *quantum mechanics*,[3] for translating the classical theory of the perturbations of a vibrating system into the corresponding quantum formulae. In particular he studied the problem of an oscillator exposed to an external field of radiation: his method was based on the correspondence-principle, the frequency belonging to a transition between states characterised by the quantum numbers

[1] *Naturwiss.* xiv (1926), p. 455
[2] *ZS. f. P.* xxvi (1924), p. 379, communicated 13 June 1924
[3] This was the first occurrence of the term in the literature of theoretical physics.

n and n' corresponding to the overtone $|n-n'|\nu$ of the classical solution: but the results he obtained were afterwards shown to be correct in the light of the later form of quantum theory.

A theory of scattering is also essentially a theory of refraction. For when (in classical theory) light is scattered by atoms, the scattered light has the same frequency as the incident light and interferes with it: the total effect produced is the same as if the primary wave alone existed, but was propagated with a different velocity: and this change of velocity is the essential feature in refraction. In the Ladenburg-Kramers formula, the terms

$$-\sum_j \frac{A_1^j}{\nu_j^2(\nu_j^2-\nu^2)}$$

affect the refraction in the opposite sense to the other terms. This was verified in 1928 by R. Ladenburg[1] and by H. Kopfermann and R. Ladenburg,[2] who studied the refractive index, in the neighbourhood of a certain spectral line of neon. When a current was passed through the gas, many of the atoms were thereby raised to an excited level, and it was found that the refractive index dropped, the refraction due to the ordinary fall in energy-level being partly counterbalanced by that due to the rise.

Even before the appearance of Kramers's paper, new possibilities in regard to scattering had been indicated by A. Smekal.[3] It may happen that the emission of scattered radiation is associated with a quantum transition in the scattering structure, in which case there will be a difference of frequency between the scattered and primary rays of the same order of magnitude as the spectral frequencies of the scattering atomic system; if ν denotes the frequency of the primary radiation, and $h\nu_k$ or $h\nu_l$ denotes the change of energy of the atom in the transition, according as this change happens in the positive or negative direction, then there may be scattered radiation (of low intensity) of frequency $\nu+\nu_k$ or $\nu-\nu_l$. At the time of Smekal's paper, his conjecture of *anomalous scattering*, as he called it, had not been verified experimentally: but in 1928 Sir Chandrasakara V. Raman[4] showed that light scattered in water and other transparent substances contains radiations of frequency quite different from that of the incident light, and it was at once seen that this was the effect predicted by Smekal: almost simultaneously G. Landsberg and L. Mandelstam[5] found the same effect experimentally, working with quartz.

In general, radiation of a definite frequency ν generates scattered

[1] *ZS. f. P.* xlviii (1928), p. 15

[2] ibid., pp. 26, 51; *ZS. f. phys. Chem.* cxxxix (1928), p. 375

[3] *Naturwiss.* xi (1923), p. 873

[4] *Ind. Journ. of Phys.* ii (1928), p. 387; C. V. Raman and K. S. Krishnan, ibid., p. 399

[5] *Naturwiss.* xvi (1928), p. 557; *ZS. f. P.* l (1928), p. 769

radiation of several modified frequencies, all of the form $\nu \pm \nu'$, where ν' is either an infra-red frequency in the absorption spectrum of the scattering material, or a difference of such frequencies.[1]

In the year following the publication of Smekal's paper, H. A. Kramers and W. Heisenberg[2] made an exhaustive study of the radiation which an atom emits when irradiated by incident light. Their method, as in Born's paper published a few months earlier, was first to study by classical theory the perturbation by incident radiation of an atom regarded as a multiply-periodic dynamical system, and then to employ the correspondence-principle in order to translate the results into their quantum-theoretic form.

According to classical theory, under the irradiation the system emits a scattered radiation, whose intensity is proportional to the intensity of the incident light: when it is analysed into harmonic components, a component involves the sum or difference of the frequency of the incident light and frequencies occurring in the undisturbed motion (as in the Smekal-Raman effect). The quantum-theoretic formulae must satisfy the condition that in the region of great quantum numbers, where successive stationary states differ comparatively little from each other, the quantum scattering must tend asymptotically to coincide with the classical scattering. This condition is satisfied by interpreting certain derivatives which occur in the classical formulae as differences of two quantities: in this way, Kramers and Heisenberg obtained equations which involved only the frequencies characteristic of transitions, while all symbols relating to the mathematical theory of multiply-periodic dynamical systems disappeared.

It was shown that when irradiated with monochromatic light, an atom emits not only spherical waves of the same frequency as the incident light, and coherent with it, but also systems of non-coherent spherical waves, whose frequencies are combinations of that frequency with other frequencies, which correspond to thinkable transitions to other stationary states. These additional systems of spherical waves occur as scattered light, but they do not contribute to the phenomena of dispersion and absorption of the incident radiation.

An interesting comment on the Kramers-Heisenberg formula was made by P. Jordan,[3] who remarked that it remained valid even when the incident radiation consisted of very long electromagnetic waves, which in the limit of zero frequency tend to a field-strength constant in time: and that in this limiting case, the formula actually yields the changes of frequency in spectral lines which are observed in the Stark effect. He drew the moral that discontinuous jumps must

[1] A review of literature on the Smekal-Raman effect to Feb. 1931 was given by K. W. F. Kohlrausch, *Phys. ZS.* xxxii (1931), p. 385. On the Smekal-Raman effect in molecules and crystals, cf. E. Fermi, *Mem. Acc. Ital. Fis.* Nr. 3 (1932), p. 1.

[2] *ZS. f. P.* xxxi (1925), p. 681

[3] P. Jordan, *Anschauliche Quantentheorie* (Berlin, 1936), p. 85

not be regarded as the essential characteristic of quantum theory: for phenomena in which they occur can be connected continuously by intermediate types with phenomena in which there is no discontinuity.

About this time much attention was given to scattering of a different type. In 1912 C. A. Sadler and P. Mesham,[1] working in L. R. Wilberforce's laboratory at Liverpool University, showed that when a homogeneous beam of X-rays is scattered by a substance of low atomic weight, the scattered rays are of a softer type (i.e. of longer wave-length).[2] Moreover, in the case of the γ-rays produced by a radium salt, it was shown by J. A. Gray[3] that the secondary or scattered γ-rays were less penetrating than the primary, that this 'softening' was due to a real change in the character of the primary rays when the secondary rays were formed, and that it increased with the angle between the primary and secondary rays (generally called the *angle of scattering*).[4] An explanation of these phenomena favoured by physicists at the time was that the primary beam was not truly homegeneous, and that its softer components were more strongly scattered than the harder ones: this, however, was explicitly denied by Sadler and Mesham. In 1917 C. G. Barkla[5] propounded the hypothesis, that there existed a new series of characteristic radiations which were of shorter wave-length than the K- and L-series which he had discovered previously, and which he named the *J-series*: and that the softening observed in the scattered radiation from light elements was due to the admixture of this J-radiation with radiation of the same hardness as the primary.[6] This explanation, however, lost credit when it was found[7] that the J-series had no critical absorption limit similar to the absorption limit of the K- and L-series, and that it showed under spectroscopic observation[8] no spectral lines such as might have been expected. When the J-series explanation was dismissed,[9] there still seemed to be a

[1] *Phil. Mag.* xxiv (1912), p. 138; cf. J. Laub, *Ann. d. Phys.* xlvi (1915), p. 785, and J. A. Gray, *J. Frank. Inst.* cxc (1920), p. 633

[2] It will be remembered that the secondary X-rays in general consist of a mixture of the characteristic rays discovered by Barkla (K, L, etc.) and of truly scattered rays. The wave-length of the former depends solely on the chemical nature of the scattering substance, but the wave-length of the latter is the same (or nearly the same) as that of the primary rays. For the heavier elements the characteristic rays predominate, but for elements of low atomic weight at that time only the truly scattered rays were normally observable.

[3] *Phil. Mag.* xxvi (1913), p. 611

[4] These results were confirmed as regards γ-rays by D. C. H. Florance (working at Manchester under Rutherford) in *Phil. Mag.* xxvii (1914), p. 225, and Arthur H. Compton, *Phil. Mag.* xli (1921), p. 749.

[5] *Phil. Trans.* ccxvii (1917), p. 315

[6] cf. C. G. Barkla and Margaret P. White, *Phil. Mag.* xxxiv (1917), p. 270. They found an abnormally great mass absorption coefficient for aluminium at 0·37Å, and regarded this as additional proof of the reality of J-radiation.

[7] cf. F. K. Richtmyer and Kerr Grant, *Phys. Rev.* xv (1920), p. 547; C. W. Hewlett, *Phys. Rev.* xvii (1921), p. 284

[8] W. Duane and T. Shimizu, *Phys. Rev.* xiii (1919), p. 289; xiv (1919), p. 389

[9] Further proof of this, depending on the polarisation of the scattered rays, was given by A. H. Compton and C. F. Hagenow, *J. Opt. Soc. Amer.* viii (1924), p. 487.

possibility of explaining the increase of wave-length in the scattered light without departing from classical theory: for[1] radiation-pressure (which had not been taken into account in the previous classical theories of scattering) might set the electrons in motion in the direction of the primary radiation, and the wave-length of the scattered light would then be increased by the Doppler effect. On performing the calculations, it was found that the increase in wave-length should follow the law actually verified by experiment, namely that it would be proportional to $\sin^2 \frac{1}{2}\theta$ where θ denotes the angle of scattering: but its magnitude as calculated was not in agreement with observation.

There seemed to be a likelihood therefore that some new type of physical explanation was required. In October 1922 a Bulletin[2] of the National Research Council of the U.S.A. appeared, written by A. H. Compton and containing a full discussion of secondary radiations produced by X-rays: in this the author suggested that when an X-ray quantum is scattered, it spends all its energy and momentum upon some particular electron. This electron in turn re-radiates the whole quantum (degraded in frequency) in some definite direction; the change in momentum of the X-ray quantum due to the change in its frequency and direction of propagation is associated with a recoil of the scattering electron. The ordinary conservation laws of energy and momentum are obeyed,[3] so that the energy of the recoil electron accounts for the difference between the energy of the incident photon and the energy of the scattered photon.

Compton's theory was communicated to a meeting of the American Physical Society on 1–2 December 1922,[4] and published more fully in May of the following year.[5] On examining the scattered rays[6] from light elements spectroscopically, he found that their spectra showed lines corresponding to those in the primary rays, and also other lines corresponding to these but displaced slightly towards the longer wave-lengths; and that the difference in wave-length increased rapidly at large angles of scattering. It is these displaced lines which represent the *Compton effect.*

In the Compton effect the electron is effectively free, i.e. it is so feebly bound to the nucleus of the atom that the binding-energy can be neglected in comparison with the energy $h\nu$ of the incident quant; this condition is realised in the scattering of hard X-rays by elements of low atomic number. When the frequency is decreased, or the atomic number is increased, the binding forces can no longer be neglected in comparison with the energy of the incident photon,

[1] cf. O. Halpern, *ZS. f. P.* xxx (1924), p. 153; G. Wentzel, *Phys. ZS.* xxvi (1925), p. 436

[2] Vol. IV, No. 20

[3] It may be noted that while the photo-electric effect shows that the *energy* of radiation is transferred in quanta, the Compton effect shows that *momentum* also is transferred in quanta.

[4] *Phys. Rev.* xxi (1923), p. 207

[5] *Phys. Rev.* xxi (1923), p. 483

[6] *Phys. Rev.* xxii (Nov. 1923), p. 409

and the phenomenon passes over into the photo-electric effect (in the case when the energy of the incident photon is transferred to the electron) or ordinary scattering (in the case when the photon retains its energy and changes only its direction).

The Compton effect may be discussed in an elementary way by light-quantum methods as follows : [1]

Let the incident light-quantum, of frequency ν, propagated in the positive direction of the axis of x, encounter at O the electron, which recoils in a direction making an angle θ with the axis Ox, with velocity v, while the light-quantum, degraded to frequency ν', is scattered in a direction making an angle $-\phi$ with Ox. Then (using the relativist formulae for energy and momentum), the equation of conservation of energy is

$$h\nu = h\nu' + mc^2\left\{\left(1-\frac{v^2}{c^2}\right)^{-\frac{1}{2}} - 1\right\},$$

while the equations of conservation of momentum are

$$\frac{h\nu}{c} = \frac{h\nu'}{c}\cos\phi + \frac{mv}{\sqrt{\left(1-\frac{v^2}{c^2}\right)}}\cos\theta,$$

$$\frac{h\nu'}{c}\sin\phi = \frac{mv}{\sqrt{\left(1-\frac{v^2}{c^2}\right)}}\sin\theta.$$

From these equations we have to calculate ν' in terms of ν and ϕ, which are supposed given. We readily find

$$\nu' = \frac{\nu}{1+\frac{h\nu}{mc^2}(1-\cos\phi)}.$$

The increment of wave-length $\Delta\lambda$ or $c/\nu' - c/\nu$ is

$$\Delta\lambda = \frac{2h}{mc}\sin^2\frac{\phi}{2},$$

a formula which was definitely confirmed by observation.

[1] In addition to the papers quoted above and below, cf. the following papers of 1923 and 1924 : P. Debye, *Phys. ZS.* xxiv (April 1923), p. 161, who discovered the theory independently. A. H. Compton, *Phil. Mag.* xlvi (1923), p. 897 ; *J. Frank. Inst.* cxcviii (1924), p 57. G. E. M. Jauncey, *Phys. Rev.* xxii (1923), p 233. P. A. Ross, *Proc. N.A.S.* ix (1923), p. 246 ; x (1924), p. 304 ; *Phys. Rev.* xxii (1923), p. 524. M. de Broglie, *Proc. Phys. Soc.* xxxvi (1924), p. 423. D. Skobelzyn, *ZS. f. P.* xxiv (1924), p. 393 ; xxviii (1924), p. 278. A. H. Compton and A. W. Simon, *Phys. Rev.* xxv (1925), p. 306

Eliminating ν' and ν, we have

$$\left(1+\frac{h\nu}{mc^2}\right)\tan\frac{\phi}{2}=\cot\theta.$$

The kinetic energy of the recoiling electron is

$$\frac{h\nu\dfrac{2h\nu}{mc^2}\cos^2\theta}{\left(1+\dfrac{h\nu}{mc^2}\right)^2-\dfrac{h^2\nu^2}{m^2c^4}\cos^2\theta}.$$

The quantity (h/mc), which has the dimensions of a length, is called the *Compton wave-length.* Its value is $0{\cdot}0242\times10^{-8}$ cm. Since the mass of a quantum is $h\nu/c^2$ or $h/\lambda c$, it is seen at once that *a quantum of radiation, whose wave-length is the Compton wave-length, has a mass equal to the mass of the electron.*

The recoil electrons of the Compton effect were studied by C. T. R. Wilson,[1] using his cloud-expansion method. He found that X-radiation of wave-length less than about 0·5Å in air produced two classes of β-ray tracks, namely (*a*), those of electrons ejected with initial kinetic energy comparable to a quantum of the incident radiation: these were photo-electrons: and (*b*), tracks of very short range, which were the recoil electrons. A. H. Compton and J. C. Hubbard,[2] discussing Wilson's results, showed that the motion of the recoil electrons corresponds precisely to Compton's theory.

We have seen that in the spectrum of scattered X-rays there are, in addition to the lines corresponding to the Compton effect, lines corresponding exactly to the primary X-rays. These unshifted lines (which are the only lines to appear when the primary rays are those of visible light) may be explained by supposing that some electrons are closely attached to the nucleus and must scatter while nearly at rest. The theory of this state of affairs was investigated by Compton,[3] who also explained certain results which had been obtained by Duane and his collaborators,[4] and which appeared to be inconsistent with the original Compton theory.

A more accurate treatment of the Compton effect, making use of later developments in general theory, was given in 1929 by

[1] *Proc. R.S.*(A), civ (1923), p. 1; cf. also W. Bothe of Charlottenburg, *ZS. f. P.* xvi (1923), p. 319; xx (1923), p. 237

[2] *Phys. Rev.* xxiii (1924), p. 439

[3] *Phys. Rev.* xxiv (1924), p. 168; *Nature*, cxiv (1924), p. 627

[4] G. L. Clark and W. Duane, *Proc. N.A.S.* ix (1923), pp. 413, 419; x (1924), pp. 41, 92; xi (1925), p. 173. G. L. Clark, W. W. Stifler and W. Duane, *Phys. Rev.* xxiii (1924), p. 551. A. H. Armstrong, W. Duane and W. W. Stifler, *Proc. N.A.S.* x (1924), p. 374. S. K. Allison, G. L. Clark and W. Duane, *Proc. N.A.S.* x (1924), p. 379. J. A. Becker, *Proc. N.A.S.* x (1924), p. 342. S. K. Allison and W. Duane, *Proc. N.A.S.* xi (1925), p. 25. cf. P. A. Ross and D. L. Webster, *Proc. N.A.S.* xi (1925), pp. 56, 61. A. H. Compton and J. A. Bearden, *Proc. N.A.S.* xi (1925), p. 117

O. Klein and Y. Nishina.[1] The formulae they obtained have been found experimentally to be accurate even with the hardest type of radiation.[2]

The Compton effect raised in an acute form the controversy regarding the reality of light-darts as contrasted with spherical waves of light: for in Compton's explanation, the incident and diffracted X-ray quanta were supposed to have definite directions of propagation. Opinion among theoretical physicists was divided: 'In a recent letter to me,' wrote A. H. Compton in 1924[3] 'Sommerfeld has expressed the opinion that this discovery of the change of wave-length of radiation, due to scattering, sounds the death knell of the wave theory. On the other hand, the truth of the spherical wave hypothesis indicated by interference experiments has led Darwin and Bohr, in conversation with me, to choose rather the abandonment of the principles of conservation of energy and momentum.' The latter policy was embodied in a hypothesis put forward in 1924 by N. Bohr, H. A. Kramers and J. C. Slater,[4] in which it was accepted, that in atomic processes, energy and momentum are only statistically conserved.

They abandoned the principle, common to all previous physical theories, that an atom which is emitting or absorbing radiation must be losing or gaining energy: in its place they introduced the notion of *virtual radiation*, which is propagated in spreading waves as in the electromagnetic theory of light, but which does not transmit energy or momentum: it has indeed no connection with physical reality except the capacity to generate in atoms a probability for the occurrence of transitions: and transitions of the atoms are the only phenomena actually observable. A transition of an atom from one state to another is accompanied by changes of energy and momentum, but is *not* accompanied by radiation: thus the part played by the atom in its relations with radiation reduces to interaction with the field of virtual radiation, while the atom remains in a stationary state. An atom in a stationary state is continually emitting virtual radiation, compounded of all the frequencies corresponding to possible transitions between this state and lower states: this radiation is emitted both spontaneously and by stimulation (in accordance with Einstein's principles of 1917). While in this state, the atom is also capable of *absorbing* radiation corresponding to transitions to states of higher energy. The absorption is performed by *virtual oscillators* situated in the atoms, the frequencies of these oscillators corresponding to the energy-differences between the state of the atom and all

[1] *ZS. f. P.* lii (1929), p. 853

[2] cf. H. C. Trueblood and D. H. Loughridge, *Phys. Rev.* liv (1938), p. 545; Z. Bay and Z. Szepesi, *ZS. f. P.* cxii (1939), p. 20

[3] *J. Frank. Inst.* cxcviii (1924), p. 57

[4] J. C. Slater, *Nature*, cxiii (1924), p. 307. N. Bohr, H. A. Kramers, and J. C. Slater, *Phil. Mag.* xlvii (1924), p. 785; *ZS. f. P.* xxiv (1924), p. 69. J. H. Van Vleck, *Phys. Rev.* xxiv (1924), p. 330. R. Becker, *ZS. f. P.* xxvii (1924), p. 173. J. C. Slater, *Phys. Rev.* xxv (1925), p. 395

higher states. When a virtual oscillator is absorbing virtual radiation, the atom to which it belongs has a certain *probability* of making a transition to the higher state corresponding to the frequency of this virtual oscillator. A transition marks the change from the continuous radiation appropriate to the old state, to the continuous radiation appropriate to the new state: simultaneously with the transition, some virtual oscillators disappear and others come into being: the transition has no other influence on the radiation. The occurrence of a transition in a given atom depends on the initial state of this atom and on the states of the atoms with which it is in communication through the field of virtual radiation, but not on transition processes in the latter atoms: so there is no *direct connection* between the transition of one atom from a higher to a lower state, and the transition of another atom from a lower to a higher state: the principles of energy and momentum are retained in a *statistical* sense, though not in individual interactions of atoms with radiation. The atoms scatter radiation which is incident on them, acting as secondary sources of virtual radiation which interferes with the incident radiation. In any transition, say between states (p) and (q), the energy of the atom changes by $h\nu_{pq}$ and its momentum by $h\nu_{pq}/c$. If the transition is a spontaneous one, the direction of this momentum is random: but if it is stimulated, i.e. induced by the surrounding virtual radiation, the direction of the momentum is the same as that of the wave propagation in this virtual field.

The Bohr-Kramers-Slater theory was wrecked when it was shown to be inconsistent with the results of more refined experiments relating to the Compton effect. One of these was performed by W. Bothe and Hans Geiger.[1] According to Compton's theory, a recoil electron is emitted simultaneously with the scattering of every quantum; while according to the Bohr-Kramers-Slater theory, the connection was much less close, the recoil electrons being emitted only occasionally, while the scattering of virtual radiation is continuous. In Bothe and Geiger's experiment two different Geiger counters counted respectively the recoil electrons, and the photo-electrons produced by the scattered photons. A great many coincidences in time were observed, so many that the probability of their occurrence on the Bohr-Kramers-Slater theory was only 1/400,000. It was therefore concluded that the conservation of energy and momentum holds in individual encounters, and the Bohr-Kramers-Slater theory could not be true.

Another experiment was performed by A. H. Compton and A. W. Simon,[2] who remarked that if in a Wilson cloud-experiment the quantum of scattered X-rays produces a photo-electron in the chamber, then a line drawn from the beginning of the recoil track to the beginning of the track of the photo-electron gives the direction

[1] *ZS. f. P.* xxvi (1924), p. 44; xxxii (1925), p. 639
[2] *Phys. Rev.* xxvi (1925), p. 289

of the quantum after scattering. It was therefore possible to test the truth of the equation

$$\left(1+\frac{h\nu}{mc^2}\right)\tan\frac{\phi}{2}=\cot\theta$$

which connects the directions of the scattered quantum and the recoil electron. If the energy of the scattered X-rays were propagated in spreading waves of the classical type, there would be no correlation whatever between the directions in which the recoil electrons proceed and the directions of the points at which the photo-electrons are ejected by the scattered photons. The results of Compton and Simon's experiment showed that the scattered photons proceed in definite directions and that the above equation connecting ϕ and θ is true.[1] And this result, like that of Bothe and Geiger, is fatal to the Bohr-Kramers-Slater hypothesis.

The discovery of the Compton effect opened a new prospect of solving a problem which had for some years baffled theoretical physicists. The thermal equilibrium between radiation and electrons in a reflecting enclosure had been investigated by H. A. Lorentz[2] and A. D. Fokker[3] on the basis of classical electrodynamics: and they had shown that Planck's law of spectral distribution of radiation could not persist in such a reflecting enclosure, if an electron were present: and, moreover, that if the Planck distribution were artificially maintained, the electron could not maintain the amount $\frac{3}{2}kT$ of mean translational kinetic energy required by the statistical theory of heat: so the classical theory failed to account for the interaction of pure-temperature radiation with free electrons. W. Pauli now,[4] basing his investigation on the work of Compton and Debye, attacked the problem of finding a quantum-theoretic mechanism for the interaction of radiation with free electrons, which should satisfy the thermodynamic requirement that electrons with the Maxwellian distribution of velocities can be in equilibrium with radiation whose spectral distribution is determined by Planck's radiation-formula. He found that the probability of a Compton interaction between a photon $h\nu$ and an electron could be represented as the sum of two expressions, one of which was proportional to the radiation-density of the primary frequency ν, while the other was proportional to the product of this radiation-density and the radiation-density of the frequency ν' which arises through the Compton process. The latter term was puzzling from the philosophic point of view, since it seemed to imply that the probability of something happening depended on something that had not yet happened. However, it was shown by W. Bothe[5] that Pauli's second term was a mistake, arising from

[1] cf. R. S. Shankland, *Phys. Rev.* lii (1937), p. 414
[2] At the Solvay conference in Brussels in 1911
[3] *Diss.*, Leiden, 1913; *Arch. Néerl*, iv (1918), p. 379
[4] *ZS. f. P.* xviii (1923), p. 272
[5] *ZS. f. P.* xxiii (1924), p. 214

Pauli's assumption that the photons scattered all have the energy $h\nu$: whereas, as we have seen earlier,[1] in pure-temperature radiation there are also definite proportions of photons having the energies $2h\nu$, $3h\nu$, etc.: if it is assumed that each of these is scattered as a whole, in exactly the same way as the photons of energy $h\nu$, then Pauli's second term falls out, and the theory becomes much simpler.[2]

In the latter half of 1923 Louis de Broglie introduced a new conception which proved to be of great importance in quantum theory. The analogy of Fermat's Principle in Optics with the Principle of Least Action in Dynamics suggested to him the desirability of studying more profoundly the parallelism between corpuscular dynamics and the propagation of waves, and attaching to it a physical meaning. He developed this idea first in a series of notes in the *Comptes Rendus*,[3] then in a doctorate thesis sustained in 1924,[4] and in other papers.[5]

In de Broglie's theory, with the motion of any electron or material particle there is associated a system of plane waves, such that the velocity of the electron is equal to the group-velocity[6] of the waves. Let m be the mass and v the velocity of the particle. It is assumed that the frequency ν of the waves is given by Planck's relation

$$E = h\nu,$$

when

$$E = \frac{mc^2}{\sqrt{\left(1 - \frac{v^2}{c^2}\right)}}$$

is the kinetic energy of the particle, so

$$h\nu = \frac{mc^2}{\sqrt{\left(1 - \frac{v^2}{c^2}\right)}}. \tag{1}$$

Since v is equal to the group-velocity of the waves, we have

$$v = \frac{d\nu}{d(1/\lambda)} \tag{2}$$

where λ is the wave-length of the waves, so that $\lambda = V/\nu$ where V is the phase-velocity of the waves.

[1] cf. p. 103

[2] A simple treatment of the equilibrium between a Maxwellian distribution of atoms, and radiation obeying Planck's law, is given by P. Jordan, *ZS. f. P.* xxx (1924), p. 297.

[3] *Comptes Rendus*, clxxvii (10 Sept. 1923), p. 507; ibid. (24 Sept. 1923), p. 548; ibid. (8 Oct. 1923), p. 630; clxxix (7 July 1924), p. 39; ibid. (13 Oct. 1924), p. 676; ibid. (17 Nov. 1924), p. 1039

[4] *Thèse*, Paris, Edit. Musson et Cie., 1924

[5] *Phil. Mag.* xlvii (Feb. 1924), p. 446; *Annales de phys.*(10) iii (1925), p. 22.

[6] cf. Vol. I, p. 253, *note* 4. For the purpose of calculating group-velocity, ν is regarded as a function of λ with c and m/h as fixed constants.

From (1) and (2) we have

$$d\left(\frac{1}{\lambda}\right)=\frac{mc^2}{hv}\,d\left(1-\frac{v^2}{c^2}\right)^{-\frac{1}{2}}=\frac{m}{h}\left(1-\frac{v^2}{c^2}\right)^{-\frac{3}{2}}dv=\frac{1}{h}dp$$

where

$$p=\frac{mv}{\sqrt{\left(1-\frac{v^2}{c^2}\right)}}$$

is the momentum of the particle. Integrating, we have $1/\lambda=p/h$, or $\lambda=h/p$. *This equation gives the wave-length of the de Broglie wave associated with a particle of momentum p.*

The phase-velocity V of the de Broglie wave is

$$V=\lambda\nu=\frac{h}{p}\cdot\frac{E}{h}=\frac{E}{p}$$

or

$$V=\frac{c^2}{v}$$

an equation which gives the phase-velocity of the de Broglie wave.[1]

Now a wave-motion of frequency $E/2\pi\hbar$ and of wave-length $2\pi\hbar/p$, where p has the components (p_x, p_y, p_z), is represented by a wave-function

$$\psi=\exp.\left\{\frac{i}{\hbar}(Et-p_xx-p_yy-p_zz)\right\}.$$

[1] The following derivation of de Broglie's result was given by Einstein, *Berlin Sitz.* (1925), p. 3.

A material particle of mass m is first correlated to a frequency ν_0 conformably to the equation

$$mc^2=h\nu_0$$

The particle now rests with respect to a Galilean system K′, in which we imagine an oscillation of frequency ν_0 everywhere synchronous. Relative to a system K′, with respect to which K′ with the mass m is moved with velocity v along the positive X-axis, there exists a wave-like process of the kind

$$\sin\left(2\pi\nu_0\frac{t-\frac{v}{c^2}x}{\sqrt{\left(1-\frac{v^2}{c^2}\right)}}\right).$$

The frequency ν and phase-velocity V of this process are thus given by

$$\nu=\frac{\nu_0}{\sqrt{\left(1-\frac{v^2}{c^2}\right)}},\qquad V=\frac{c^2}{v}.$$

The relation

$$\text{(phase velocity)}\times\text{(group velocity)}=c^2$$

was shown by R. W. Ditchburn, *Revue optique* xxvii (1948), p. 4, and J. L. Synge, *Rev. Opt.* xxxi (1952), p. 121, to be a necessary consequence of relativity theory.

This is the analytical expression of the de Broglie wave associated with a particle whose kinetic energy is E *and whose momentum is* (p_x, p_y, p_z).

But a wave of ordinary light of frequency ν in the direction (l, m, n) is represented by a wave-function

$$\psi = \exp.\left[2\pi i\nu\left\{t - \left(\frac{1}{c}\right)(lx + my + nz)\right\}\right]$$

or

$$\psi = \exp.\left\{\frac{i}{\hbar}(\mathrm{E}t - p_x x - p_y y - p_z z)\right\}$$

when E denotes the energy $h\nu$ and (p_x, p_y, p_z) the momentum $h\nu/c$ of the corresponding photon. Comparing this with the above expression for the de Broglie wave, we see that *an ordinary wave of light is simply the de Broglie wave belonging to the associated photon.* It follows that if a de Broglie wave is regarded as a quantum effect, then *the interference and diffraction of light must be regarded as essentially quantum effects.* It is, in fact, a mistake to speak of the wave-theory of light as the 'classical' theory: that it is usually so called is due to the historical accident that the wave-theory of light happened to be discovered before the photon theory, which is the corpuscular theory. When interference is treated by the corpuscular theory (Duane's method, p. 142), then its quantum character is shown by the fact that quantum jumps of momentum make their appearance. Moreover, the interference and diffraction of light are evidently of the same nature as the interference and diffraction of electron beams and of molecular rays, and the latter phenomena are undoubtedly quantum effects.

The principle of Fermat applied to the wave may be shown to be identical with the principle of Least Action applied to the particle; in fact, $\delta\int ds/\lambda = 0$ is equivalent to $\delta\int p\,ds = 0$ if p is a constant multiple of $1/\lambda$.

De Broglie now showed that his theory provided a very simple interpretation of Bohr's quantum condition for stationary states of the hydrogen atom. That condition was, that the angular momentum of the atom should be a whole-number multiple of $\hbar$ or $h/2\pi$, say $nh/2\pi$. But if r denotes the radius of the orbit and p the linear momentum of the electron, the angular momentum is rp: so the condition becomes

$$2\pi rp = nh$$

or

$$2\pi r = n\lambda$$

where λ is the wave-length of the de Broglie wave associated with the electron. This equation, however, means simply that *the circumference of the orbit of the electron must be a whole-number multiple of the wave-length of the de Broglie wave.*

More generally, the Wilson-Sommerfeld quantum condition is that the Action must be a multiple of h. But the Action is $\int p\,ds$ where p denotes the momentum and ds the element of the path: so $h\int ds/\lambda$ must be a multiple of h, where λ is the de Broglie wave-length: or, $\int ds/\lambda$ must be a whole number. We can express this by saying that *the de Broglie wave must return to the same phase when the electron completes one revolution of its orbit.*

The connection between the Wilson-Sommerfeld condition and the wave-theory can be seen also in the case of the diffraction of light or electron beams or corpuscular rays by an infinite reflecting plane grating. The solution of this problem by the corpuscular theory (Duane's method) depends on the Wilson-Sommerfeld condition, which yields (cf. p. 143)

$$(pd/h)\ (\sin i - \sin r) = \text{a whole number},$$

where p now denotes the total momentum of the quantum of light. The solution by the wave-theory of Young and Fresnel, on the other hand, depends on the principle that $(1/2\pi)$ times the difference in phase between the rays reflected from adjacent spacing-intervals must be a whole number, and this at once gives

$$(d/\lambda)\ (\sin i - \sin r) = \text{a whole number}.$$

These two equations are identical if $(p/h) = (1/\lambda)$, which is assured by de Broglie's relation.

In July 1925 a research student named Walter Elsasser, working under James Franck at Göttingen, made an important contribution[1] to the theory. Franck had been told by his colleague Max Born of an investigation made in America by Clinton J. Davisson and C. H. Kunsman,[2] who had studied the angular distribution of electrons reflected at a platinum plate, and had found at certain angles strong maxima of the intensity of the electronic beam. Born, who knew of de Broglie's theory, mentioned it in this connection, and Franck proposed to Elsasser that he should examine the question whether Davisson's maxima could be explained in some way by de Broglie's waves. Elsasser showed that they could in fact be interpreted as an effect due to interference of the waves. There were strong maxima, which with increasing electron-velocity approximated to the positions of the maxima which would be observed if light of the wave-length given by de Broglie's law, namely, $\lambda = h/mv$ were diffracted at an optical plane grating, the constants of the grating being those of platinum crystals. This was the first confirmation of de Broglie's theory which was based on comparison

[1] *Naturwiss*, xiii (1925), p. 711 [2] *Phys. Rev.* xxii (1923), p. 242

with experiment. Elsasser remarked further that de Broglie's theory provided solutions for some other puzzles of current physics. He discussed an effect discovered by C. Ramsauer[1] in Germany, and independently but later by J. S. Townsend and V. A. Bailey[2] at Oxford, and by R. N. Chaudhuri[3] working in O. W. Richardson's laboratory in London, namely, that the mean free path of an electron in the inert gases becomes exceedingly long when the velocity of the electron is reduced: thus when an electron moving with a velocity of about 10^8 cm./sec. collides with a molecule of a non-inert gas, in general it loses more than one per cent of its energy, but when it collides with a molecule of argon, it loses only about one ten-thousandth part of its energy: and the mean free path of such an electron in argon is about ten times as long as that calculated from the kinetic theory. Elsasser explained this by showing that when slow electrons are scattered by atoms of the inert gases, the effect follows the same laws as the classical scattering of radiation, of the associated de Broglie wave-length, by small spheres whose radius is the same as that of the atom.

Elsasser further suggested that electrons reflected at a single large crystal of some substance might show the diffraction-effect of the de Broglie waves decisively.[4] The phenomenon thus predicted was found experimentally in 1927 by Clinton J. Davisson and Lester H. Germer[5] of the Bell Telephone Co., who found that a beam of slow electrons, reflected from the face of a target cut from a single crystal of nickel, gave well-defined beams of scattered electrons in various directions in front of the target: in fact, diffraction phenomena were observed precisely similar to those obtained with X-rays, of a wave-length connected with the momentum of the electrons by de Broglie's formula.

In Nov.–Dec. 1927, George Paget Thomson[6] examined the scattering of cathode rays by a very thin metallic film, which could be regarded as a microcrystallic aggregate, and confirmed the fact that a beam of electrons behaves like a wave: from the size of the rings in the diffraction-pattern it was possible to deduce the wave-length of the waves causing them, and in all cases he obtained the value $\lambda = h/p$. If a stream of electrons is directed at a screen in which

[1] *Ann. d. Phys.* lxiv (1921), p. 451; lxvi (1921), p. 546; lxxii (1923), p. 345

[2] *Phil. Mag.* xliii (1922), p. 593 [3] *Phil. Mag.* xlvi (1923), p. 461

[4] Another experimental investigation which could be explained by de Broglie's theory was that of E. G. Dymond, *Nature*, cxviii (1926), p. 336, on the scattering of electrons in helium; the moving electrons could be associated with plane de Broglie waves, whose interference governed the scattering. I. Langmuir, *Phys. Rev.* xxvii (1906), p. 806 had shown that inelastic collisions in several gases lead to very small angles of scattering.

[5] *Phys. Rev.* xxx (1927), p. 707; *Nature*, cxix (16 April 1927), p. 558; *Proc. N.A.S.* xiv (1928), p. 317; cf. K. Schaposchnikow, *ZS. f. P.* lii (1928), p. 451; E. Rupp, *Ann. d. Phys.* i (1929), p. 801; C. J. Davisson, *J. Frank. Inst.*, ccv (1928), p. 597

[6] G. P. Thomson and A. Reid, *Nature*, cxix (1927), p. 890. G. P. Thomson, *Proc. R.S.*(A), cxvii (1928), p. 600; cxxviii (1930), p. 641. A. Reid, *Proc. R.S.*(A), cxix (1928), p. 663. R. Ironside, ibid., p. 668. S. Kikuchi, *Proc. Tokyo Ac.* iv (1928), pp. 271, 354, 471

there are two small holes close together, an interference-pattern is produced just as with light.

The experimental verification of de Broglie's formula was further extended by E. Rupp,[1] who succeeded in obtaining electronic diffraction by a ruled grating.

The electron-energies for which the formula has been verified range from 50 to 10^6 electron-volts.[2]

These effects have been observed not only with streams of electrons (cathode rays and the β-rays from radio-active sources), but also with streams of material particles: I. Estermann and O. Stern[3] in 1930 found that molecular rays of hydrogen and helium, impinging on a crystal face of lithium fluoride, were diffracted, giving a distribution of intensity corresponding to the spectra formed by a crossed grating. The wave-length, calculated from the known constants of the crystal, was in agreement with de Broglie's formula.

Theoretical papers on the diffraction of electrons at crystals were published not long after the Davisson-Germer experiment by Hans Bethe,[4] a pupil of Sommerfeld, and C. G. Darwin.[5]

The next notable advance in physical theory was made by Satyandra Nath Bose[6] of Dacca University, in a short paper giving a new derivation of Planck's formula of radiation: Einstein, who seems to have translated it into German from an English manuscript sent to him by Bose, recognised at once its importance and its connection with de Broglie's theory.

Bose regarded the radiation as composed of photons, which for statistical purposes could be treated like the particles of a gas, but with the important difference that photons are indistinguishable from each other, so that instead of considering the allocation of individual distinguishable photons among a set of states, he fixes attention on the number of states that contain a given number of photons. He assumes that the total energy E of the photons is given, and that they are contained in an enclosure of unit volume.

A photon $h\nu$ may be specified by its co-ordinates (x, y, z) and the three components of its momentum (p_x, p_y, p_z). Since the total momentum is $h\nu/c$, we have

$$p_x^2 + p_y^2 + p_z^2 = r^2 \quad \text{where} \quad r = h\nu/c.$$

Let volume in the six-dimensional space of (x, y, z, p_x, p_y, p_z) be

[1] *ZS. f. P.* lii (1928), p. 8; *Phys. ZS.* xxix (1928), p. 837

[2] J. V. Hughes, *Phil. Mag.* xix (1935), p. 129

[3] *ZS. f. P.* lxi (1930), p. 95: cf. T. H. Johnson, *Phys. Rev.* xxxi (1928), p. 103: F. Knauer and O. Stern, *ZS. f. P.* ciii (1929), p. 779

[4] *Ann. d. Phys.* lxxxvii (1928), p. 55

[5] *Proc. R.S.*(A), cxx (1928), p. 631. This paper is concerned with the polarisation of electron-waves, on which, see also C. Davisson and L. H. Germer, *Phys. Rev.* xxxiii (1929), p. 760, and E. Rupp, *ZS. f. P.* liii (1929), p. 548.

[6] *ZS. f. P.* xxvi (1924), p. 178

called *phase-space.* Then to the frequency-range from ν to $\nu+d\nu$ there corresponds the phase-space

$$\int dx\,dy\,dz\,dp_x\,dp_y\,dp_z = 4\pi r^2 dr = 4\pi\,\frac{h^3\nu^2}{c^3}\,d\nu.$$

Bose now assumes that this phase-space is partitioned into cells of volume h^3, so there are $4\pi\nu^2 d\nu/c^3$ cells. In order to take account of polarisation we must double the number, so we obtain $8\pi\nu^2 d\nu/c^3$ cells.

Now let N_s be the number of photons in the frequency-range $d\nu_s$, and consider the number of ways in which these can be allocated among the cells belonging to $d\nu_s$. Let p_0^s be the number of vacant cells, p_1^s the number of cells that contain one photon, p_2^s the number of cells that contain two photons etc. Then the number of possible ways of choosing a set of p_0^s cells, a set of p_1^s cells, etc. out of a total of $8\pi\nu^2 d\nu/c^3$ cells is

$$\frac{A_s!}{p_0^s\,!\,p_1^s!\,p_2^s\,!\,\ldots}, \quad \text{where} \quad A_s = \frac{8\pi\nu_s^2}{c^3}\,d\nu_s,$$

and we have

$$N_s = \sum_r r p_r^s.$$

As the fundamental assumption of his statistics, Bose assumes that if a particular quantum state is considered, then all values for the number of particles in that state are equally likely, so the probability of any distribution specified by the p_r^s is measured by the number of different ways in which it can be realised. Hence the probability of the state specified by the p_r^s (now taking into account the whole range of frequencies) is

$$W = \prod_s \frac{A_s!}{p_0^s!\,p_1^s!\,p_2^s!\,\ldots}.$$

Since the p_r^s are large, we can use Stirling's approximation

$$\log n! = n \log n - n$$

so $$\log W = \sum_s A_s \log A_s - \sum_s \sum_r p_r^s \log p_r^s, \quad \text{since} \quad A_s = \sum_r p_r^s.$$

This expression is to be a maximum subject to the condition

$$E = \sum_s N_s h\nu_s, \quad \text{when} \quad N_s = \sum_r r p_r^s.$$

The usual conditions become in this case

$$\sum_s \sum_r \delta p_r^s (1 + \log p_r^s) = 0, \qquad \sum_s \delta N_s . h\nu_s = 0,$$

where

$$\delta N_s = \sum_r r\delta p_r^s, \qquad \sum_r \delta p_r^s = 0,$$

which give

$$\sum_s \sum_r \delta p_r^s \left\{ (1 + \log p_r^s + \lambda^s) + \frac{rh\nu_s}{\beta} \right\} = 0$$

where β and the λ^s are constants : so

$$p_r^s = B_s e^{-\frac{rh\nu_s}{\beta}}$$

where the B_s are constants.

Therefore

$$A_s = \sum_r p_r^s = \sum_r B_s e^{-\frac{rh\nu_s}{\beta}} = B_s \left(1 - e^{-\frac{h\nu_s}{\beta}}\right)^{-1}$$

or

$$B_s = A_s \left(1 - e^{-\frac{h\nu_s}{\beta}}\right)$$

while $N_s = \sum_r r p_r^s = A_s \sum_r r e^{-\frac{rh\nu_s}{\beta}} \left(1 - e^{-\frac{h\nu_s}{\beta}}\right) = A_s e^{-\frac{h\nu_s}{\beta}} \left(1 - e^{-\frac{h\nu_s}{\beta}}\right)^{-1}.$

Thus

$$E = \sum_s N_s h\nu_s = \sum_s \frac{8\pi h\nu_s^3}{c^3} \; e^{-\frac{h\nu_s}{\beta}} \left(1 - e^{-\frac{h\nu_s}{\beta}}\right)^{-1} d\nu_s.$$

The entropy is

$$S = k \log W$$

which gives

$$S = k \left\{ \frac{E}{\beta} - \sum_s A_s \log \left(1 - e^{\frac{h\nu_s}{\beta}}\right) \right\}.$$

Since $\partial S / \partial E = 1/T$ where T denotes the absolute temperature, we have

$$\beta = kT$$

so

$$E = \sum_s \frac{8\pi h\nu_s^3}{c^3} \; \frac{1}{e^{\frac{h\nu_s}{kT}} - 1} d\nu_s$$

which is equivalent to Planck's formula.

Bose's paper therefore showed that in order to obtain Planck's law of radiation, we must assume that photons obey a particular kind of statistics. This point may be illustrated by an analogy, as follows. Consider an empty railway train standing at a platform, with passengers on the platform who get into the train ; and suppose that when they have all taken their places, p_0 compartments are vacant, p_1 compartments have one passenger apiece, p_2 compartments have two passengers apiece, and so on : so that if A is the number of compartments in the train, we have

$$A = p_0 + p_1 + p_2 + - \ldots,$$

and if N is the number of passengers, we have

$$N = 0 \cdot p_0 + 1 \cdot p_1 + 2 \cdot p_2 + \ldots .$$

For simplicity, we shall assume that comparatively few people are travelling, so that the number of compartments is greater than the number of passengers. Let us inquire, what is the probability of this particular distribution specified by the numbers $(p_0, p_1, p_2, \ldots)$. Evidently the probability depends on the assumption that we make regarding the motives which influence passengers in their choice of a compartment. Three such assumptions are as follows :

(i) We might assume that each passenger chooses his compartment at random, without regard to whether there are already any other passengers in it, or not.

(ii) We might assume that each passenger likes to have a compartment to himself, so he refuses to enter any compartment which already has an occupant.

(iii) We might assume that among the passengers there are small family parties whose members wish to be together in the same compartment, so that if we know that at least one place in a compartment is occupied, there is a certain probability (arising from this fact) that other seats in it will also be occupied. We may regard each family party or unattached traveller as a unit, and assume that each unit chooses its compartment at random, without regard to whether there are already other passengers in it, or not.

It is evident that these three different assumptions will give quite different values for the probability of any particular distribution $(p_0, p_1, p_2, \ldots)$: this we express by saying that they give rise to *different statistics*. The difference between *classical statistics* and the different kinds of *quantum statistics* may be illustrated by this analogy.

Bose's discovery was immediately extended by Einstein,[1] to the study of a monatomic ideal gas. The difference between Bose's

[1] *Berlin Sitz.* (1924), p. 261 ; (1925), pp. 3, 18

photons and Einstein's gas-particles is that for photons the energy is c times the momentum, whereas for particles a different equation holds: and, moreover, in Bose's problem the total energy is fixed but the number of photons is not fixed, whereas in Einstein's problem the total number of particles is definite. These differences, however, do not affect the general plan of the investigation. The analysis leads to the following conclusions: the average number of particles of mass m in unit volume with energies in the range ϵ to $\epsilon + d\epsilon$ is

$$\frac{2\pi}{h^3}\frac{(2m)^{\frac{3}{2}}\epsilon^{\frac{1}{2}}d\epsilon}{e^{\epsilon/kT+\mu}-1} \tag{1}$$

where μ is a constant: whence *the total number of particles in unit volume is*

$$n=\frac{2\pi}{h^3}(2m)^{\frac{3}{2}}\int_0^\infty\frac{\epsilon^{\frac{1}{2}}d\epsilon}{e^{\epsilon/kT+\mu}-1}=\frac{2\pi}{h^3}(2mkT)^{\frac{3}{2}}\int_0^\infty\frac{x^{\frac{1}{2}}dx}{e^{x+\mu}-1} \tag{2}$$

and *the total energy in unit volume is*

$$E=\frac{2\pi}{h^3}(2m)^{\frac{3}{2}}\int_0^\infty\frac{\epsilon^{\frac{3}{2}}d\epsilon}{e^{\epsilon/kT+\mu}-1}=\frac{2\pi}{h^3}(2m)^{\frac{3}{2}}(kT)^{\frac{5}{2}}\int_0^\infty\frac{x^{\frac{3}{2}}dx}{e^{x+\mu}-1}. \tag{3}$$

These are the fundamental formulae of what is generally called *Bose-Einstein* statistics.

Now consider the relation between these formulae and the formulae of the classical (Maxwellian) kinetic theory of gases. We should expect the classical formulae to be obtained as the limiting case when $h\to 0$, in which case it is evident from (1) that μ must tend to infinity in such a way that h^3e^μ has a finite value, say λ. From (2) we then have, in this limiting case,

$$n=\frac{2\pi}{\lambda}(2mkT)^{\frac{3}{2}}\int_0^\infty x^{\frac{1}{2}}e^{-x}\,dx=\frac{\pi^{\frac{3}{2}}}{\lambda}(2mkT)^{\frac{3}{2}} \tag{4}$$

while (1) states in the limiting case that the total number of particles in unit volume with energies between ϵ and $\epsilon + d\epsilon$ is

$$\frac{2\pi}{\lambda}(2m)^{\frac{3}{2}}e^{-\epsilon/kT}\epsilon^{\frac{1}{2}}\,d\epsilon$$

or, by (4), out of a total of n particles, the number with energies between ϵ and $\epsilon + d\epsilon$ is

$$\frac{2n}{\pi^{\frac{1}{2}}(kT)^{\frac{3}{2}}}e^{-\epsilon/kT}\epsilon^{\frac{1}{2}}\,d\epsilon$$

which is precisely the Maxwellian formula.

Moreover (3) becomes in this limiting case

$$E = \frac{2\pi}{\lambda} . (2m)^{\frac{3}{2}} . (kT)^{\frac{5}{2}} . T(\tfrac{5}{2}) = \tfrac{3}{2} nkT$$

which is the Maxwellian formula for the total energy.

We see that in Bose statistics, the slower molecules are more numerous, as compared with the faster ones, than is the case in Maxwell's theory.

The similarity in statistical behaviour between Bose's photons and the particles of a gas, revealed by this investigation, was examined further by Einstein. He pointed out [1] that de Broglie's discovery made it possible to correlate to any system of material particles a (scalar) wave-field: and he showed the close connection between the fluctuations of energy in systems of waves and in systems of particles. We have seen [2] that the mean-square of the fluctuations of energy per unit volume in the frequency-range from ν to $\nu + d\nu$ in the radiation in an enclosure at temperature T is

$$\overline{\epsilon^2} = h\nu E + \frac{c^3 E^2}{8\pi\nu^2 d\nu},$$

the first term representing the fluctuation in the number of molecules in unit volume of an ideal gas on the classical theory, when each molecule has energy $h\nu$. Einstein now showed that when the gas-particles are assumed to satisfy the Bose-Einstein statistics, *both* terms appear. In other words, a Bose-Einstein gas differs from a Maxwellian gas in precisely the same way as radiation obeying Planck's formula differs from radiation obeying the law of Wien.

The next advance in the theory of quantum statistics was made by Enrico Fermi [3] (*b.* 1901). He remarked that in the Maxwellian kinetic theory of gases, the average kinetic energy per molecule is $\frac{3}{2}kT$, and hence the molecular heat at constant volume (i.e. the heat that must be communicated to one gramme-molecule in order that its temperature may be raised one degree, the volume remaining unchanged), calculated from this theory, is $c_v = \frac{3}{2}R$, where R is the gas-constant. If, however, Nernst's thermodynamical law, which requires $(dE/dT) \to 0$ as $T \to 0$, is applicable to an ideal gas, then c_v must vanish in the limit when $T \to 0$, and therefore (as Einstein had remarked in 1906) the Maxwellian theory cannot be true at very low temperatures. The reason for this must, he argued, be sought in the quantification of molecular motions, and this quantification be made to depend on Pauli's exclusion principle, that one system can

[1] At page 9 of the first paper in *Berlin Sitz.* (1925). [2] cf. p. 101 *supra*

[3] *Lincei Rend.* iii (7 Feb. 1926), p. 145; *ZS. f. P.* xxxvi (1926), p. 902. A contribution of great importance was made by P.A.M. Dirac somewhat later in the year, *Proc. R.S.*(A), cxii (1926), p. 661, on account of which the type of statistics introduced by Fermi is often called the Fermi-Dirac statistics; but as Dirac's paper involves the ideas of wave-mechanics, its description is postponed for the present.

never contain two elements of equal value with exactly the same set of quantum numbers. Thus he asserted that at most one molecule with specified quantum numbers could be present in an ideal gas : where by quantum numbers he understood not only those which relate to the internal motions of the molecules, but also those which specify the molecule's motion of translation.

The quantification may be performed as follows : suppose the gas is contained in a cubical vessel whose edge is of length l, so that the possible quantum values of the components of momentum of the particle are $p_x = (s_1 h/l)$, $p_y = (s_2 h/l)$, $p_z = (s_3 h/l)$, where s_1, s_2, s_3 are whole numbers or zero ; then Pauli's principle asserts that in the whole gas there can be at most one particle with specified quantum numbers s_1, s_2, s_3.

The energy of the particle is

$$\epsilon = \frac{h^2}{2l^2 m} (s_1^2 + s_2^2 + s_3^2)$$

and as in the case of Bose statistics, the number of quantum states of the particle which correspond to kinetic energy between ϵ_s and $\epsilon_s + d\epsilon_s$ is

$$R_s = \frac{2\pi V}{h^3} (2m)^{\frac{3}{2}} \epsilon_s^{\frac{1}{2}} d\epsilon_s.$$

Now suppose that N_s particles have energies between ϵ_s and $\epsilon_s + d\epsilon_s$, so of the R_s states, N_s are occupied (by one particle each) and the rest unoccupied. The number of ways in which this can occur is

$$\frac{R_s\,!}{N_s\,!\,(R_s - N_s)\,!}$$

so the total number of ways in which the allocation specified by the N_s can be realised is

$$W = \prod_s \frac{R_s\,!}{N_s\,!\,(R_s - N_s)\,!}$$

and we assume that this is proportional to the probability of this allocation. Thus

$$\log W = \sum_s \left\{ \log R_s\,! - \log N_s\,! - \log (R_s - N_s)\,! \right\}.$$

Using Stirling's approximation to the log. of a factorial, this gives

$$\log W = \sum_s \left\{ R_s \log R_s - N_s \log N_s - (R_s - N_s) \log (R_s - N_s) \right\}.$$

This is to be made a maximum, subject to

$$\sum_s N_s = n, \text{ where } n \text{ is the total number of particles}$$

and

$$\sum_s \epsilon_s N_s = E, \text{ where E is the total energy.}$$

The usual procedure yields

$$N_s = \frac{R_s}{e^{\beta\epsilon_s + \mu} + 1}$$

where β and μ are independent of s. Now the entropy is

$$S = k \log W$$

and

$$\frac{dS}{dE} = \frac{1}{T},$$

whence we find

$$\frac{1}{T} = k\beta.$$

Hence *the number of particles per unit volume with kinetic energy between* ϵ *and* $\epsilon + d\epsilon$ is

$$n(\epsilon)d\epsilon = \frac{2\pi(2m)^{\frac{3}{2}}}{h^3} \cdot \frac{\epsilon^{\frac{1}{2}}d\epsilon}{e^{\epsilon/kT + \mu} + 1}$$

This is the fundamental equation of the Fermi statistics.[1]

The *total density of particles* is therefore

$$n = \int_0^\infty n(\epsilon)d\epsilon = \frac{(2mkT)^{\frac{3}{2}}2\pi}{h^3}\int_0^\infty \frac{x^{\frac{1}{2}}dx}{e^{x+\mu}+1}$$

an equation which determines μ as a function of the density and temperature.

The parameter μ has a thermodynamical significance. If for a finite mass of gas U is the total energy, S the entropy, T the temperature, p the pressure and v the volume then

$$G = U - TS + pv$$

is called the *Gibbs's thermodynamical potential*; and the *Gibbs's thermodynamical potential per molecule* is defined as $\psi = \partial G/\partial n$, where n denotes the number of molecules, and the temperature and pressure are kept constant in the differentiation. The *Fermi constant* μ is now given [2] by

$$\mu = -\frac{\psi}{kT}.$$

[1] This may be modified, e.g. when the particles concerned have different possibilities of spin.

[2] W. Pauli, *ZS. f. P.* xli (1927), p. 81, *at* p. 91

The *total kinetic energy of the particles in unit volume* is

$$E=\int_0^\infty \epsilon n(\epsilon)d\epsilon=\frac{2\pi}{h^3}(2m)^{\frac{3}{2}}(kT)^{\frac{5}{2}}\int_0^\infty \frac{x^{\frac{3}{2}}dx}{e^{x+\mu}+1}.$$

The close similarity of these equations to the corresponding equations in the Bose statistics is evident, the only difference being in the occurrence of $+1$ instead of -1 in the denominator. In exactly the same way as in the case of Bose statistics, we see that in the limit when $h\to 0$, $\mu\to+\infty$, Fermi statistics pass into Maxwellian statistics: speaking physically, a Fermi gas approximates to a classical gas at sufficiently high temperatures and low pressures. The deviation of Fermi statistics from classical behaviour is in the opposite direction to the deviation in the case of Bose statistics.[1]

The pressure of a gas in Fermi statistics is related to the energy-density by the same equation as in classical theory,

$$p=\tfrac{2}{3}\frac{E}{V}.$$

Classical statistics is not the only limiting case of Fermi statistics: there is another limiting case at the opposite extreme, namely when μ is very large and *negative*. If we write $-\alpha$ for μ, the integrals occurring in the theory are of the form

$$\int_0^\infty \frac{x^p dx}{e^{x-\alpha}+1}$$

and it can be shown that

$$\lim_{\alpha\to\infty}\left\{\alpha^{-p-1}\int_0^\infty \frac{x^p dx}{e^{x-\alpha}+1}\right\}=\frac{1}{p+1}$$

so the total density of particles becomes

$$n=\frac{4\pi}{3h^3}(2mkT\alpha)^{\frac{3}{2}}$$

and the total kinetic energy of the particles in unit volume is

$$E=\frac{4\pi}{5h^3}(2m)^{\frac{3}{2}}(kT\alpha)^{\frac{5}{2}}.$$

Eliminating α, we have

$$E=\frac{3}{40}\left(\frac{6}{\pi}\right)^{\frac{2}{3}}\frac{h^2}{m}n^{\frac{5}{3}}\text{ approximately,}$$

the neglected terms involving T^2 and higher powers [2] of T.

[1] The thermodynamical functions for a gas with Fermi statistics were studied by E. C. Stoner, *Phil. Mag.* xxviii (1939), p. 257.

[2] For an electron-gas, e.g. in metals, there is an extra factor 2 arising from the two possible values of the spin of the electron.

It is evident from this equation that at the zero of absolute temperature the particles of a gas with Fermi statistics are not at rest, but have a definite zero-point energy.

In this limiting case we see from the above value of n that

$$nh^3(mkT)^{-\frac{3}{2}} \gg 1,$$

which is realised if the temperature is low, the density n is large and the mass m of the individual particles is small. The deviation from classical behaviour is due fundamentally to the circumstance that the quantum states of very small momentum are all occupied by particles. In fact, since (in accordance with Nernst's heat theorem) the entropy S vanishes at the absolute zero of temperature, and since $S = k \log W$, it follows that $W = 1$ when $T = 0$, i.e. there is only one way of distributing the particles: they occupy every quantum state in the neighbourhood of the state of zero momentum.

This fact has an interesting application in astrophysics. In 1844 Bessel concluded from irregularities in the motion of Sirius that it was one component of a double star, the other member of the pair being invisible with the telescopes then available. Some years later the companion was observed telescopically, and found to be a star between the 8th and 9th magnitude. Its mass is not much less than that of the sun, and it is a 'white' star, so that its surface-brightness must be greater than the sun's; but its total radiation is only about $\frac{1}{360}$th of that of the sun: hence the area of its surfaces must be very much smaller than the sun's, and in fact not much larger than the earth's. Its density must therefore be very great—about 60,000 times that of water. This surprising inference was confirmed in 1925, when Walter S. Adams [1] of Mount Wilson found in the spectrum of the companion of Sirius a displacement of the lines which might well be the decrease of frequency that is to be expected in lines that have been emitted in an intense gravitational field.

The explanation [2] of this abnormal density is that matter can exist in such a dense state if it has so much energy that the electrons escape from the nuclei, so they are no longer bound in ordinary atomic orbits but are, in the main, free. The density of the matter is then limited no longer by the size of atoms, but only by the sizes of the electrons and atomic nuclei: and as the volumes of these are perhaps 10^{-14} of the volumes of atoms, we may conclude that densities of as much as 10^{14} times that of terrestrial matter may be possible: this is much greater than that of any of the 'white dwarfs,' as stars like the companion of Sirius are called.

According to the classical theory of the relation between energy and temperature, such a star, having excessively great energy, would have a very high temperature, and would therefore radiate intensely. It was, however, shown by R. H. Fowler [3] that this is

[1] *Proc. N.A.S.* xi (1925), p. 382

[2] Due chiefly to Eddington

[3] *Mon. Not. R.A.S.* lxxxvii (1926), p. 114

not the case. The radiation depends on the temperature, and depends on the energy only in so far as the energy determines the temperature: the apparent difficulty was due to the use of the classical correlation between energy and temperature: the correct relation, for such dense stellar matter, is that of the above-mentioned limiting case of Fermi's statistics, when μ is very large and negative: the energy is still very great, but the temperature is not correspondingly great, and indeed ultimately approaches zero: so radiation, which depends on temperature, stops when the dense matter still has ample energy.

The absolutely final state (the 'black dwarf') is one in which there is only one possible configuration left: the star is then analogous to one gigantic molecule in its lowest quantum state, and the temperature (which ceases to have any meaning) may be said to be zero.

It was shown in 1927 by L. S. Ornstein and H. A. Kramers[1] of Utrecht that the formulae of Fermi statistics may be derived in a totally different way by considering the kinetics of reactions.

Let the possible values of the energy of a gas-molecule in an enclosure be $\epsilon_1, \epsilon_2, \epsilon_3, \ldots$; for simplicity we suppose these values all different. Consider two molecules in the states k' and l' respectively, which by their interaction are changed to the states k'' and l'' respectively. Let the *a priori* probability that this transition should take place in unit time be denoted by $a_{\prime\prime}{}'$. In order that the transition may actually take place, it is necessary that there should exist a molecule in the k' state and one in the l' state, and, moreover (by the Pauli principle), that neither a molecule in the k'' state nor one in the l'' state is present. Denoting by $\bar{n}_k$ the probability that a molecule is in the k state, the mean frequency for the process in which k', l' tend to k'', l'', is

$$A_{\prime\prime}{}' = a_{\prime\prime}{}' \, \bar{n}_{k'} \bar{n}_{l'} (1 - \bar{n}_{k''}) (1 - \bar{n}_{l''})$$

while the frequency of the reverse process is

$$A_{\prime}{}'' = a_{\prime}{}'' \, \bar{n}_{k''} \bar{n}_{l''} (1 - \bar{n}_{k'}) (1 - \bar{n}_{l'}).$$

In thermodynamic equilibrium we must have

$$A_{\prime\prime}{}' = A_{\prime}{}''$$

and if we postulate further, that the *a priori* probability for reverse processes is equal, i.e.

$$a_{\prime\prime}{}' = a_{\prime}{}'',$$

[1] *ZS. f. P.* xlii (1927), p. 481. On the kinetic relations of quantum statistics, cf. also: P. Jordan, *ZS. f. P.* xxxiii (1925), p. 649; xli (1927), p. 711. W. Bothe, *ZS. f. P.* xlvi (1928), p. 327. L. Nordheim, *Proc. R.S.*(A), cxix (1928), p. 689. S. Kikuchi and L. Nordheim, *ZS. f. P.* lx (1930), p. 652. S. Flügge, *ZS. f. P.* xciii (1935), p. 804

then if we write

$$\frac{\bar{n}_k}{1-\bar{n}_k}=m_k,$$

the above equations give

$$m_{k'}\, m_{l'}=m_{k''}\, m_{l''}.$$

This equation must hold for all the values of k', l', k'', l'', for which the equation of conservation of energy

$$\epsilon_{k'}+\epsilon_{l'}=\epsilon_{k''}+\epsilon_{l''}$$

holds: whence it follows that the only possible reasonable distribution is given by

$$m_k=e^{-\alpha-\beta\epsilon_k}$$

where α and β are constants: whence we have

$$\bar{n}_k=\frac{e^{-\alpha-\beta\epsilon_k}}{1+e^{-\alpha-\beta\epsilon_k}}=\frac{1}{e^{\alpha+\beta\epsilon_k}+1}$$

which is Fermi's law of distribution.

Another application of the Fermi statistics was to the theory of electrons in metals.

In 1927 Pauli [1] succeeded in accounting for the observed character of the paramagnetism of the alkali metals, which is feeble and nearly independent of the temperature, by assuming that the conducting electrons in a metal may be regarded as an electron-gas which is degenerate in the sense of Fermi's statistics (i.e. μ is large and negative). In the following year Sommerfeld [2] discussed the main problems of the electron-theory of metals on the same assumption: the conducting electrons may be supposed to have free paths of the order of 100 times the atomic distances.

To see that the assumption is justified, we have to show that

$$nh^3(mkT)^{-\frac{3}{2}}\gg 1.$$

Now we may suppose that the number n of free electrons is of the same order as the number of atoms, say 10^{22} per cm^3. The mass of

[1] *ZS. f. P.* xli (1927), p. 81; communicated 16 Dec. 1926; published 10 Feb. 1927

[2] Preliminary note in *Naturwiss*, xv (14 Oct. 1927), p. 825; xvi (1928), p. 374; *ZS. f. P.* xlvii (1928), p. 1. His associate C. Eckart, ibid. p. 38, discussed the Volta-effect, and his associate W. V. Houston, *ZS. f. P.* xlviii (1928), p. 449, discussed electric conduction. Sommerfeld's theory has been developed and improved by: E. Kretschmann, *ZS. f. P.* xlviii (1928), p. 739. F. Block, *ZS. f. P.* lii (1928), p. 555; lix (1930), p. 208. L. W. Nordheim, *Proc. R.S.*(A), cxix (1928), p. 689. R. Peierls, *ZS. f. P.* liii (1929), p. 255; *Ann. d. Phys.* iv (1930), p. 121; v (1930), p. 244. L. Brillouin, *Les Statistiques quantiques* (Paris, 1930), Tome II. L. Nordheim, *Ann. d. Phys.* ix (1931), p. 607. A. H. Wilson, *Proc. R.S.*(A), cxxxviii (1932), p. 594. On semi-conductors, cf. A. H. Wilson, *Proc. R.S.*(A), cxxxiii (1931), p. 458; cxxxiv (1931), p. 277.

the electron m is 9×10^{-28} gr., Planck's constant h is $6{\cdot}6\times10^{-27}$ erg. sec., Boltzmann's constant k is $1{\cdot}4\times10^{-16}$ erg./deg. Thus the condition is roughly

$$10^{-57}\,(10^{-43}\,\mathrm{T})^{-\frac{3}{2}}\gg1$$

or

$$\mathrm{T}\ll10^{5}$$

which is satisfied at all ordinary temperatures. Thus *the electron-gas inside a metal has a degenerate Fermi distribution.* At the zero of absolute temperature, there is a finite energy given by the equation

$$\mathrm{E}=\frac{3}{40}\left(\frac{3}{\pi}\right)^{\frac{2}{3}}\frac{h^2}{m}\,n^{\frac{5}{3}},$$

and the derivative $d\mathrm{E}/d\mathrm{T}$ is zero in accordance with Nernst's theorem: the specific heat of the electron-gas is in fact proportional to T for low temperatures, whereas the classical statistics made it a constant. Thus it is understood why the specific heat of the electrons in metals is exceedingly small: the specific heat of a metal is usually what would be expected if it were due to the metallic atoms alone.

The replacement of classical by Fermi statistics does not appreciably change the theoretical ratio of the thermal and electrical conductivities.

We have seen that in Fermi statistics, the probability that a quantum state whose energy is ϵ is occupied

is
$$(e^{\epsilon/k\mathrm{T}+\mu}+1)^{-1},$$

and that $\mu=-\psi/k\mathrm{T}$ where ψ is Gibbs's thermodynamical potential per molecule. For the electron-gas inside a metal, this quantity ψ is called the *electrochemical potential.* It has the property that the electrochemical potentials of electrons in any two regions which are in thermal equilibrium (e.g. electrons in different metals which are in contact with each other at the same temperature) must be equal.

In a second paper [1] in the *Zeitschrift für Physik*, Sommerfeld discussed thermo-electric phenomena, obtaining new formulae for the Peltier and Thomson effects, and showing that the new expression for the Thomson heat agreed with the experimental values much better than the old. Shortly before this a new thermo-electric effect had been discovered by P. W. Bridgman,[2] namely, that when an electric current passes across an interface where the crystal orientation changes, an 'internal Peltier heat' is developed. A theoretical discussion of this and other thermoelectric phenomena was given in 1929 by P. Ehrenfest and A. J. Rutgens.[3]

[1] *ZS. f. P.* xlvii (1928), p. 43
[2] *Proc. N.A.S.* xi (1925), p. 608; *Phys. Rev.* xxxi (1928), p. 221
[3] *Proc. Amst. Ac.* xxxii (1929), p. 698

The twentieth century brought considerable developments in the subject of thermionics. Richardson had deduced his original formula[1] from the assumption that the free electrons in a metal have the same energy as is attributed in the kinetic theory of gases to gas-molecules at the same temperature as the metal. H. A. Wilson[2] showed that the phenomena could not be explained completely by this theory, and proposed to replace Richardson's treatment by one analogous to the theory of evaporation. Now the latent heat of vaporisation of a liquid may be shown by thermodynamical methods to be given by what is called the *equation of Clapeyron and Clausius*, which may be written in the form

$$\chi = \mathrm{RT}^2 . \frac{1}{p} \frac{dp}{d\mathrm{T}}$$

where χ is the latent heat of evaporation per gramme-molecule at the absolute temperature T, R is the gas-constant per gramme-molecule (= 1·987 cal./deg.), and p is the vapour-pressure at temperature T : and Wilson proposed to apply this equation. Under the influence of these ideas, Richardson[3] suggested as an alternative to his earlier formula that the saturation emission per unit area, in ampères per cm.^{2}, should be represented by an equation

$$\mathrm{I} = \mathrm{AT}^2 e^{-\frac{\chi}{k\mathrm{T}}}$$

where A is a constant, k is Boltzmann's constant, and χ represents the energy required to get an electron through the surface : χ, which is called the *work-function*, is analogous to the latent heat of evaporation of a monatomic gas : it is usually reckoned in electron-volts and is often written $e\phi$ where ϕ is expressed in volts, and so is in fact the potential in volts necessary to impart to an electron the kinetic energy required for evaporation.

It may be remarked that although the conducting electrons inside a metal have a degenerate Fermi distribution, the external electrons produced by thermionic emission have a distribution which is practically Maxwellian : this is explained by their much smaller concentration.

In 1923 Saul Dushman,[4] by use of the Clapeyron-Clausius formula, and on the assumption that the electrons within the metal obey Maxwellian statistics, obtained Richardson's second formula, showing, however, that it is not rigorously true if χ is a function of temperature (though for clean metals the deviations are small) ;

[1] cf. Vol. I, pp. 425–8

[2] *Phil. Trans.*(A), ccii (1903), p. 243

[3] *Phil. Mag.* xxviii (1914), p. 633

[4] *Phys. Rev.* xxi (1923), p. 623 ; cf. earlier papers by : W. Schottky, *Phys. ZS.* xv (1914), p. 872 ; xx (1919), pp. 49, 220. M. von Laue, *Ann. d. Phys.* lviii (1919), p. 695. R. C. Tolman, *J. Am. Chem. Soc.* xlii (1920), p. 1185 ; xliii (1921), p. 866. And cf. later papers by : P. W. Bridgman, *Phys. Rev.* xxvii (1926), p. 173 ; xxxi (1928), p. 90. L. Tonks and I. Langmuir, *Phys. Rev.* xxix (1927), p. 524. L. Tonks, *Phys. Rev.* xxxii (1928), p. 284. K. F. Herzfeld, *Phys. Rev.* xxxv (1930), p. 248

in fact χ should be understood as relating to the work-function for the zero of absolute temperature: and he showed that A is a universal constant, having the value

$$A = 2\frac{\pi m e k^2}{h^3} = 60{\cdot}2 \text{ amp./cm.}^2 \text{ deg}^2.$$

Sommerfeld's discovery that the electrons in a metal have degenerate Fermi statistics naturally led [1] to a new treatment of thermionics, but the formula finally obtained had the same general character as Richardson's formula of 1914. L. W. Nordheim [2] found

$$i = A(1-r)T^2 e^{-\frac{\chi}{kT}}$$

the quantities in this equation being defined as follows: A has twice the value of Dushman's universal constant A, so

$$A = \frac{4\pi m e k^2}{h^3} = 120 \text{ amp./cm.}^2 \text{ deg.}^2:$$

the factor 2 arises when we take account of the two possible values of the electron-spin. It might be thought easy to discriminate experimentally between Dushman's and Nordheim's formulae for A: but small uncertainties in the values of χ and T, which occur in the exponential, affect the value of i so much that the experimental determination of A is very uncertain.[3] r is the reflection coefficient, i.e. the ratio of the number of electrons reflected internally at the surface to the total number reaching it, only those electrons being considered that have velocity components normal to the surface sufficient to allow them to escape. r is small and need not usually be considered, since the experimental value of A is so doubtful. χ corresponds to Richardson's work-function: but χ now has the form

$$\chi = W_a - W_i$$

where W_a is e times the difference in electrostatic potential between the inside and outside of the metal, and W_i is an energy which depends on the pressure of the electron-gas, and which assists the escape of the electrons from the metal: we may take W_i to be the kinetic energy of the highest filled level at the absolute zero of temperature.

Now we saw that in a Fermi distribution, the number of quantum

[1] A. Sommerfeld, *ZS. f. P.* xlvii (1928), p. 1

[2] *ZS. f. P.* xlvi (1928), p. 833; *Phys. ZS.* xxx (1929), p. 177

[3] L. A. du Bridge, *Phys. Rev.* xxxi (1928), pp. 236, 912, found that for clean platinum the thermionic constant A has the value 14,000 amp./cm.^{2} deg^2, which is 230 times Dushman's theoretical value 60·2; cf. L. A. du Bridge, *Proc. N.A.S.* xiv (1928), p. 788, and R. H. Fowler, *Proc. R.S.*(A), cxxii (1929), p. 36.

states of a particle (in unit volume) which correspond to kinetic energy between ϵ and $\epsilon + d\epsilon$ is

$$\frac{2\pi}{h^3}(2m)^{\frac{3}{2}}\epsilon^{\frac{1}{2}}d\epsilon$$

or inserting a factor 2 on account of the two values of the electron-spin, it is

$$\frac{4\pi}{h^3}(2m)^{\frac{3}{2}}\epsilon^{\frac{1}{2}}d\epsilon.$$

At zero temperature, all the quantum states are occupied up to a certain level of kinetic energy, say W_i : so if n be the number of free electrons per unit volume, we have

$$n = \frac{4\pi}{h^3}(2m)^{\frac{3}{2}}\int_0^{W_i} \epsilon^{\frac{1}{2}}d\epsilon = \frac{8\pi}{3h^3}(2mW_i)^{\frac{3}{2}}$$

so

$$W_i = \frac{h^2}{2m}\left(\frac{3}{8\pi}\right)^{\frac{2}{3}}n^{\frac{2}{3}}.$$

It can be shown that

$$W_i = \mu + e\Phi_i$$

where μ is the electrochemical potential of the electrons just inside the metal, and Φ_i is the electrostatic potential just inside the metal.[1]

The function W_a, the true total energy necessary to liberate an electron from the metal, may be determined independently by a method which depends on the diffraction of beams of electrons by metallic crystals.[2] For although the experiments of Davisson and Germer, already referred to, were explained qualitatively by de Broglie's theory of the wave-behaviour of electrons, there was a quantitative discrepancy, which was resolved by assuming that the wave-length of the electron-waves in the metal was different from the wave-length *in vacuo*, or in other words, that the metal had a refractive index for the electron-waves : this refractive index, which could be determined from the diffraction-experiments, determined the grating-potential and thence the function W_a : W_a was found to be considerably greater than the work-function χ.

In a complete treatment of thermionic and photoelectric phenomena, it is necessary to take account of factors that cannot be fully discussed here, e.g. the fact that an electron just outside a metal is

[1] The use of the electrochemical potential in thermionics is due to W. Schottky and H. Rothe, *Handbuch d. Experimentalphysik* (Leipzig, 1928), XII/2.

[2] cf. C. Eckart, *Proc. N.A.S.* xii (1927), p. 460 ; H. Bethe, *Naturwiss*, xv (1927), p. 787 ; L. Rosenfeld and E. E. Witmer, *ZS. f. P.* xlix (1928), p. 534

influenced by a force due to its own image in the metal :[1] in the case of a flat perfectly-conducting surface this force is of amount $e^2/4z^2$, where z is the distance of the electron from the surface.

The constants and functions that occur in thermionics occur also in related branches of physics. If two metals A and B at the same temperature are in contact, electrons flow from one to the other until the electrochemical potentials of the electrons in the two metals are equal, and there is a difference of potential (the *Volta effect* or *contact potential-difference* [2]) between a point just outside A and a point just outside B. Since the work-function χ is

$$W_a - W_i \quad \text{or} \quad -e\Phi_0 + e\Phi_i - (\mu + e\Phi_i)$$

where Φ_0 is the electrostatic potential at a point just outside the metal, we see that *the Volta effect* $(\Phi_0)_A - (\Phi_0)_B$ *is equal to* $(\chi_B - \chi_A)/e$, *where* χ_A *and* χ_B *are the thermionic work-functions of the two metals*: so if i_A and i_B are the thermionic electric saturation currents per unit area at the same temperature, the contact potential-difference is [3]

$$\frac{kT}{e} \log \frac{i_B}{i_A}.$$

Thermionics is closely connected also with photoelectricity, as we have seen in Chapter III.[4] The frequency of light which will just eject electrons photoelectrically, but with zero velocity, is called the *threshold frequency*, and the corresponding wave-length is called the *photoelectric long wave-length limit.* R. A. Millikan [5] verified, by a series of careful experiments, that if ν_0 is the threshold frequency, the photoelectric quantity $h\nu_0$ is equal to the thermionic work-function $e\phi$ measured at the same temperature. L. A. Du Bridge [6] found that for clean platinum the photoelectric long wave-length limit is 1962Å, which by the above equation corresponds to 6·30 volts, while the thermionic work-function is 6·35 volts: the two are thus in agreement within the limits of error.

The theory of the connection between photoelectric threshold frequency and thermionic work-function was studied in the light of Sommerfeld's electron theory of metals by R. H. Fowler.[7]

[1] The theory of electric images is due to William Thomson (Kelvin). This effect had been considered in P. Lenard's early work on the photoelectric effect, *Ann. d. Phys.* viii (1902), p. 149 ; cf. P. Debye, *Ann. d. Phys.* xxxiii (1910), p. 441, and Walter Schottky, *Phys. ZS.* xv (1914), p. 872.

[2] cf. Vol. I, p. 71

[3] O. W. Richardson, *Phil. Mag.* xxiii (1912), p. 265 ; verified by O. W. Richardson and F. S. Robertson, *Phil. Mag.* xliii (1922), p. 557

[4] cf. p. 90 *supra*

[5] *Phys. Rev.* xviii (1921), p. 236

[6] *Phys. Rev.* xxxi (1928), pp. 236, 912 ; cf. also A. H. Warner, *Proc. N.A.S.* xiii (1927), p. 56, who worked with tungsten.

[7] *Proc. R.S.*(A), cxviii (March, 1928), p. 229; cf. G. Wentzel, *Sommerfeld Festschrift* (1928), p. 79

The effect of temperature on the photoelectric sensibility of a clean metal near the threshold was studied by J. A. Becker and D. W. Mueller[1] in 1928, by E. O. Lawrence and L. B. Linford[2] in 1930 and by R. H. Fowler in 1931.[3] It was shown that the photoelectric long wave-length limit is shifted towards the red.

Some interesting experiments on the photoelectric properties of thin films were carried out in 1929 by H. E. Ives and A. R. Olpin.[4] They found that the long-wave limit of photoelectric action in the case of films of the alkali metals on platinum varies with the thickness of the film. As the film accumulates, the long-wave limit moves towards the red end of the spectrum, reaches an extreme position and then recedes again to the final position characteristic of the metal in bulk. The wave-length of the maximum excursion of the long-wave limit was found in every case to coincide with the first line of the principal series of the metal in the form of vapour, i.e. the resonance potential. This seemed to suggest that photoelectric emission is caused when sufficient energy is given to the atom to produce its first stage of excitation.

Richardson[5] in 1912 studied the photoelectric emission from a surface exposed to black-body radiation corresponding to some high temperature T: this has been called the *complete photoelectric effect.* It was shown by A. Becker[6] that the relative distribution of the velocities of the electrons released from platinum photoelectrically under these conditions is identical with that of electrons released thermionically. If there is equilibrium, the surface must itself be at temperature T, and then the total number of electrons leaving the surface, whether as a result of thermionic or photoelectric action, must be given by the Richardson equation.

Sommerfeld's discovery that the conducting electrons in metals obey Fermi statistics made possible a satisfactory theory of a phenomenon which had long been known[7] but which, in some of its features, had not been explained: namely, the extraction of electrons from cold metals by intense electric fields.[8] Experimental work by R. A. Millikan and his coadjutors[9] had shown that the currents obtained at a given voltage are independent of the temperature, provided the latter is not so high as to approach the temperature at which thermionic emission becomes appreciable: whence Millikan

[1] *Phys. Rev.* xxxi (1928), p. 431

[2] *Phys. Rev.* xxxvi (1920), p. 482

[3] *Phys. Rev.* xxxviii (1931), p. 45; cf. J. A. Becker and W. H. Brittain, *Phys. Rev.* xlv. (1934), p. 694

[4] *Phys. Rev.* xxxiv (1929), p. 117; cf. Hughes and du Bridge, *Photoelectric Phenomena* (New York, 1932), p. 178

[5] *Phil. Mag.* xxiii (1912), p. 594

[6] *Ann. d. Phys.* lx (1919), p. 30

[7] Knowledge of it seems to have evolved gradually from R. F. Earhart's experiments on very short sparks. *Phil. Mag.* i (1901), p. 147.

[8] A first approximate theory was given by W. Schottky, *ZS. f. P.* xiv (1923), p. 63, following on J. E. Lilienfeld, *Phys. ZS.* xxiii (1922), p. 306.

[9] R. A. Millikan and C. F. Eyring, *Phys. Rev.* xxvii (1926), p. 51; R. A. Millikan and C. C. Lauritsen, *Proc. N.A.S.* xiv (1928), p. 45; cf. also N. A. de Bruyne, *Proc. Camb. Phil. Soc.* xxiv (1928), p. 518

concluded that the conduction electrons extracted in this way do not share in the thermal agitation, whereas the thermions, which are expelled at high temperatures, do.

Millikan and Lauritsen showed that if I is the current obtained by an applied electric force F, then log I when plotted against I/F yields approximately a straight line. In 1928 the subject was attacked in several theoretical papers,[1] most completely by R. H. Fowler and L. Nordheim.[2] They calculated the emission coefficient (i.e. the ratio of the number of electrons going through the surface, to the number of electrons incident from inside the metal) and the reflection coefficient, at the surface, and integrated over all incident electrons according to Sommerfeld's electron theory of metals; and they showed that it is the electrons with small energies that are pulled out by strong fields. Now these electrons with small energies have Fermi statistics, and that is why the intensity of the emitted current is, at ordinary temperatures, independent of the temperature. The formula obtained theoretically by Fowler and Nordheim for the current I was

$$I = \frac{\epsilon}{2\pi h} \frac{\psi^{\frac{1}{2}}}{(\chi+\psi)\chi^{\frac{1}{2}}} F^2 e^{-4\kappa\chi^{\frac{3}{2}}/3F}$$

where ϵ is the electron-charge, $\kappa^2 = 8\pi^2 m/h^2$, ψ is Gibbs's thermodynamical potential per electron, χ is the thermionic work-function and F is the applied electric force. This formula agrees with the experimental results.[3]

[1] R. H. Fowler, *Proc. R.S.*(A), cxvii (1928), p. 549; L. Nordheim, *ZS. f. P.* xlvi (1928), p. 833; W. V. Houston, *ZS. f. P.* xlvii (1928), p. 33 (working with Sommerfeld); O. W. Richardson, *Proc. R.S.*(A), cxvii (1928), p. 719; W. S. Pforte, *ZS. f. P.* xlix (1928), p. 46

[2] *Proc. R.S.*(A), cxix (1928), p. 173; cf. also J. R. Oppenheimer, *Phys. Rev.* xxxi (1928), p. 66

[3] cf. also: O. W. Richardson, *Proc. R.S.*(A), cxix (1928), p. 531; N. A. de Bruyne, *Proc. R.S.*(A), cxx (1928), p. 423; A. T. Waterman, *Proc. R.S.*(A), cxxi (1928), p. 28; L. W. Nordheim, *Proc. R.S.*(A), cxxi (1928), p. 626; T. E. Stern, B. S. Gossling and R. H. Fowler, *Proc. R.S.*(A), cxxiv (1929), p. 699; W. V. Houston, *Phys. Rev.* xxxiii (1929), p. 361

Chapter VII

MAGNETISM AND ELECTROMAGNETISM, 1900–26

At the end of the nineteenth century the classical theory of electrons was well established, and one magnetic phenomenon, namely, the Zeeman effect, had been explained in terms of it by Lorentz and Larmor. The time was evidently ripe for the consideration of diamagnetic, paramagnetic and ferromagnetic phenomena in the light of electron-theory.

It will be remembered[1] that Weber had explained diamagnetism (the magnetic polarisation of bodies induced in a direction opposite to that of the magnetising field) by postulating the existence in molecules of circuits whose electric resistance is zero (but whose self-induction is not zero), so that the creation of an external field causes induced currents in them: the total magnetic flux through the circuits remains zero, and therefore by Lenz's law the direction of the induced currents corresponds to diamagnetism. Paramagnetism was explained by postulating Ampèrean electric currents in the molecules, whose planes were orientated by the magnetising field: and Ewing had developed on this basis an explanation of ferromagnetism.

The re-statement of these ideas in terms of the theory of electrons was undertaken in 1901–3 by W. Voigt[2] and J. J. Thomson.[3] They studied the effect of an external magnetic field on the motion of a number of electrons, which are situated at equal intervals round the circumference of a circle, and are rotating in its plane with uniform velocity round its centre; and they found that if a substance contained a uniform distribution of such systems, the coefficient of magnetisation of the substance would be zero; so that it would be impossible to explain the magnetic or diamagnetic properties of bodies by supposing that the atoms contain charged particles circulating in closed periodic orbits under the action of central forces.

The origin of the difference between the effects produced by charged particles freely describing orbits, and those produced by constant electric currents flowing in circular circuits, as in Ampère's theory of magnetism, is that in the case of the particles describing their orbits we get, in addition to the effects due to the constant electric currents, effects of the same character as those due to the induction of currents in conductors by the variation of the magnetic field: these induced currents tend to make the body diamagnetic,

[1] cf. Vol. I, pp. 208–11 [2] *Gött. Nach.* (1901), p. 169; *Ann. d. Phys.* ix (1902), p. 115
[3] *Phil. Mag.*(6) vi (1903), p. 673

while the Ampèrean currents tend to make it magnetic ; and in the case of the particles describing free orbits, these tendencies balance each other.

Voigt showed that spinning electrons, obstructed in their motion by continual impacts, would lead to paramagnetism or diamagnetism according as they possessed, immediately after the impact, an average excess of potential or kinetic energy. However, besides the complexity of this representation, there is the objection that it attributes to the same cause phenomena so different as paramagnetism and diamagnetism, and that it offers no interpretation of certain laws, established experimentally by P. Curie,[1] namely, that the paramagnetic susceptibility varies inversely as the absolute temperature, whereas diamagnetism is in all observed cases except bismuth, rigorously independent of T.

The first successful application of electron-theory to the general problem of magnetism was made in 1905 by Paul Langevin.[2] He accepted Weber's view that diamagnetism is really a property possessed by all bodies, and that the so-called paramagnetic and ferromagnetic substances are those in which the diamagnetism is masked by vastly greater paramagnetic and ferromagnetic effects. The condition for the absence of paramagnetism and ferromagnetism is that the molecules of the substance should have no magnetic moment, so that they have no tendency to orient themselves in an external magnetic field.

In order to explain diamagnetism, we observe that the external applied magnetic field H creates a Larmor precession,[3] each electron acquiring an additional angular velocity $eH/2mc$ in its orbit[4]; whence it follows, as Langevin showed, that the increase of the magnetic moment of a molecule due to one particular electron circulating in it is

$$\Delta M = -\frac{He^2}{4mc^2}\overline{r^2}$$

where r is the distance of the electron from the atomic nucleus, projected in a plane perpendicular to H, and $\overline{r^2}$ is an average extended over the duration of several revolutions. The negative sign shows that the effect is diamagnetic. In order to obtain the total effect on a molecule we must sum over all the electrons in it. Thus if N denotes Avogadro's number, i.e. the number of molecules in a gramme-molecule (which is the same for all elements and compounds), then *the diamagnetic susceptibility per gramme-molecule is*

$$-\frac{e^2N}{4mc^2}\sum\overline{r^2}$$

[1] *Ann. chim. phys.* v (1895), p. 289

[2] *Comptes Rendus*, cxxxix (26 Dec. 1904), p. 1204 ; *Soc. Française de phys., Résumés* (1905), p. 13 * ; *Ann. chim. phys.*(8) v (1905), p. 70

[3] cf. Vol. I, pp. 415–6

[4] As usual, H and M are in electromagnetic units, and e in electrostatic units.

where the summation is taken over all the electrons in the atom or molecule, and (r, z, ϕ) are the cylindrical co-ordinates of the electron when the z-axis is taken in the direction of the magnetic field H.[1] Since atoms which exhibit diamagnetism without paramagnetism have spherical symmetry, $\sum \overline{r^2}$ can be replaced by $\frac{2}{3}\sum R^2$, where R denotes the distance of the electron from the centre of the atom.

The smallness of the diamagnetic effects is accounted for by the smallness of the radius r, which is necessarily less than the molecular dimensions.

The diamagnetic property is acquired instantaneously, at the moment of the creation of the external field.

The intramolecular motions of the electrons depend very little on the temperature, as is shown by the fixity of spectral lines : so the diamagnetic susceptibility should vary very little with the temperature, in agreement with Curie's experimental result. It varies also very little with the physical or chemical state : *diamagnetism is an atomic property.*

The exception presented by solid bismuth, whose diamagnetic susceptibility diminishes approximately linearly when the temperature rises, was attributed by Langevin (following J. J. Thomson) to the presence in the metal of free conduction-electrons.

The fact that diamagnetism and the Zeeman effect both depend on the Larmor precession shows that the two phenomena (at any rate when considered classically) are very closely connected, and indeed may be regarded as different aspects of the same phenomenon.

If the intrinsic magnetic moment of a molecule is not null, there is superposed on the diamagnetic effect another phenomenon, due to the orientation of the molecular magnets by the external field; this *paramagnetic* effect, when it exists, is large compared with the diamagnetic, and completely masks it. In discussing it we shall assume that the molecular magnets are not associated in aggregates within which their mutual actions are of importance (this is the case with *ferromagnetic* substances, which will be considered later) : and in fact we shall suppose that they can be treated in the same way as the molecules in the kinetic theory of gases. In an external magnetic field H, a molecule whose magnetic moment is M has a potential energy $-\text{MH} \cos \alpha$, where α is the angle between M and H ; the potential energy being thus least when the molecular magnet is parallel to the external field. The situation may now be compared with that of a gas composed of heavy molecules and acted on by gravity, where the ascent of a molecule causes an increase in its gravitational potential energy and a decrease in its kinetic energy ; similarly the magnetic molecule experiences a change of kinetic energy when α changes, and the partition of kinetic energy between the molecules becomes incompatible with thermal

[1] W. Pauli, *ZS. f. P.* ii (1920), p. 201

equilibrium. Through the mediation of collisions, a rearrangement is brought about, by which the mean kinetic energy of a molecule is made independent of its orientation: the temperature becomes uniform, and the molecular magnets are directed preferentially in the direction of H; though this orientation is not universal, on account of the thermal agitation and the collisions. Unlike diamagnetism, paramagnetism does not appear instantaneously, since collisions are required to create it.

Langevin found that if μ denotes the intrinsic magnetic moment of a molecule, T the absolute temperature, k Boltzmann's constant, N Avogadro's number, M the magnetisation in one gramme-molecule, then

$$\mathrm{M} = \mathrm{N}\mu\left(\coth \mathrm{X} - \frac{1}{\mathrm{X}}\right) \quad \text{where} \quad \mathrm{X} = \frac{\mu \mathrm{H}}{k\mathrm{T}}.$$

When X is small, so that we need retain only the first term in (coth $\mathrm{X} - 1/\mathrm{X}$), this gives for the molecular susceptibility χ,

$$\chi = \frac{\mathrm{M}}{\mathrm{H}} = \frac{\mathrm{N}\mu^2}{3k\mathrm{T}};$$

this formula involves Curie's law, that the paramagnetic susceptibility varies inversely as the absolute temperature.[1]

From the point of view of the Rutherford-Bohr theory of atomic structure, paramagnetism, being due to the possession of intrinsic magnetic moment, is associated with *incomplete shells* of electrons[2]; for every closed shell has spherical symmetry. This explains why the alkaline and alkaline-earth elements in the metallic state are paramagnetic, while their salts are diamagnetic: for the electrons which do not belong to closed shells in the metals are taken into gaps in the shells of the elements with which they are combined. In the case of the incomplete inner shells of the rare earths, however, this kind of compensation does not take place, so the salts are paramagnetic.[3]

In 1907 Langevin's theory was extended so as to give an account of ferromagnetism, by Pierre Weiss[4] (*b.* 1865). Ferromagnetism is a property of *molecular aggregates* (crystals), so Weiss took into consideration the *internal* magnetic field, assuming that each molecule is acted on by the surrounding molecules with a force equal to that which it would experience in a uniform field proportional to the intensity of the magnetisation and in the same direction. On this

[1] On the magnetic susceptibility of oxygen, hydrogen and helium, cf. A. P. Wills and L. G. Hector, *Phys. Rev.*, xxiii (1924), p. 209; for helium, neon, argon and nitrogen, cf. L. G. Hector, *Phys. Rev.*, xxiv (1924), p. 418. For theoretical predictions regarding these gases, cf. G. Joos, *ZS. f. P.* xix (1923), p. 347.

[2] cf. N. W. Taylor and G. N. Lewis, *Proc. N.A.S.*, xi (1925), p. 456

[3] F. Hund, *ZS. f. P.* xxxiii (1925), p. 855

[4] *Bull. des séances de la soc. fr. de phys., Année* 1907, p. 95; *Journ. de phys.* vi (1907), p. 661

assumption Weiss showed that a ferromagnetic body when heated to a certain critical temperature, the *Curie point*,[1] ceases to be ferromagnetic, and that for higher temperatures the inverse of the susceptibility is a linear function of the excess of the temperature above the Curie point.

It follows from Weiss's work that in a ferromagnetic substance there must be *domains* (i.e. regions larger than a single atom or molecule) which are inherently magnetic, although if the magnetic moments of the domains are not oriented in any preferential direction, a finite block of the substance may show no magnetisation.

Weiss found [2] that the magnetic moments of all known molecules were multiples of a certain greatest common divisor, which he called a *magneton.* This is called *Weiss's magneton* in order to distinguish it from the natural unit of magnetic moment to which W. Pauli [3] in 1920 gave the name *Bohr magneton,* and which has the value

$$\frac{he}{4\pi mc}.$$

The Bohr magneton is nearly five times the Weiss magneton. For the vapours of the elements in the first column of the Newlands-Mendeléev table (i.e. the alkalis and Cu, Ag, Au), it was shown by W. Gerlach and O. Stern,[4] by W. Gerlach and A. C. Cilliers,[5] and by J. B. Taylor,[6] that the intrinsic magnetic moment is one Bohr magneton. As we have seen,[7] the Bohr magneton was found in 1925 to be the magnetic moment of an electron ; and from this time the Weiss magneton ceased to figure in physical theory.[8]

The researches of Langevin and Weiss represented notable advances in the theory of magnetism, and the agreement of their results with experimental data was striking: yet it was shown in Niels Bohr's inaugural-dissertation [9] in 1911 and by Miss H. J. van Leeuwen,[10] a pupil of Lorentz's, in 1919, that the validity of these results could be explained only by supposing that Langevin and Weiss had not consistently applied classical statistics to all the degrees of freedom concerned : in other words, they had made assumptions of a quantistic character. The difficulty was eventually removed only by the development of quantum mechanics.

An account must now be given of some questions which had been

[1] The name was introduced by P. Weiss and H. Kamerlingh Onnes, *Comm. phys. Labor. Leiden,* No. 114 (1910), p. 3.

[2] *Journ. de phys.* i (1911), pp. 900, 965 ; *Phys. ZS.* xii (1911), p. 935 ; cf. B. Cabrera, *Journ. de phys.* iii (1922), p. 443

[3] *Phys. ZS.* xxi (1920), p. 615

[4] *Ann. d. Phys.* lxxiv (1924), p. 673

[5] *ZS. f. P.* xxvi (1924), p. 106

[6] *Phys. Rev.* xxviii (1926), p. 576

[7] cf. p. 136 *supra*

[8] cf. W. Gerlach, *Phys. ZS.* xxiv (1923), p. 275

[9] N. Bohr, *Studier over Metallernes Elektronteori* (120 pp.), Copenhagen, 1911

[10] *Inaugural dissertation,* Leiden, 1919 : her argument, which was based on Boltzmann's H-theorem, was substantially reproduced in her paper *J. de phys. et le radium,* ii (1921), p. 361 ; cf. also J. N. Kroo, *Ann. d. Phys.* xlii (1913), p. 1354, who argued that the electron theory could explain only diamagnetism ; E. Holm, *Ann. d. Phys.* xliv (1914), p. 241 ; R. Gans, *Ann. d. Phys.* xlix (1916), p. 149.

left unsettled in the nineteenth century and were definitely answered in the twentieth.

Maxwell had suggested[1] the possibility that an electromotive force might be produced by simply altering the velocity of a conductor. This effect was not known to exist until 1916, when it was observed by Richard C. Tolman and T. Dale Stewart.[2] They rotated a coil of copper wire about its axis at a high speed and then suddenly brought it to rest, so that a pulse of current was produced at the instant of stopping by the tendency of the free electrons to continue in motion. The ends of the coil were connected to a sensitive ballistic galvanometer, which enabled the experimenters to measure the pulse of current thus produced. Denoting by R the total resistance in the circuit, by l the length of the rotating coil, by v the rim speed of the coil, by Q the pulse of electricity which passes through the galvanometer at the instant of stopping, by F the charge carried in electrolysis by one gramme-ion, that is Ne/c electromagnetic units, where N is Avogadro's number, and by M the effective mass of the carrier of the current (the electron), they found the equation

$$Q = \frac{Mvl}{FR},$$

by use of which they inferred from the experiments that the carriers of electric current in metals have approximately the same ratio of mass to charge as an electron.

Maxwell[3] mentioned another possible effect which might be caused by the carriers of electricity in conductors. He proposed to take a circular coil of a great many windings and suspend it by a fine vertical wire, so that the windings are horizontal and the coil is capable of rotating about a vertical axis. A current is supposed to be conveyed into the coil by means of the suspending wire, and, after passing round the windings, to complete its circuit by passing downwards through a wire which is in the same line with the suspending wire and dips into a cup of mercury. If a current is sent through the coil, then at the moment of starting it, a force would require to be supplied in order to produce the angular momentum of the carriers of electricity passing round the coil; and as this must be supplied by the elasticity of the suspending wire, the coil must rotate in the opposite direction.

Maxwell failed to detect this phenomenon experimentally, but it was successfully observed in 1931 by S. J. Barnett[4] (*b.* 1873). Like the preceding effects of electron-inertia, it provides a measure of m/e.

[1] In § 577 of his *Treatise*

[2] *Phys. Rev.*(2) viii (1916), p. 97; ix (1917), p. 164; cf. R. C. Tolman, S. Karrer and E. W. Guernsey, *Phys. Rev.* xxi (1923), p. 525; R. C. Tolman and L. M. Mott-Smith, *Phys. Rev.* xxviii (1926), p. 794

[3] *Treatise*, § 574

[4] *Phil. Mag.* xii (1931), p. 349

In 1908 O. W. Richardson[1] suggested the existence of a mechanical effect accompanying the act of magnetisation. He imagined a long thin cylindrical bar of iron suspended by a fibre, so that it is capable of small rotations about a vertical axis. When the bar is not magnetised, the electrons which are moving in closed orbits in the molecules (Ampère's molecular currents) will not possess any resultant angular momentum, since one azimuth is as probable as another for the orbits. Now consider the effect of suddenly applying a vertical magnetic field: the orbits will orient themselves so as to leave a balance in favour of the plane perpendicular to the direction of the field, and thus an angular momentum of electrons will be created about the axis of suspension. This must be balanced by an equal reaction elsewhere, and therefore a twisting of the suspended system as a whole is to be expected.

Richardson himself did not succeed in obtaining the effect, which was first observed by A. Einstein and W. J. de Haas[2] in 1915. They were followed by many other experimenters, particularly J. Q. Stewart[3] of Princeton and W. Sucksmith and L. F. Bates,[4] working with Professor A. P. Chattock in Bristol. Sucksmith and Bates concluded that the *gyromagnetic ratio*, i.e. the ratio of the angular momentum of an elementary magnet to its magnetic moment, has a value only slightly greater than mc/e. This is only half the value of the ratio for an electron circulating in an orbit, and has been taken to imply that the magnetic elements responsible for the phenomenon are chiefly not orbital electrons but are electrons spinning on their own diameters.[5]

The converse of the Richardson effect has also been observed, namely, it has been found possible to magnetise iron rods by spinning them about their axes. John Perry[6] said in 1890 'Rotating a large mass of iron rapidly in one direction and then in the other in the neighbourhood of a delicately-suspended magnetic needle ought, I think, to give rise to magnetic phenomena. I have hitherto failed to obtain any trace of magnetic action, but I attribute my failure to the comparatively slow speed of rotation which I have employed, and to the want of delicacy of my magnetometer.'

The effect predicted by Perry was anticipated also by Schuster[7] in 1912, but was first observed in 1914–15 by S. J. Barnett[8]: the Ampèrean molecular currents, since they possess angular momentum,

[1] *Phys. Rev.* xxvi (1908), p. 248

[2] *Verh. d. deutsch. phys. Ges.* xvii (1915), p. 152; cf. A. Einstein, ibid. xviii (1916), p. 173 and W. J. de Haas, ibid. xviii (1916), p. 423

[3] *Phys. Rev.* xi (1918), p. 100

[4] *Proc. R.S.*(A), civ (1923), p. 499; cf. W. Sucksmith, *Proc. R.S.*(A), cviii (1925), p. 638; also Emil Beck, *Ann. d. Phys.* lx (1919), p. 109; and G. Arvidsson, *Phys. ZS.* xxi (1920), p. 88

[5] On the gyromagnetic effect for paramagnetic substances, cf. W. Sucksmith, *Proc. R.S.*(A), cxxviii (1930), p. 276; cxxxv (1932), p. 276

[6] J. Perry, *Spinning Tops*, p. 112

[7] *Proc. Phys. Soc.* xxiv (1911–12), p. 121

[8] *Phys. Rev.*(2) vi (1915), p. 239; *Bull. Nat. Res. Council*, iii (1922), p. 235; cf. S. J. Barnett and L. J. H. Barnett, *Phys. Rev.* xx (1922), p. 90; *Nature*, cxii (1923), p. 186

behave like the wheels of gyroscopes, changing their orientation, with a tendency to make their rotations become parallel to the impressed rotation. Thus a preponderance of magnetic moment is caused along the axis of the impressed rotation.

From time to time papers appeared on questions belonging to the same class as the problem of *unipolar induction*. This problem, which had first been considered by Faraday,[1] may be stated as follows. A magnet, symmetrical with regard to an axis about which it rotates, has sliding contact with the ends A and B of a stationary wire ACB, at two points A and B not in the same equatorial plane of the magnet. As Faraday found, a steady current flows through the wire. The question is, do the lines of magnetic induction rotate with the magnet, so that the electromotive force is produced when they cut the stationary wire ACB: or do the lines of magnetic induction remain fixed, so that the electromotive force is produced when the moving part of the circuit (the magnet) rotates through them? Faraday believed the latter explanation to be the true one: but Weber, who introduced the name *unipolar induction*, took the contrary view.[2] It was known in the nineteenth century that the experimental results could be explained equally well on either hypothesis.

More generally, we can consider a magnet which is capable of being rotated about its axis of symmetry (whether it is actually so rotated or not) and a conducting circuit composed of two parts, one of which is rotated about the axis of the magnet (in the case considered above it is the magnet itself) while the other, ACB, remains fixed. It is found that if the moving part of the circuit is distinct from the magnet, the electromotive force is independent of whether the magnet is rotating or not. The electromotive force round the circuit is in all cases, if ω denotes the angular velocity of the moving part of the circuit, $(\omega/2\pi c)$ times the flux of magnetic induction through any cylindrical surface having as boundaries the circles described round the axis of revolution by the two contact points A and B of the fixed with the rotating part of the circuit.

In the twentieth century some new types of experiment were devised. In 1913 Marjorie Wilson and H. A. Wilson[3] constructed a non-conducting magnet by embedding a large number of small steel spheres in a matrix of wax, and rotated this in a magnetic field whose direction was parallel to the axis of rotation. Their measures of the induced electromotive force were in satisfactory agreement with the predictions of the electromagnetic theory of moving bodies published by A. Einstein and J. Laub[4] in 1908.

Further contributions to problems of this class were made by

[1] cf. Vol. I, pp. 173–4

[2] loc. cit., Vol. I

[3] *Proc. R.S.*(A), lxxxix (1913), p. 99

[4] *Ann. d. Phys.* xxvi (1908), p. 532

S. J. Barnett,[1] E. H. Kennard,[2] G. B. Pegram,[3] and W. F. G. Swann,[4] and in 1922 a comprehensive review of the whole subject was published by J. T. Tate.[5]

The problems of classical electromagnetic theory which were studied in the early years of the twentieth century related chiefly to the expression of the field due to a moving electron, and the rate at which energy is radiated from it outwards. Formulae for the electric and magnetic vectors of the field were given by many writers.[6] The most interesting terms are those which involve the acceleration of the electron. Denote by (v_x, v_y, v_z) the velocity and by $(w_x, w_y\ w_z)$ the acceleration of the electron at the instant $\bar{t}$ when it emits the radiation which reaches the point (x, y, z) at the instant t: let the co-ordinates of the electron at time t' be (x', y', z'), and let $x'(\bar{t}), y'(\bar{t}), z'(\bar{t})$, be denoted by $(\bar{x}', \bar{y}', \bar{z}')$; and let

$$\bar{r}^2 = (\bar{x}' - x)^2 + (\bar{y}' - y)^2 + (\bar{z}' - z)^2,$$

so

$$\bar{t} = t - \bar{r}/c.$$

Then the terms in the x-component of the electric force which involve the acceleration are

$$\mathrm{D}_x = -\frac{ew_x}{\bar{r}s^2} + \frac{e\{(\bar{x}' - x)w_x + (\bar{y}' - y)w_y + (\bar{z}' - z)w_z\}}{\bar{r}^2\, s^3}\left\{\frac{\bar{x}' - x}{\bar{r}} + \frac{v_x}{c}\right\}$$

where

$$s = 1 + \frac{(\bar{x}' - x)v_x + (\bar{y}' - y)v_y + (\bar{z}' - z)v_z}{c\bar{r}}.$$

From this we have at once

$$\mathrm{D}_x\,(\bar{x}' - x) + \mathrm{D}_y\,(\bar{y}' - y) + \mathrm{D}_z\,(\bar{z}' - z) = 0$$

so *the vector* $(\mathrm{D}_x, \mathrm{D}_y, \mathrm{D}_z)$ *is perpendicular to* $\bar{\mathbf{r}}$. Moreover, the part of the magnetic vector which depends on the acceleration of the electron (call it **H**) *is perpendicular to both* **D** *and* $\bar{\mathbf{r}}$, *and equal in magnitude to* **D**. So the *wave of acceleration*, as Langevin [7] called the field specified by **D** and **H**, *has all the characters of a wave of light.*

At great distances from the electron, the intensity of **D** and **H** decreases like $1/r$, whereas the intensity of the terms in the electric

[1] *Phys. ZS.* xiii (1912), p. 803; *Phys. Rev.* xxxv (1912), p. 323
[2] *Phys. ZS.* xiii (1912), p. 1155; *Phys. Rev.*(2) i (1913), p. 355
[3] *Phys. Rev.* x (1917), p. 591
[4] *Phys. Rev.* xv (1920), p. 365
[5] *Bull. Nat. Res. Council*, iv, Part 6 (1922), p. 75
[6] K. Schwarzschild, *Gött. Nach.* (1903), p. 132. G. Herglotz, *Gött. Nach.* (1903), p. 357. H. A. Lorentz, *Proc. Amst. Acad.* v (1903), p. 608. A. W. Conway, *Proc. Lond. M.S.*(2) i (1903), p. 154. A. Sommerfeld, *Gött. Nach.* (1904), p. 99. P. Langevin, *J. de phys.* iv (1905), p. 165. H. Poincarè, *Palermo Rend.* xxi (1906), p. 129. G. A. Schott, *Ann. d. Phys.* xxiv (1907), p. 637; xxv (1908), p. 63. F. R. Sharpe, *Bull. Amer. Math. Soc.* xiv (1908), p. 330. A. Sommerfeld, *Munich Sitz.* (1911), p. 51. A. W. Conway, *Proc. R.I.A.* xxix, A (1911), p. 1; *Proc. R.S.*(A), xciv (1918), p. 436
[7] loc. cit.

and magnetic vectors which do not involve w decrease like $1/r^2$: so *at great distances the field consists solely of the wave of acceleration*, which represents the *radiation* emitted by the electron.

The rate of loss of energy by radiation from a charge e moving with an acceleration w and a velocity small compared with c is, as we have seen.[1]

$$\frac{2}{3}\frac{e^2w^2}{c^3}.$$

When v/c is no longer neglected, it was shown by Heaviside[2] that *the energy radiated per second is*

$$\frac{2}{3}\frac{e^2w^2}{c^3}\cdot\frac{1-(v^2/c^2)\sin^2 \widehat{vw}}{(1-v^2/c^2)^3}.$$

The question as to a reaction or 'back pressure' experienced by a moving mass on account of its own emission of radiant energy was discussed by O. Heaviside,[3] M. Abraham,[4] J. H. Poynting,[5] A. W. Conway,[6] J. Larmor[7] and Leigh Page.[8]

A striking unification of electromagnetic theory was published in 1912 by Leigh Page[9] (*b.* 1884). It had been realised long before by Priestley[10] that from the experimental fact that there is no electric force in the space inside a charged closed hollow conductor, it is possible to deduce the law of the inverse square between electric charges, and so the whole science of electrostatics. It was now shown by Page that if a knowledge of the relativity theory of Poincaré and Lorentz is assumed, the effect of electric charges in motion can be deduced from a knowledge of their behaviour when at rest, and thus the existence of magnetic force may be inferred from electrostatics: magnetic force is in fact merely a name introduced in order to describe those terms in the ponderomotive force on an electron which depend on its velocity. In this way Page showed that Ampère's law for the force between current-elements, Faraday's law of the induction of currents and the whole of the Maxwellian electromagnetic theory, can be derived from the simple assertion of the absence of electric effects within a charged closed hollow conductor.

In 1914 two new representations of electromagnetic actions were introduced, both of which were evidently inspired by Maxwell's *Encyclopaedia Britannica* article on the aether, in which it was regarded as composed of corpuscles, moving in all directions with the velocity

[1] cf. Vol. I, p. 396
[2] *Nature*, lxvii (1902), p. 6
[3] *Nature*, lxvii (1902), p. 6
[4] *Ann. d. Phys.* x (1903), p. 156; *Brit. Ass. Rep., Cambridge* 1904, p. 436
[5] *Phil. Trans.* ccii (1904), p. 525
[6] *Proc. R.I.A.* xxvii (1908), p. 169
[7] *Proc. Int. Cong. Math., Cambridge* 1912, Vol. I, p. 213; *Nature*, xcix (1917), p. 404
[8] *Phys. Rev.* xi (1918), p. 376
[9] *Amer. J. Sci.* xxxiv (1912), p. 57; *Phys. ZS.* xiii (1912), p. 609
[10] cf. Vol. I, p. 53

of light, never colliding with each other, and possessing some vector quality such as rotation.

The first of these representations was due to Ebenezer Cunningham of Cambridge[1] (*b.* 1881). Cunningham remarked that if the aether is supposed to be at rest, as it is in Lorentz's theory, then the transfer of energy represented by the Poynting vector cannot be identified with the rate of work of the stress in the aether. This, indeed, can be done only if the aether is supposed to be in motion ; for a stress on a stationary element of area does not transmit any energy across that element. He therefore proposed to assign a velocity to the aether at every point, such that a state of stress in the medium would account both for the transference of momentum and for the flow of energy.

He found that the component of the aether-velocity which is in the direction of the Poynting vector must be the smaller root of the quadratic equation $(c^2+x^2)g=2Wx$ where g is the value of the momentum $(1/c)$ $[\mathbf{E}\,.\,\mathbf{H}]$, and W is the density of energy : that the other component lies in a definite direction in the plane of **E** and **H** : and that the *total* velocity of the aether at every point must be equal to c : so the direction and magnitude of the aethereal velocity are completely determined. The scheme is relativistically invariant : so that an objective aether is not necessarily foreign to the point of view of the principle of relativity : and the mechanical categories of momentum, energy, and stress can also be maintained in their entirety. The only essential modification of the ordinary theory of material media is that (as in relativity generally) we no longer take momentum to be in the direction of, or proportional to, the velocity.

The *true stress* in the aether may be defined as differing from the Faraday-Maxwell stress by an amount representing the rate at which momentum is *convected* by the aether across stationary elements of area. This 'true stress' consists of a tension P, defined by the equation

$$P^2=W^2-c^2g^2$$

in the direction of the component velocity of the aether in the plane of **E** and **H**, together with an equal pressure P in all directions at right angles to this. It is this 'true stress' which does work by acting on moving elements, and so transfers energy.

Cunningham gave examples of the determination of the aether-velocity in some simple cases. For a train of plane waves of light the stress P vanishes, and the system is one of pure convection : the aether moves as a whole in the direction of propagation of the waves. For a moving point charge, the aether moves as if continually emitted from the charge with velocity c, every element travelling uniformly in a straight line after emission.

[1] E. Cunningham, *The Principle of Relativity*, Cambridge, 1914 ; *Relativity and the Electron Theory*, London, 1915

The continual emission from an electron of corpuscles moving with velocity c in all directions is the chief feature also of an *emission theory of electromagnetism* published in the same year (1914) by Leigh Page.[1] Page's work is in some ways reminiscent of ideas which had been put forward by J. J. Thomson in an Adamson lecture[2] delivered in Manchester University in 1907, when the concept of aether in motion with velocity c was brought into relation with Thomson's favourite concept of moving lines of electric force.

Page proposed to regard an electron, when viewed in an inertial system in which it is at rest, as being like a sphere whose surface is studded with *emittors* distributed uniformly over it : each emittor is continually projecting into the surrounding space a stream of *corpuscles*, each corpuscle moving radially in a straight line with the velocity of light : it is assumed that the emittors have no rotation relative to the inertial system. When the electron is in motion in any way, the stream of corpuscles that have been ejected from any one emittor at successive instants form a curve in space which, as Page showed, is a *line of electric force* in the field due to the moving electron.

As in Thomson's theory,[3] each electron is supposed to possess its own system of lines of force, independently of other electrons, so that in general there will be as many lines of force crossing at a point as there are electrons in the field.

Owing to the motion of the electron, the direction of the line of force at a point of space does not generally coincide with the direction of motion of the corpuscles at the point. The component of the electric vector **d** in any direction is measured by the number of lines of force which cross unit area at right angles to this direction. The magnetic vector **h** is defined as in Thomson's papers, by the equation

$$\mathbf{h}=\frac{1}{c}\left[\mathbf{c}.\,\mathbf{d}\right],$$

where **c** is the (vector) velocity of the corpuscles at the point. Page worked out the consequences of these assumptions for an electron of given velocity and acceleration, and found expressions for **d** and **h** at any point of the field at any time, which agreed exactly with the values deduced by previous investigators from the Maxwell-Lorentz equations.

In 1913 the existence of a hitherto unknown phenomenon was deduced by John Gaston Leathem[4] (1871–1923) of Cambridge,

[1] *Amer. J. Sci.* xxxviii (1914), p. 169 ; L. Page, *An Introduction to Electrodynamics*, Boston, 1922 ; L. Page and N. I. Adams, *Electrodynamics*, London, 1941

[2] *Manchester Univ. Lectures*, No. 8 (Manchester Univ. Press, 1908) : reprinted in the *Smithsonian Report* for 1908, p. 233 ; cf. also J. J. Thomson, *Phil. Mag.* xxxix (1920), p. 679, and *Mem. and Proc. Manchester Lit. and Phil. Soc.* lxxv (1930–1), p. 77

[3] cf. J. J. Thomson, *Proc. Camb. Phil. Soc.* xv (1909), p. 65

[4] *Proc. R.S.*(A), lxxxix (1913), p. 31 ; cf. Larmor's *Collected Papers*, Vol. II, p. 72

from classical electrodynamics, namely, a minute mechanical force exerted by a varying electric field on a magnetic dipole. On the assumption that the magnetism is due to molecular electric currents, he found that the ponderomotive force[1] on the dipole is

$$\left(m_x\frac{\partial}{\partial x}+m_y\frac{\partial}{\partial y}+m_z\frac{\partial}{\partial z}\right)\mathbf{H}+4\pi\left[\mathbf{m},\frac{\partial\mathbf{D}}{\partial t}\right]$$

where **m** is the magnetic moment of the dipole, **H** is the magnetic force, and **D** is the electric displacement. The first term is the ordinary formula for the force exerted on a magnetic dipole, regarded as a polarised combination of positive and negative magnetism, by a field of magnetic force. The second term was unexpected: it represents a mechanical force exerted on a magnet by a displacement-current, at right angles to the displacement-current and to the magnetic moment, and proportional to the product of the two and the sine of the angle between them.

It might seem possible to test the matter by hanging a small magnet horizontally between the horizontal plates of a charged condenser and then effecting a non-oscillatory discharge of the condenser. If the upper plate were originally charged with positive electricity, the displacement-current on discharge would be upwards, and an eastwards impulse on the magnet might be looked for: the effect would, however, be too small to be observed.

In the first quarter of the twentieth century several interesting results in classical electrodynamics were discovered by Richard Hargreaves (1853–1939). In 1908 he proved[2] that Lorentz's fundamental equations (with $c=1$) can be replaced by two integral-equations

$$\iint(\mathrm{H}_x dydz+\mathrm{H}_y dzdx+\mathrm{H}_z dxdy+\mathrm{E}_x dxdt+\mathrm{E}_y dydt+\mathrm{E}_z dzdt)=0 \qquad \text{(I)}$$

and

$$\iint(\mathrm{E}_x dydz+\mathrm{E}_y dzdx+\mathrm{E}_z dxdy-\mathrm{H}_x dxdt-\mathrm{H}_y dydt-\mathrm{H}_z dzdt)$$

$$=-\iiint(\rho w_x dydzdt+\rho w_y dzdxdt+\rho w_z dxdydt-\rho dxdydz). \qquad \text{(II)}$$

Here (t, x, y, z) denotes the co-ordinates of a point in space-time, **E** is the electric and **H** the magnetic vector, and **w** is the velocity of the charge-density ρ. Let any closed two-dimensional manifold S_2 in the four-dimensional space-time be assigned, and let S_2 be the boundary of a three-dimensional manifold S_3 in which the co-ordinates t, x, y, z are functions of three parameters α, β, γ, of which $\gamma=0$ on S_2 and $\gamma<0$ in S_3: so that on S_2 the co-ordinates

[1] In electromagnetic units

[2] *Trans. Camb. Phil. Soc.* xxi (1908), p. 107

are functions of α, β only. Then any term such as $\iint H_x dydz$ may be interpreted to mean $\iint H_x\, \partial(y, z)/\partial(\alpha, \beta)\, d\alpha d\beta$ taken over S_2, and any term such as $\iiint \rho dxdydz$ may be interpreted to mean

$$\iiint \rho \frac{\partial(x, y, z)}{\partial(\alpha, \beta, \gamma)} d\alpha d\beta d\gamma$$

taken over S_3. In the general case the quantities occurring in these equations are evaluated at different points of space *at different times*: the integrals are thus more general than the usual surface and volume integrals.

Let S be an arbitrary closed surface in the (x, y, z) space, and let t be expressed in terms of (x, y, z) by an arbitrary law $t = t(x, y, z)$. Each particle is supposed to be within S at the moment when its charge is evaluated, but since the charges on the particles are evaluated at different times, the particles need not all be within the closed surface at a given time. This explains why the *total charge on the particles* is not $\iiint \rho dxdydz$, but is the triple integral on the right-hand side of equation (II).

Hargreaves showed further that the equations

$$E_x = -\frac{\partial \Phi}{\partial x} - \frac{\partial A_x}{\partial t}, \qquad H_x = \frac{\partial A_z}{\partial y} - \frac{\partial A_y}{\partial z} \quad \text{etc.,}$$

which express the electric and magnetic vectors in terms of the scalar and vector potentials, are equivalent to the single integral equation

$$\int (A_x dx + A_y dy + A_z dz - \Phi dt)$$
$$= \iint (H_x dydz + H_y dzdx + H_z dxdy + E_x dxdt + E_y dydt + E_z dzdt).$$

Here we are considering a closed curve, and a surface bounded by this curve, and we can suppose that t is expressed in terms of (x, y, z) by an arbitrary known law: then the line-integral may be understood to mean

$$\int \left[\left(A_x - \Phi \frac{\partial t}{\partial x}\right) dx + \left(A_y - \Phi \frac{\partial t}{\partial y}\right) dy + \left(A_z - \Phi \frac{\partial t}{\partial z}\right) dz \right].$$

In 1920 Hargreaves[1] generalised Maxwell's theory of light-pressure[2] by showing that when a general electromagnetic field is present near the surface of any perfectly-reflecting (i.e. perfectly-conducting) body, which is in motion in any way, the pressure is normal, and is measured by the difference between the magnetic and electric energies (per unit volume) at the surface.

[1] *Phil. Mag.* xxxix (1920), p. 662 [2] cf. Vol. I, pp. 274–5

In 1922 Hargreaves[1] discovered remarkable expressions for the scalar and vector potentials of a moving electron. Let $x'(t)$, $y'(t)$, $z'(t)$ specify the position of the electron at time $\bar{t}$: let (t, x, y, z) be the world-point for which the potentials are to be calculated: let $\bar{t}$ be the value of t for the electron at the instant when it emits actions which reach the point (x, y, z) at the instant t: let $\bar{x}'$ denote $x'(\bar{t})$, let $\bar{y}'$ denote $y'(\bar{t})$ and let $\bar{z}'$ denote $z'(\bar{t})$. Then Hargreaves showed that *the scalar potential has the value*

$$\Phi = -\tfrac{1}{2}\, ec \left(\frac{\partial^2}{\partial x^2} + \frac{\partial^2}{\partial y^2} + \frac{\partial^2}{\partial z^2} - \frac{1}{c^2}\frac{\partial^2}{\partial t^2}\right)(\bar{t} - t),$$

and *the x-component of the vector-potential* is

$$A_x = -\tfrac{1}{2}e\left(\frac{\partial^2}{\partial x^2} + \frac{\partial^2}{\partial y^2} + \frac{\partial^2}{\partial z^2} - \frac{1}{c^2}\frac{\partial^2}{\partial t^2}\right)(\bar{x}' - x).$$

Other properties found by him were:

$$\left(\frac{\partial \bar{t}}{\partial x}\right)^2 + \left(\frac{\partial \bar{t}}{\partial y}\right)^2 + \left(\frac{\partial \bar{t}}{\partial z}\right)^2 - \frac{1}{c^2}\left(\frac{\partial \bar{t}}{\partial t}\right)^2 = 0$$

and

$$\left(\frac{\partial^2}{\partial x^2} + \frac{\partial^2}{\partial y^2} + \frac{\partial^2}{\partial z^2} - \frac{1}{c^2}\frac{\partial^2}{\partial t^2}\right)^2 (\bar{t} - t) = 0:$$

in fact, the harmonic operator applied twice to any function of $\bar{t}$ yields a zero result, except at the source.

[1] *Mess. of Math.* lii (1922), p. 34

Chapter VIII

THE DISCOVERY OF MATRIX-MECHANICS

A young German named Werner Heisenberg (*b.* 1901), shortly after taking his doctor's degree in 1923 under Sommerfeld at Munich, moved to Niels Bohr's research school at Copenhagen. Here he became closely associated with H. A. Kramers, who in 1924 made important contributions to the theory of dispersion,[1] in the development of which Heisenberg took part.

The concepts employed in this theory suggested to Heisenberg a new approach to the general problems of atomic theory. It will be remembered that the quantum theory of dispersion had originated in Ladenburg's successful translation into quantum language of the analysis that was used in the classical theory. In place of classical electrons in motion within the atom, Ladenburg introduced into the formulae transitions between stationary states: so that instead of the atom being regarded as a Rutherford planetary system of nucleus and electrons obeying the laws of classical dynamics, its behaviour with respect to incident radiation was predicted by means of calculations based on the 'virtual orchestra.'[2] The great advantage thereby gained depended on the circumstance that the motions of the electrons in the Rutherford planetary system were completely unobservable, and did not yield directly the frequencies of the spectral lines emitted by the atom: whereas the virtual orchestra emitted radiations of the frequencies that were actually observed, and thus was much more closely related to physical experiments.

Heisenberg saw that this idea of replacing the classical dynamics of the Rutherford atom by formulae based on the virtual orchestra could be applied in a far wider connection. He took as his primary aim to lay the foundations of a quantum-theoretic mechanics which should be based exclusively on relations between quantities that are actually observable. Previous investigators had found integrals of the classical equations of motion of the atomic system, and so had obtained formulae for the co-ordinates and velocities of the electrons as functions of the time. These formulae Heisenberg now abandoned, on the ground that they do not represent anything that is accessible to direct observation: and in their place he proposed to make the virtual orchestra the central feature of atomic theory.

By taking this step, he made it no longer necessary to find first the classical solution of a problem and then translate it into quantum language: he proved, in fact, that it is possible to translate the

[1] cf. p. 203 *supra* [2] cf. p. 204 *supra*

classical problem into a quantum problem at the very beginning, before solving it: that is to say, he translated the fundamental laws of classical dynamics into a system of fundamental quantum laws constituting what Born [1] had adumbrated under the name *quantum mechanics.*

He considered problems defined by differential equations, such as, for instance, the problem of the anharmonic oscillator, which is defined by the differential equation

$$\frac{d^2x}{dt^2} + \omega_0{}^2 x + \lambda x^2 = 0$$

and inquired how a solution of this equation can be obtained which, like a virtual orchestra, refers to a doubly-infinite aggregate of transitions between stationary states. Let us suppose that x can be *represented* by an aggregate of terms x_{mn}, where x_{mn} is associated with the transition between the stationary states m and n: and let these terms be arranged in a double array thus:

$$\begin{matrix} x_{11} & x_{12} & x_{13} & x_{14} & \cdot & \cdot \\ x_{21} & x_{22} & x_{23} & x_{24} & \cdot & \cdot \\ x_{31} & x_{32} & x_{33} & x_{34} & \cdot & \cdot \\ \cdot & \cdot & \cdot & \cdot & \cdot & \cdot \\ \cdot & \cdot & \cdot & \cdot & \cdot & \cdot \end{matrix}$$

The differential equation involves x^2, and therefore Heisenberg suggested that x^2 should be capable of being represented by a similar array

$$\begin{matrix} (x^2)_{11} & (x^2)_{12} & (x^2)_{13} & \cdot\ \cdot\ \cdot\ \cdot \\ (x^2)_{21} & (x^2)_{22} & (x^2)_{23} & \cdot\ \cdot\ \cdot\ \cdot \\ (x^2)_{31} & (x^2)_{32} & (x^2)_{33} & \cdot\ \cdot\ \cdot\ \cdot \\ \cdot & \cdot & \cdot & \cdot\ \cdot\ \cdot\ \cdot \\ \cdot & \cdot & \cdot & \cdot\ \cdot\ \cdot\ \cdot \end{matrix}$$

The question is, what is the relation between the elements $(x^2)_{mn}$ and the elements x_{mn}? Now since x_{mn} refers to the transition between the stationary states m and n, we may expect it to have the time-factor $e^{2\pi i\nu(m,\,n)t}$, where $\nu(m, n) = (W_m - W_n)/h$ is the frequency of

[1] cf. p. 204 *supra*

the spectral line associated with this transition. Guided by the principle that the time-factor of $(x^2)_{mn}$ must be the same as the time-factor of x_{mn}, Heisenberg suggested that the element $(x^2)_{mn}$ could be expressed in terms of the elements of the x-array by the equation

$$(x^2)_{mn} = \sum_r x_{mr}\, x_{rn}.$$

He went further and proposed that if x and y are two different co-ordinates, then

$$(xy)_{mn} = \sum_r x_{mr}\, y_{rn}.$$

Since this involves the equation

$$(yx)_{mn} = \sum_r y_{mr}\, x_{rn},$$

it is evident that in general the products xy and yx are represented by different arrays: *multiplication of the quantum-theoretic x and y is not commutative.*

During the month of June 1925 Heisenberg, then on holiday in the island of Heligoland, worked out the solution of the anharmonic oscillator equation according to these ideas, and was delighted to find that principles such as the conservation of energy could be fitted into his system. In the first week of July 1925,[1] he wrote a paper embodying his new theory. Being then on Professor Max Born's staff at Göttingen, he took the MS to Born and asked him to read it, at the same time asking for leave of absence for the rest of the term (which ended about 1 August), as he had been invited to lecture at the Cavendish Laboratory in Cambridge. Born, who granted the leave of absence and then read the MS., at once recognised and identified the law of multiplication which Heisenberg had introduced for the virtual-orchestra arrays: for having attended lectures on non-commutative algebras by Rosanes in Breslau and then discussed the subject with Otto Toeplitz in Göttingen, Born saw that Heisenberg's law was simply the law of multiplication of *matrices.* For the benefit of readers who are not acquainted with matrix-theory, some simple explanations may be given here. Consider any square array

$$\begin{pmatrix} a_{11} & a_{12} & a_{13} & \dots & a_{1n} \\ a_{21} & a_{22} & a_{23} & \dots & a_{2n} \\ \cdot & \cdot & \cdot & \cdot & \cdot \\ \cdot & \cdot & \cdot & \cdot & \cdot \\ a_{n1} & a_{n2} & a_{n3} & \dots & a_{nn} \end{pmatrix}$$

[1] *ZS. f. P.* xxxiii (1925), p. 879 (received 29 July 1925)

formed of ordinary (real or complex) numbers a_{11}, a_{12}, . . ., which we shall call the *elements*. This array we shall call a *matrix*,[1] and we shall regard it as capable of undergoing operations such as addition and multiplication—in fact, as a kind of generalised *number*, which can be represented by a single letter. The number n of rows or columns is called the *order*. Two matrices of the same order

$$A \equiv \begin{pmatrix} a_{11} & a_{12} & \cdots & a_{1n} \\ a_{21} & a_{22} & \cdots & a_{2n} \\ \cdot & \cdot & \cdots & \cdot \\ \cdot & \cdot & \cdots & \cdot \\ a_{n1} & a_{n2} & \cdots & a_{nn} \end{pmatrix} \qquad B \equiv \begin{pmatrix} b_{11} & b_{12} & \cdots & b_{1n} \\ b_{21} & b_{22} & \cdots & b_{2n} \\ \cdot & \cdot & \cdots & \cdot \\ \cdot & \cdot & \cdots & \cdot \\ b_{n1} & b_{n2} & \cdots & b_{nn} \end{pmatrix}$$

are said to be *equal* when their elements are equal, each to each: that is

$$a_{pq} = b_{pq} \qquad (p, q = 1, 2, \ldots n).$$

The *sum* of A and B is defined to be the matrix

$$\begin{pmatrix} a_{11}+b_{11} & a_{12}+b_{12} & \cdots & a_{1n}+b_{1n} \\ a_{21}+b_{21} & a_{22}+b_{22} & \cdots & a_{2n}+b_{2n} \\ \cdot & \cdot & \cdots & \cdot \\ \cdot & \cdot & \cdots & \cdot \\ a_{n1}+b_{n1} & a_{n2}+b_{n2} & \cdots & a_{nn}+b_{nn} \end{pmatrix}$$

It is denoted by A + B. Evidently addition so defined is *commutative* (that is, A + B = B + A) and *associative* (that is, [A + B] + C = A + [B + C]). The *null matrix* is defined to be a matrix all of whose elements are zero: and the *unit matrix* is defined to be the matrix

$$\begin{pmatrix} 1 & 0 & 0 & \cdots & 0 \\ 0 & 1 & 0 & \cdots & 0 \\ 0 & 0 & 1 & \cdots & 0 \\ \cdot & \cdot & \cdot & \cdots & \cdot \\ \cdot & \cdot & \cdot & \cdots & \cdot \\ 0 & 0 & 0 & \cdots & 1 \end{pmatrix}$$

If k denotes an ordinary real or complex number, the matrix

$$\begin{pmatrix} ka_{11} & ka_{12} & \cdots & ka_{1n} \\ ka_{21} & ka_{22} & \cdots & ka_{2n} \\ \cdot & \cdot & \cdots & \cdot \\ \cdot & \cdot & \cdots & \cdot \\ ka_{n1} & ka_{n2} & \cdots & ka_{nn} \end{pmatrix}$$

[1] The term *matrix* was introduced in 1850 by J. J. Sylvester (Papers I, p. 145); but the theory was really founded by A. Cayley, *J. für Math.* 1 (1855), p. 282 and *Phil. Trans.* cxlviii (1858), p. 17. Under the name *linear vector operator* the same idea had been developed previously by Sir W. R. Hamilton: some of the more important theorems are given in his *Lectures on Quaternions* (1852).

which is obtained by multiplying every element of the matrix A by k, is called the *product* of A by k, and denoted by kA.

We shall now define the multiplication of two matrices. With the above matrices A and B we can associate two linear substitutions

$$\begin{cases} y_1 = a_{11}x_1 + a_{12}x_2 + \quad . \; . \; . \quad + a_{1n}x_n \\ y_2 = a_{21}x_1 + a_{22}x_2 + \quad . \; . \; . \quad + a_{2n}x_n \\ . \; . \; . \; . \; . \; . \; . \; . \; . \; . \; . \\ . \; . \; . \; . \; . \; . \; . \; . \; . \; . \; . \\ y_n = a_{n1}x_1 + a_{n2}x_2 + \; . \; . \; . \quad + a_{nn}x_n \end{cases}$$

and

$$\begin{cases} u_1 = b_{11}y_1 + b_{12}y_2 + \quad . \; . \; . \quad + b_{1n}y_n \\ u_2 = b_{21}y_1 + b_{22}y_2 + \quad . \; . \; . \quad + b_{2n}y_n \\ . \; . \; . \; . \; . \; . \; . \; . \; . \; . \; . \\ . \; . \; . \; . \; . \; . \; . \; . \; . \; . \; . \\ u_n = b_{n1}y_1 + b_{n2}y_2 + \; . \; . \; . \quad + b_{nn}y_n. \end{cases}$$

A matrix may in fact be regarded as a symbol of linear operation, representing the associated substitution.

The effect of performing these substitutions in succession is represented by the equations

$$\begin{cases} u_1 = (b_{11}a_{11} + b_{12}a_{21} + \; . \; . \; .)x_1 + (b_{11}a_{12} + b_{12}a_{22} + \; . \; . \; .)x_2 + \; . \; . \; . \; . \\ . \; . \; . \; . \; . \; . \; . \; . \\ u_n = (b_{n1}a_{11} + b_{n2}a_{21} + \; . \; . \; .)x_1 + (b_{n1}a_{12} + b_{n2}a_{22} + \; . \; . \; .)x_2 + \; . \; . \; . \; . \end{cases}$$

This suggests that the *product* BA of the two matrices should be defined to be the matrix which has for its element in the pth row and q^{th} column the quantity

$$b_{p1}a_{1q} + b_{p2}a_{2q} + \; . \; . \; . \; + b_{pn}a_{nq},$$

so in multiplying matrices we multiply the *rows* of the first factor, element by element, into the *columns* of the second factor. Multiplication so defined is always possible and unique, and satisfies the associative law

$$A(BC) = (AB)C.$$

It does not, however, satisfy the commutative law: that is, AB and BA are in general two different matrices. The distributive law connecting multiplication with addition, namely,

$$A(B + C) = AB + AC$$
$$(B + C)A = BA + CA$$

is always satisfied.

The derivation, with respect to the time, of a matrix whose elements are A_{mn} is the matrix whose elements are $\partial A_{mn}/\partial t$. The derivative of a function of a matrix with respect to the matrix of which it is a function may be defined by the equation

$$\frac{df(\mathrm{A})}{d\mathrm{A}} = \lim_{k \to 0} \frac{1}{k}\{f(\mathrm{A}+k1) - f(\mathrm{A})\}$$

where 1 denotes the unit matrix: whence we readily see that matrix differentiation obeys the same formal laws as ordinary differentiation.

The above elementary account refers to matrices of *finite* order. The matrices which represent the co-ordinates in quantum theory are of *infinite* order, and differ in some of their properties from finite matrices: in particular, the multiplication of infinite matrices is not necessarily associative (though the associative law holds for row-finite matrices, i.e. those in which each row has only a finite number of non-zero elements).

Let us now return to Born's train of thought. He considered dynamical systems with one degree of freedom, represented classically by the differential equations

$$\frac{dq}{dt} = \frac{\partial \mathrm{H}}{\partial p}, \qquad \frac{dp}{dt} = -\frac{\partial \mathrm{H}}{\partial q},$$

where q denotes the co-ordinate, p the momentum, and H the Hamiltonian function: and following Heisenberg, he retained the form of these equations, but assumed that q and p can be represented by matrices, (representing a matrix by the element in its nth row and m^{th} column)

$$\mathbf{q} = \{q(n, m)e^{2\pi i\nu(n, m) t}\}$$

$$\mathbf{p} = \{p(n, m)e^{2\pi i\nu(n, m) t}\}$$

where $\nu(n, m)$ denotes the frequency belonging to the transition between the stationary states with the quantum numbers n and m, so

$$h\nu(n, m) = \mathrm{W}_n - \mathrm{W}_m$$

where W_n and W_m denote the values of the energy in the states.[1]

Since $\nu(n, n) = 0$, the elements in the leading diagonals of the matrices do not involve the time. This suggests the question, what is the physical meaning of these diagonal-elements? Now in the classical Fourier expansion of a variable, the constant term is equal to the average value (with respect to the time) of that variable in

[1] Since $\nu(m, n) = -\nu(n, m)$, it is natural to assume that the matrices are *Hermitean*, i.e. the elements obtained by interchanging rows and columns are conjugate complex quantities.

It is obvious that if two matrices which have equal frequencies in corresponding elements (m, n) are multiplied together, the frequency of the corresponding element (m, n) in the resulting product-matrix will again be the same as in the factor matrices.

the type of motion considered: so by the correspondence-principle we infer that *a non-temporal element x_{nn} of the matrix representing a variable x is to be interpreted as the average value of the variable x in the stationary state corresponding to the quantum number n, when all phases are equally probable.* The non-diagonal elements have not in general an equally direct physical interpretation: but it is obvious that a knowledge of the non-diagonal elements of the matrix representing a variable x would be necessary in order to calculate the diagonal elements of (say) x^2: and, moreover, we must not lose sight of the connection of the non-diagonal element x_{mn} with the transition between the stationary states m and n: this will be referred to later.

Conversely, if the derivative of a matrix with respect to the time vanishes, the matrix must be a diagonal matrix. Hence the equation of conservation of energy $(d\text{H}/dt) = 0$, has in general[1] the consequence, that *the matrix which represents the energy* H *is a diagonal matrix.* The diagonal element H_{nn} of this matrix represents the average value of H in the stationary state of the system for which the energy is H_n, i.e. it is H_n itself: that is, $\text{H}_{nn} = \text{H}_n$. So Bohr's frequency-condition may be written

$$\nu(n, m) = \frac{\text{H}(n, n) - \text{H}(m, m)}{h}.$$

As might be expected in matrix-calculus, the products **pq** and **qp** are not equal. Now Heisenberg had given a formula, derived originally by W. Kuhn[2] of Copenhagen and W. Thomas[3] of Breslau, which constituted a translation, into the new quantum theory, of the Wilson-Sommerfeld relation

$$\int p\,dq = nh:$$

and from this formula Born deduced that the terms in the leading diagonal of the matrix

$$\mathbf{pq} - \mathbf{qp}$$

must all be equal and must each have the value $h/2\pi i$. He could not, however, obtain the values of the non-diagonal elements in the matrix, though he suspected that they might all be zero.

At this stage (about the middle of July 1925) he called in the help of his other assistant, Pascual Jordan (*b.* 1902), who in a few days succeeded in showing from the canonical equations of motion

$$\frac{d\mathbf{q}}{dt} = \frac{\partial \mathbf{H}}{\partial \mathbf{p}}, \qquad \frac{d\mathbf{p}}{dt} = -\frac{\partial \mathbf{H}}{\partial \mathbf{q}}$$

[1] The case of degenerate systems requires further consideration.
[2] *ZS. f. P.* xxxiii (1925), p. 408
[3] *Naturwiss,* xiii (1925), p. 627

that the derivative of $\mathbf{pq}-\mathbf{qp}$ with respect to the time must vanish : $\mathbf{pq}-\mathbf{qp}$ must therefore be a diagonal matrix, and Born's guess was correct. *The equation thus arrived at, namely,*

$$\mathbf{pq}-\mathbf{qp}=\frac{h}{2\pi i}.\mathbf{1}$$

where $\mathbf{1}$ *denotes the unit matrix, corresponds in the matrix-mechanics to the Wilson-Sommerfeld quantum condition*

$$\int pdq=nh$$

of the older quantum theory. It is known as the *commutation rule.* A proof of it, much simpler than the method by which Born and Jordan established it, is as follows : from the equation

$$\mathbf{q}=\{q(n,\ m)e^{2\pi i\nu(n,\ m)\,t}\}$$

we have

$$\frac{\partial\mathbf{q}}{\partial t}=\frac{i}{\hbar}\{(\mathrm{W}_n-\mathrm{W}_m)q(n,\ m)e^{2\pi i\nu(n,\ m)\,t}\}$$

$$=\frac{i}{\hbar}(\mathbf{Hq}-\mathbf{qH})$$

or

$$\mathbf{Hq}-\mathbf{qH}=\frac{\hbar}{i}\frac{\partial\mathbf{H}}{\partial\mathbf{p}}$$

and therefore the equation

$$\mathbf{fq}-\mathbf{qf}=\frac{\hbar}{i}\frac{\partial\mathbf{f}}{\partial\mathbf{p}} \qquad (1)$$

is valid for $\mathbf{f}=\mathbf{H}$ and $\mathbf{f}=\mathbf{q}$. Now suppose that (1) is valid for any two values of $\mathbf{f}$, say $\mathbf{f}=\mathbf{a}$ and $\mathbf{f}=\mathbf{b}$: then we can show easily that it must be valid for $\mathbf{f}=\mathbf{a}+\mathbf{b}$ and for $\mathbf{f}=\mathbf{ab}$. Since all matrix functions depend on repeated additions and multiplications, we conclude that equation (1) is valid when $\mathbf{f}$ is any function of $\mathbf{H}$ and $\mathbf{q}$. Now the equation $\mathbf{H}=\mathbf{H}(\mathbf{q},\ \mathbf{p})$ determines $\mathbf{p}$ as a function of $\mathbf{H}$ and $\mathbf{q}$: and therefore equation (1) is valid for $\mathbf{f}=\mathbf{p}$: that is, we have[1]

$$\mathbf{pq}-\mathbf{qp}=\frac{\hbar}{i}\mathbf{1}$$

It may be remarked that this relation could not hold if $\mathbf{p}$ and $\mathbf{q}$ were finite matrices, since in that case the sum of the diagonal elements of $\mathbf{pq}-\mathbf{qp}$ would be zero.

Born and Jordan published their discoveries in a paper which was received by the editor of the *Zeitschrift für Physik* on 27 September.[2]

[1] On the physical meaning of the commutation-rule, cf. H. A. Kramers, *Physika*, v (1925), p. 369

[2] *ZS. f. P.* xxxiv (1925), p. 858

In it they applied the theory to the case of the harmonic oscillator. For the harmonic oscillator the Hamiltonian function is

$$\mathrm{H} = \frac{1}{2m}(p^2 + m^2\omega^2 q^2) \tag{1}$$

so the matrix equations of motion are

$$\frac{d\mathbf{q}}{dt} = \frac{\partial \mathbf{H}}{\partial \mathbf{p}} = \frac{1}{m}\mathbf{p}, \quad \frac{d\mathbf{p}}{dt} = -\frac{\partial \mathbf{H}}{\partial \mathbf{q}} = -m\omega^2\mathbf{q}. \tag{2}$$

Hence

$$\frac{d^2\mathbf{q}}{dt^2} + \omega^2\mathbf{q} = 0$$

of which the solution is

$$\mathbf{q} = \mathbf{A}e^{i\omega t} + \mathbf{B}e^{-i\omega t} \tag{3}$$

where **A** and **B** are matrices not involving the time.

Now in a matrix which has the time-factor $e^{i\omega t}$, in the system of reference in which **H** is represented by a diagonal-matrix, the only possible non-zero elements are those to which the time-factor $e^{i\omega t}$ belongs, i.e. those in the sub-diagonal immediately below the principal diagonal. Thus

$$\mathbf{A} = \begin{pmatrix} 0 & 0 & 0 & 0 \ldots \\ \alpha_1 & 0 & 0 & 0 \ldots \\ 0 & \alpha_2 & 0 & 0 \ldots \\ 0 & 0 & \alpha_3 & 0 \ldots \\ \cdot & \cdot & \cdot & \cdot \\ \cdot & \cdot & \cdot & \cdot \end{pmatrix} \text{ and } \mathbf{B} = \begin{pmatrix} 0 & \beta_1 & 0 & 0 \ldots \\ 0 & 0 & \beta_2 & 0 \ldots \\ 0 & 0 & 0 & \beta_3 \ldots \\ \cdot & \cdot & \cdot & \cdot \\ \cdot & \cdot & \cdot & \cdot \end{pmatrix} \tag{4}$$

From (2) and (3)

$$\mathbf{p} = m\frac{d\mathbf{q}}{dt} = im\omega(\mathbf{A}e^{i\omega t} - \mathbf{B}e^{-i\omega t}). \tag{5}$$

From (3) and (5),

$$\mathbf{pq} - \mathbf{qp} = 2im\omega(\mathbf{AB} - \mathbf{BA})$$

or

$$\frac{\hbar}{i} = 2im\omega \begin{pmatrix} -\alpha_1\beta_1 & 0 & 0 & \ldots \\ 0 & \alpha_1\beta_1 - \alpha_2\beta_2 & 0 & \ldots \\ 0 & 0 & \alpha_2\beta_2 - \alpha_3\beta_3 & \ldots \\ \cdot & \cdot & \cdot & \cdot \\ \cdot & \cdot & \cdot & \cdot \end{pmatrix}$$

whence

$$\alpha_1\beta_1 = \frac{\hbar}{2m\omega}, \quad \alpha_2\beta_2 = \frac{2\hbar}{2m\omega}, \quad \alpha_3\beta_3 = \frac{3\hbar}{2m\omega}, \quad \text{etc.} \tag{6}$$

Also $$\mathbf{H} = \frac{1}{2m}(\mathbf{p}^2 + m^2\omega^2\mathbf{q}^2) = m\omega^2(\mathbf{A}\,\mathbf{B} + \mathbf{B}\,\mathbf{A})$$

$$= m\omega^2 \begin{pmatrix} \alpha_1\beta_1 & 0 & 0 & 0\ldots \\ 0 & \alpha_1\beta_1 + \alpha_2\beta_2 & 0 & 0\ldots \\ 0 & 0 & \alpha_2\beta_2 + \alpha_3\beta_3 & 0\ldots \\ \cdots & \cdots & \cdots & \cdots \\ \cdots & \cdots & \cdots & \cdots \end{pmatrix}$$

$$= \frac{\omega\hbar}{2} \begin{pmatrix} 1 & 0 & 0 & 0\ldots \\ 0 & 3 & 0 & 0\ldots \\ 0 & 0 & 5 & 0\ldots \\ \cdots & \cdots & \cdots & \cdots \\ \cdots & \cdots & \cdots & \cdots \end{pmatrix} \tag{7}$$

so $\mathbf{H}$ is a diagonal matrix, as it should be.

Moreover, $\mathbf{q}$ must be a Hermitean matrix, so α_r and β_r must be complex-conjugates : and therefore from (6)

$$\alpha_1 = \left(\frac{\hbar}{2m\omega}\right)^{\frac{1}{2}} e^{i\delta_1}, \quad \alpha_2 = \left(\frac{2\hbar}{2m\omega}\right)^{\frac{1}{2}} e^{i\delta_2}, \quad \alpha_3 = \left(\frac{3\hbar}{2m\omega}\right)^{\frac{1}{2}} e^{i\delta_3}, \ \ldots$$

$$\beta_1 = \left(\frac{\hbar}{2m\omega}\right)^{\frac{1}{2}} e^{-i\delta_1}, \quad \beta_2 = \left(\frac{2\hbar}{2m\omega}\right)^{\frac{1}{2}} e^{-i\delta_2}, \quad \beta_3 = \left(\frac{3\hbar}{2m\omega}\right)^{\frac{1}{2}} e^{-i\delta_3}, \ \ldots$$

where δ_1 δ_2, δ_3, . . . are arbitrary real numbers. If we write $\delta_r = \gamma_r - \pi/2$, from (3), (4) and (8) we have

$$\mathbf{q} = i\left(\frac{\hbar}{2m\omega}\right)^{\frac{1}{2}} \begin{pmatrix} 0 & e^{-i(\omega t+\gamma_1)} & 0 & 0 & \ldots \\ e^{-i(\omega t+\gamma_1)} & 0 & \sqrt{2}.e^{-i(\omega t+\gamma_2)} & 0 & \ldots \\ 0 & -\sqrt{2}e^{i(\omega t+\gamma_2)} & 0 & \sqrt{3}e^{-i(\omega t+\gamma_3)} & \ldots \\ 0 & 0 & -\sqrt{3}e^{i(\omega t+\gamma_3)} & 0 & \ldots \\ \cdots & \cdots & \cdots & \cdots & \cdots \\ \cdots & \cdots & \cdots & \cdots & \cdots \end{pmatrix}$$

and then

$$\mathbf{p} = \left(\frac{m\omega\hbar}{2}\right)^{\frac{1}{2}} \cdot \begin{pmatrix} 0 & e^{-i(\omega t+\gamma_1)} & 0 & 0 & \ldots \\ e^{i(\omega t+\gamma_1)} & 0 & \sqrt{2}e^{-i(\omega t+\gamma_2)} & 0 & \ldots \\ 0 & \sqrt{2}e^{i(\omega t+\gamma_2)} & 0 & \sqrt{3}e^{-i(\omega t+\gamma_3)} & \ldots \\ 0 & 0 & \sqrt{3}e^{i(\omega t+\gamma_3)} & 0 & \ldots \\ \cdots & \cdots & \cdots & \cdots & \cdots \\ \cdots & \cdots & \cdots & \cdots & \cdots \end{pmatrix}$$

Thus *the matrices which represent the co-ordinate and the momentum in the problem of the harmonic oscillator are determined.* Equation (7) shows that *the values of the energy, corresponding to the stationary states of the oscillator,* are

$$\frac{\omega\hbar}{2}, \quad \frac{3\omega\hbar}{2}, \quad \frac{5\omega\hbar}{2} \ldots$$

or in general

$$(n+\tfrac{1}{2})\omega\hbar.$$

The squares of the moduli of the elements of the matrix which represents the electric moment of an atom are in general the measure of its *transition-probabilities.* A connection is thus set up with Einstein's coefficients A_m^n and with Planck's theory of radiation.

Born and Jordan's paper represented a great advance: it contained the formulation of matrix mechanics, the discovery of the commutation law, some simple applications to the harmonic and anharmonic oscillator and (in its last section) a discussion of the quantification of the electromagnetic field.

When the paper had been sent off to the *Zeitschrift für Physik*, Born went for a holiday with his family to the Engadine. On his return in September to Göttingen, there began a hectic time of collaboration with Jordan, and also by correspondence with Heisenberg, who was now in Copenhagen. Born had received and accepted an invitation to lecture during the winter at the Massachusetts Institute of Technology, and was bound to leave at the end of October; but the joint paper was finished in time before his departure, and was received by the editor of the *Zeitschrift für Physik* on 16 November.[1]

The first important result in it was a general method for the solution of quantum-mechanical problems, analogous to the general Hamiltonian theory of classical dynamics. A *canonical transformation* of the variables **p**, **q** to new variables **P**, **Q** was defined to be a transformation for which

$$\mathbf{pq}-\mathbf{qp}=\mathbf{PQ}-\mathbf{QP}=-i\hbar\,;$$

when this equation is satisfied then the canonical equations

$$\frac{d\mathbf{q}}{dt}=\frac{\partial \mathbf{H}}{\partial \mathbf{p}}, \qquad \frac{d\mathbf{p}}{dt}=-\frac{\partial \mathbf{H}}{\partial \mathbf{q}}$$

transform into

$$\frac{d\mathbf{Q}}{dt}=\frac{\partial \mathbf{H}}{\partial \mathbf{P}}, \qquad \frac{d\mathbf{P}}{dt}=-\frac{\partial \mathbf{H}}{\partial \mathbf{Q}}.$$

[1] M. Born, W. Heisenberg and P. Jordan, *ZS. f. P.* xxxv (1926), p. 557

A general transformation which satisfies this condition is

$$\mathbf{P} = \mathbf{SpS}^{-1}$$
$$\mathbf{Q} = \mathbf{SqS}^{-1}$$

where **S** is an arbitrary quantum-theoretic quantity. We then have for any function $f(\mathbf{P}, \mathbf{Q})$.

$$f(\mathbf{P}, \mathbf{Q}) = \mathbf{S}f(\mathbf{p}, \mathbf{q})\mathbf{S}^{-1}.$$

The importance of canonical transformations depends on the following theorem : *matrices* **p** *and* **q** *which satisfy the commutation-rule*

$$\mathbf{pq} - \mathbf{pq} = -i\hbar$$

and which, when substituted in $\mathbf{H}(\mathbf{p}, \mathbf{q})$ *make it a diagonal matrix, represent solutions of the equations*

$$\frac{d\mathbf{q}}{dt} = \frac{\partial \mathbf{H}}{\partial \mathbf{p}}, \qquad \frac{d\mathbf{p}}{dt} = -\frac{\partial \mathbf{H}}{\partial \mathbf{q}}.$$

Therefore if we take any pair of matrices $\mathbf{p}_0$, $\mathbf{q}_0$, which satisfy the commutation-rule (for instance, we might take $\mathbf{p}_0$, $\mathbf{q}_0$, to be the solution of the problem of the harmonic oscillator), then we can reduce the problem of the integration of the canonical equations for a Hamiltonian $\mathbf{H}(\mathbf{p}, \mathbf{q})$ to the following problem : *To determine a matrix* **S** *such that when*

$$\mathbf{p} = \mathbf{Sp}_0\mathbf{S}^{-1}. \qquad \mathbf{q} = \mathbf{Sq}_0\mathbf{S}^{-1},$$

the function

$$\mathbf{H}(\mathbf{p}, \mathbf{q}) \equiv \mathbf{SH}(\mathbf{p}_0, \mathbf{q}_0)\mathbf{S}^{-1}$$

is a diagonal matrix. This last equation is analogous to the Hamilton's partial differential equation of classical dynamics, and **S** corresponds in some measure to the Action-function.

The next question taken up in the memoir of the three authors was the theory of perturbations. Let a problem be defined by a Hamiltonian function

$$\mathbf{H} = \mathbf{H}_0(\mathbf{p}, \mathbf{q}) + \lambda\mathbf{H}_1(\mathbf{p}, \mathbf{q}) + \lambda^2\mathbf{H}_2(\mathbf{p}, \mathbf{q}) + \ldots$$

and suppose that the solution is known of the problem defined by the Hamiltonian function $\mathbf{H}_0(\mathbf{p}, \mathbf{q})$, so that matrices $\mathbf{p}_0$, $\mathbf{q}_0$ are known which satisfy the commutation-rule and make $\mathbf{H}_0(\mathbf{p}_0, \mathbf{q}_0)$ a diagonal matrix. Then it is required to find a transformation-matrix **S** such that if

$$\mathbf{p} = \mathbf{Sp}_0\mathbf{S}^{-1} \quad \text{and} \quad \mathbf{q} = \mathbf{Sq}_0\mathbf{S}^{-1}$$

then

$$\mathbf{H}(\mathbf{p}, \mathbf{q}) = \mathbf{SH}(\mathbf{p}_0, \mathbf{q}_0)\mathbf{S}^{-1}$$

will be a diagonal-matrix. It was shown how to solve this problem by successive approximations, and one of the formulae of Kramers' theory of dispersion was derived.

The theory was then extended to systems with more than one degree of freedom. If the Hamiltonian equations are

$$\frac{d\mathbf{q}_k}{dt}=\frac{\partial \mathbf{H}}{\partial \mathbf{p}_k}, \qquad \frac{d\mathbf{p}_k}{dt}=-\frac{\partial \mathbf{H}}{\partial \mathbf{q}_k},$$

then the commutation-rules are

$$\begin{cases} \mathbf{p}_k\mathbf{q}_l-\mathbf{q}_l\mathbf{p}_k=-i\hbar\delta_{kl}, \text{ where } \delta_{kl}=1 \text{ or } 0 \text{ according as } k=l \text{ or } k\neq l \\ \mathbf{p}_k\mathbf{p}_l-\mathbf{p}_l\mathbf{p}_k=0 \\ \mathbf{q}_k\mathbf{q}_l-\mathbf{q}_l\mathbf{q}_k=0. \end{cases}$$

Among those who listened to Heisenberg's lectures at Cambridge in the summer of 1925 was a young research student named Paul Adrien Maurice Dirac (*b.* 1902), who by a different approach arrived at a theory essentially equivalent to that devised by Born and Jordan. On 7 November he sent to the Royal Society a paper [1] in which Heisenberg's ideas were developed in an original way.

Dirac investigated the form of a quantum operation (denote it by d/dv) that satisfies the laws

$$\frac{d}{dv}\cdot(x+y)=\frac{d}{dv}\cdot x+\frac{d}{dv}\cdot y$$

and

$$\frac{d}{dv}(xy)=\left(\frac{d}{dv}\cdot x\right)y+x\cdot\left(\frac{d}{dv}\cdot y\right).$$

It was found that the most general operation satisfying these laws is

$$\frac{d}{dv}\cdot x=xy-yx$$

where y is some other quantum variable. By considering the limit, when for large quantum numbers the quantum theory passes into the classical theory, Dirac showed that the corresponding expression in classical physics is

$$i\hbar\sum_r\left(\frac{\partial x}{\partial q_r}\frac{\partial y}{\partial p_r}-\frac{\partial y}{\partial q_r}\frac{\partial x}{\partial p_r}\right)$$

where the p's and q's are a set of canonical variables of the system. This is $i\hbar$ multiplied by the well-known *Poisson-bracket expression* [3]

[1] *Proc. R.S.*(A), cix (1925), p. 642
[2] Note that the order of x and y is preserved in the second equation.
[3] cf. Whittaker, *Analytical Dynamics*, § 130

of the functions x and y. Thus, if in quantum theory we define a quantity $[x, y]$ by the equation

$$xy - yx = i\hbar[x, y],$$

then $[x, y]$ is analogous to the Poisson-bracket expressions of classical theory.

Now the Hamiltonian equations of classical theory,

$$\frac{dq_r}{dt} = \frac{\partial H}{\partial p_r}, \qquad \frac{dp_r}{dt} = -\frac{\partial H}{\partial q_r} \qquad (r = 1, 2, \ldots n)$$

may be written in terms of the classical Poisson-brackets (x, y)

$$\frac{dq_r}{dt} = (q_r, H), \qquad \frac{dp_r}{dt} = (p_r, H).$$

The fundamental postulate on which Dirac built his theory was, that *the whole of classical dynamics, so far as it can be expressed in terms of Poisson-brackets instead of derivatives, may be taken over immediately into quantum theory.* Thus, for any quantum-theoretic quantity $\mathbf{x}$ the equation of motion is

$$\frac{d\mathbf{x}}{dt} = [\mathbf{x}, \mathbf{H}] = \frac{1}{i\hbar}(\mathbf{xH} - \mathbf{Hx}).$$

Moreover, if

$$\mathbf{p}_r, \mathbf{q}_r \qquad (r = 1, 2, \ldots n)$$

are any set of canonical variables of the classical system, then we have for the classical Poisson-brackets the values

$$(q_r, q_s) = 0, \qquad (p_r, p_s) = 0, \qquad (q_r, p_s) = \delta_{rs}$$

and therefore in quantum theory we must have

$$\mathbf{q}_r\mathbf{q}_s - \mathbf{q}_s\mathbf{q}_r = 0, \qquad \mathbf{p}_r\mathbf{p}_s - \mathbf{p}_s\mathbf{p}_r = 0, \qquad \mathbf{q}_r\mathbf{p}_s - \mathbf{p}_s\mathbf{q}_r = i\hbar\delta_{rs}\mathbf{1}.$$

Thus Dirac arrived at all the fundamental equations of Heisenberg, Born and Jordan, without explicitly introducing matrices. He introduced the name *q-numbers* for the quantum-mechanical quantities whose multiplication is not in general commutative, and *c-numbers* for ordinary numbers.

In a second paper[1] he applied the theory to the hydrogen atom, and obtained the formula for the Balmer spectrum. This was done at about the same time by Pauli,[2] who observed, moreover, the automatic disappearance of certain difficulties, which had been

[1] *Proc. R.S.*(A), cx (1926), p. 561

[2] *ZS. f. P.* xxxvi (1926), p. 336

found [1] to occur in examining the simultaneous action on the hydrogen atom of crossed electric and magnetic fields. Pauli also gave a matrix-mechanical derivation of the Stark effect.

In the same year Heisenberg and Jordan [2] applied the theory (using the hypothesis of the rotating electron) to the problem of the anomalous Zeeman effect, and obtained Landé's *g*-formula.[3] The fine-structure of spectral doublets in the absence of an external field was also completely explained. Dirac [4] treated the Compton effect by quantum mechanics, and obtained formulae for the angular distribution of the recoil electrons and the scattered radiation which agreed completely with experiment.

Meanwhile Born had been in America since November 1925, and had there met Norbert Wiener (*b.* 1894), who worked with him at the problems of continuous spectra and aperiodic phenomena. In February 1926 they wrote a joint paper [5] on the formulation of the laws of quantification. As they pointed out, the representation of the quantum laws by matrices incurs serious difficulties in the case of aperiodic phenomena. For example, in uniform rectilinear motion, since no periods are present, the matrix which represents the co-ordinate can have no element outside the principal diagonal. They therefore tried to generalise the quantum rules in such a way as to cover these cases, and this they effected by developing a theory of *operators*, representing a co-ordinate by a linear operator instead of by a matrix; and they enunciated the general principle that *to every physical quantity there corresponds an operator.* The case of unaccelerated motion in one dimension, which has no periodic components, was shown to be as amenable to their methods as a periodic motion.

The development of matrix mechanics in the year following Heisenberg's first paper was amazingly rapid: and, only eight months from the date of his discovery, it was supplemented by a parallel theory, which will be described in the next chapter.[6]

[1] O. Klein, *ZS. f. P.* xxii (1924), p. 109; W. Lenz, *ZS. f. P.* xxiv (1924), p. 197
[2] *ZS. f. P.* xxxvii (1926), p. 263
[3] On the Zeeman effect, cf. also S. Goudsmit and G. E. Uhlenbeck, *ZS. f. P.* xxxv (1926), p. 618; L. H. Thomas, *Phil. Mag.* iii (1927), p. 13.
[4] *Proc. R.S.*(A), cxi (1926), p. 405
[5] *J. Math. Phys. Mass. Inst. Tech.* v (1926), p. 84; *ZS. f. P.* xxxvi (1926), p. 174
[6] On the state of matrix-mechanics at the end of 1926, cf. Dirac, *Physical interpretation of quantum dynamics, Proc. R.S.*(A), cxiii (1 Jan. 1927), p. 621

Chapter IX

THE DISCOVERY OF WAVE-MECHANICS

In 1926 a new movement began, which developed from de Broglie's principle,[1] that with any particle of (relativistic) energy E and momentum (p_x, p_y, p_z) there is associated a wave represented by a wave-function

$$\psi = e^{(i/\hbar)\,(Et - p_x x - p_y y - p_z z)}.$$

De Broglie had not hitherto extended his theory to the point of introducing a medium—a kind of aether—whose vibrations might be regarded as constituting the wave, and whose behaviour could be specified by partial differential equations. A step equivalent to this was now taken.

With the above value of ψ, we have

$$\frac{\partial^2\psi}{\partial x^2} = -\frac{p_x{}^2}{\hbar^2}\psi,$$

so

$$\frac{\partial^2\psi}{\partial x^2} + \frac{\partial^2\psi}{\partial y^2} + \frac{\partial^2\psi}{\partial z^2} = -\frac{p^2}{\hbar^2}\psi$$

and

$$\frac{1}{c^2}\frac{\partial^2\psi}{\partial t^2} = -\frac{E^2}{c^2\hbar^2}\psi,$$

so

$$\frac{1}{c^2}\frac{\partial^2\psi}{\partial t^2} - \frac{\partial^2\psi}{\partial x^2} - \frac{\partial^2\psi}{\partial y^2} - \frac{\partial^2\psi}{\partial z^2} = -\left(\frac{E^2}{c^2} - p^2\right)\frac{\psi}{\hbar^2} = -\frac{m^2c^2}{\hbar^2}\psi,$$

where m denotes the mass of the particle. Thus *the de Broglie wave-function satisfies the partial differential equation*

$$\frac{1}{c^2}\frac{\partial^2\psi}{\partial t^2} = \frac{\partial^2\psi}{\partial x^2} + \frac{\partial^2\psi}{\partial y^2} + \frac{\partial^2\psi}{\partial z^2} - \frac{m^2c^2}{\hbar^2}\psi.$$

This equation was discovered by L. de Broglie[2] and others.[3] It is satisfied by all possible de Broglie waves belonging to the particle, but it does not specify their passage into each other as the particle moves—that is, it yields no information as to the variation from moment to moment of the energy and momentum of

[1] cf. p. 214 [2] *Comptes Rendus*, clxxxiii (26 July 1926), p. 272

[3] E. Schrödinger, *Ann. d. Phys.* lxxxi (1926), p. 109; O. Klein, *ZS. f. P.* xxxvii (1926), p. 895, *at* p. 904; cf. also W. Gordon, *ZS. f. P.* xl (1926), p. 117

the particle. What was really wanted was a partial differential equation for the de Broglie waves associated with a particle, which would yield the equations of motion of the particle by a limiting process similar to that by which geometrical optics is derived from physical optics : a partial differential equation, in fact, whose *filiform solutions* (i.e. solutions which are zero everywhere except at points very close to some *curve* in space-time) are the trajectories of the particle. Such an equation can be obtained, at any rate, in the non-relativistic approximation. For from the above equations we have at once

$$\frac{\partial^2\psi}{\partial x^2}+\frac{\partial^2\psi}{\partial y^2}+\frac{\partial^2\psi}{\partial z^2}=\frac{p^2}{\mathrm{E}^2}\,\frac{\partial^2\psi}{\partial t^2}.$$

Substituting the values for p and E in terms of the velocity v of the particle, this becomes

$$\frac{\partial^2\psi}{\partial x^2}+\frac{\partial^2\psi}{\partial y^2}+\frac{\partial^2\psi}{\partial z^2}=\frac{v^2}{c^4}\,\frac{\partial^2\psi}{\partial t^2}. \tag{A}$$

This equation is not relativistically invariant, so we shall restrict ourselves to the non-relativistic case of a particle m moving in a field of potential V, whose equation of energy is

$$\tfrac{1}{2}mv^2+\mathrm{V}=\epsilon$$

where ϵ is its total non-relativistic energy. The equation (A) has filiform solutions, each of which is[1] a null geodesic of the metric for which the square of the element of interval is

$$(c^4/v^2)\,(dt)^2-(dx)^2-(dy)^2-(dz)^2.$$

It can be shown that the null geodesics of this metric are the curves which satisfy the differential equations

$$m\frac{d^2x}{dt^2}=-\frac{\partial\mathrm{V}}{\partial x}\qquad\text{and similar equations in } y \text{ and } z\,;$$

but these are the ordinary Newtonian equations of motion of the particle. Thus *the filiform solutions of equation* (A) *are precisely the trajectories of the particle in the given potential field.* The solution ψ of equation (A) is therefore the wave-function we are seeking.

The time factor in ψ is $\exp\{(i/\hbar)\mathrm{E}t\}$, so if ψ now denotes the wave-function deprived of its time-factor, we have from (A)

$$\begin{aligned}\frac{\partial^2\psi}{\partial x^2}+\frac{\partial^2\psi}{\partial y^2}+\frac{\partial^2\psi}{\partial z^2}&=-\frac{v^2}{c^4}\,\frac{E^2}{\hbar^2}\psi\\&=-\frac{m^2v^2}{\hbar^2(1-v^2/c^2)}\psi.\end{aligned}$$

[1] E. T. Whittaker, *Proc. Camb. Phil. Soc.* xxiv (1928), p. 32

Since we are considering only the non-relativistic approximation, we can replace the factor $(1-v^2/c^2)$ in the denominator by unity, and thus obtain

$$\frac{\partial^2\psi}{\partial x^2}+\frac{\partial^2\psi}{\partial y^2}+\frac{\partial^2\psi}{\partial z^2}=-\frac{m^2v^2}{h^2}\psi$$

or

$$\frac{\partial^2\psi}{\partial x^2}+\frac{\partial^2\psi}{\partial y^2}+\frac{\partial^2\psi}{\partial z^2}+\frac{2m}{\hbar^2}(\epsilon-\mathrm{V})\psi=0.$$

This equation, which was published by Erwin Schrödinger[1] in March 1926, gave the first impetus to the study of *wave-mechanics*. Schrödinger's own approach, which was different from that given above, laid stress on a connection with the theory of Hamilton's Principal Function in dynamics. This will now be considered.

He considered a particle of mass m with momentum p and total energy ϵ in a field of force of potential $\mathrm{V}(x, y, z)$, so that the (non-relativist) equation of energy is

$$\frac{1}{2m}p^2+\mathrm{V}=\epsilon, \quad \text{or} \quad p=\sqrt{\{2m(\epsilon-\mathrm{V})\}}.$$

If we associate with this moving particle a frequency ν given by $\epsilon=h\nu$, and a de Broglie wave-length λ given by $\lambda=h/p$, then the phase-velocity of the de Broglie wave is

$$\varpi=\nu\lambda=\frac{\epsilon}{h}\cdot\frac{h}{p}=\frac{\epsilon}{\sqrt{\{2m(\epsilon-\mathrm{V})\}}}.$$

Take any *phase-surface*, or surface of constant phase, at the instant $t=0$; and suppose that the equation of the phase-surface at the instant t, which has been derived (as in Huygens' Principle) by wave-propagation with the phase-velocity ϖ from this original phase-surface, has the equation

$$\tau(x, y, z)=t.$$

Then by elementary analytical geometry we have

$$\left(\frac{\partial\tau}{\partial x}\right)^2+\left(\frac{\partial\tau}{\partial y}\right)^2+\left(\frac{\partial\tau}{\partial z}\right)^2=\frac{1}{\varpi^2}$$

and therefore

$$\left(\frac{\partial\tau}{\partial x}\right)^2+\left(\frac{\partial\tau}{\partial y}\right)^2+\left(\frac{\partial\tau}{\partial z}\right)^2=\frac{2m(\epsilon-\mathrm{V})}{\epsilon^2}. \tag{1}$$

This equation can, however, be obtained in a very different way.

[1] *Ann. d. Phys.*(4) lxxix (1926), pp. 361, 489

The Hamilton's partial differential equation associated with the particle[1] is

$$\frac{\partial W}{\partial t}+H\left(x, y, z, \frac{\partial W}{\partial x}, \frac{\partial W}{\partial y}, \frac{\partial W}{\partial z}\right)=0,$$

where H is the Hamiltonian function for the particle, namely,

$$H=\frac{1}{2m}(p_x{}^2+p_y{}^2+p_z{}^2)+V(x, y, z)$$

and W is *Hamilton's Principal Function.* The Hamilton's equation is therefore in this case

$$\frac{\partial W}{\partial t}+\frac{1}{2m}\left\{\left(\frac{\partial W}{\partial x}\right)^2+\left(\frac{\partial W}{\partial y}\right)^2+\left(\frac{\partial W}{\partial z}\right)^2\right\}+V(x, y, z)=0.$$

The Principal Function may be written

$$W=-\epsilon t+S(x, y, z)$$

where ϵ denotes the total energy and S is *Hamilton's Characteristic Function.* The equation for S now becomes

$$-\epsilon+\frac{1}{2m}\left\{\left(\frac{\partial S}{\partial x}\right)^2+\left(\frac{\partial S}{\partial y}\right)^2+\left(\frac{\partial S}{\partial z}\right)^2\right\}+V(x, y, z)=0$$

or

$$\left(\frac{\partial S}{\partial x}\right)^2+\left(\frac{\partial S}{\partial y}\right)^2+\left(\frac{\partial S}{\partial z}\right)^2=2m(\epsilon-V).$$

Comparing this with equation (1), we see that *the equation of the phase-surfaces of the de Broglie waves associated with a particle, namely,*

$$\tau(x, y, z)=t$$

is obtained by equating to zero the Hamilton's Principal Function of the particle. Thus *the theory of Hamilton's Principal Function in Dynamics corresponds to Huygens' Principle in Optics.*

This investigation so far belongs to what may be called the ‘geometrical optics’ of the de Broglie waves; a similar equation exists in ordinary optics, the surfaces $\tau(x, y, z)=$ constant being the wave-fronts, whose normals are the ‘rays.’ Schrödinger now put forward the idea that the failure of classical physics to account for quantum phenomena is analogous to the failure of geometrical optics to account for interference and diffraction: and he proposed to create in connection with de Broglie waves a theory analogous to Physical Optics.

Considering the stationary states of an atom, we have seen[2] that the de Broglie wave associated with an electron in an atom

[1] cf. Whittaker, *Analytical Dynamics*, § 142 [2] cf. p. 216 *supra*

returns to the same phase when the electron completes one revolution of an orbit belonging to a stationary state. Therefore at any one point of space the de Broglie disturbance in a stationary state is purely periodic, with the frequency $\nu = \epsilon/h$, where ϵ is the energy of the stationary state; and it can be represented by a *wave-function* of the form

$$\psi = e^{(i\epsilon/\hbar)[-t+\tau(x,y,z)]}.$$

Differentiating this twice, we have

$$\frac{\partial^2\psi}{\partial x^2} = -\frac{\epsilon^2}{\hbar^2}\left(\frac{\partial\tau}{\partial x}\right)^2\psi + \frac{i\epsilon}{\hbar}\frac{\partial^2\tau}{\partial x^2}\psi,$$

which gives, using (1),

$$\frac{\partial^2\psi}{\partial x^2} + \frac{\partial^2\psi}{\partial y^2} + \frac{\partial^2\psi}{\partial z^2} = -\frac{2m(\epsilon - V)}{\hbar^2}\psi + \frac{i\epsilon}{\hbar}\psi\left(\frac{\partial^2\tau}{\partial x^2} + \frac{\partial^2\tau}{\partial y^2} + \frac{\partial^2\tau}{\partial z^2}\right).$$

The ratio of the second term to the first term on the right-hand side of this equation is excessively small if $\hbar$ is excessively small, as it is: so we may neglect the second term, and write

$$\frac{\partial^2\psi}{\partial x^2} + \frac{\partial^2\psi}{\partial y^2} + \frac{\partial^2\psi}{\partial z^2} = -\frac{2m(\epsilon - V)}{\hbar^2}\psi,$$

which is again *Schrödinger's equation for the wave-function* ψ *of the particle.*

The potential function V which occurs in this equation will possess singularities at certain points or at infinity, and these points (with others) will in general be singular points of the solution ψ of the differential equation. But if ψ is to represent the de Broglie wave of a stationary state, it must be free from singularities; and therefore Schrödinger laid down the condition that *the solution of the wave-equation corresponding to a stationary state must be one-valued, finite and free from singularities even at the singularities of* $V(x, y, z)$.[1]

Now it is known that *the partial differential equation possesses a solution of this character only for certain special values of the constant* ϵ, *which are known as the proper-values.*[2] These proper-values will be the only values of the energy ϵ which the atom can have when it is in a stationary state; and thus was justified the title which Schrödinger gave to his first paper, *Quantification as a Problem of Proper-values.*[3]

[1] The conditions laid down in Schrödinger's earlier papers were unnecessarily stringent: on this, cf. G. Jaffé, *ZS. f. P.* lxvi (1930), p. 770; R. H. Langer and N. Rosen, *Phys. Rev.* xxxvii (1931), p. 658; E. H. Kennard, *Nature*, cxxvii (1931), p. 892.

[2] The undesirable hybrid word *eigenvalues* has often been used, but *proper-values*, which is in every way preferable, is used in the English translation of Schrödinger's *Collected Papers on Wave Mechanics* (Blackie and Son, 1928).

[3] 'In the winter of 1926, Born and Jordan having just announced a new development in quantum mechanics, I found more than twenty Americans in Göttingen at this fount of quantum wisdom. A year later they were at Zürich, with Schrödinger. A couple of years later, Heisenberg at Leipzig, and then Dirac at Cambridge, held the Elijah mantle of quantum theory.' (K. T. Compton, *Nature*, cxxxix (1937), p. 222.)

The first problem that Schrödinger investigated by his new method was that of the hydrogen atom, consisting of a proton and an electron; denoting their distance apart by r, so that the potential energy is $-e^2/r$, and neglecting the motion of the nucleus, the wave-equation is (now writing E for the total energy)

$$\frac{\partial^2\psi}{\partial x^2}+\frac{\partial^2\psi}{\partial y^2}+\frac{\partial^2\psi}{\partial z^2}+\frac{2m}{\hbar^2}\left(\mathrm{E}+\frac{e^2}{r}\right)\psi=0.$$

We want the values of E for which solutions of this equation exist that are one-valued and everywhere finite.[1] To find these, we introduce spherical-polar co-ordinates defined by

$$x=r\sin\theta\cos\phi, \qquad y=r\sin\theta\sin\phi, \qquad z=r\cos\theta,$$

and try to obtain a solution of the form

$$\psi=\mathrm{RY}$$

where R is a function of r alone, and Y is a function of θ and ϕ only. The wave-equation now becomes

$$\frac{1}{r^2\mathrm{R}}\frac{d}{dr}\left(r^2\frac{d\mathrm{R}}{dr}\right)+\frac{1}{r^2\mathrm{Y}}\left[\frac{1}{\sin\theta}\frac{\partial}{\partial\theta}\left(\sin\theta\frac{\partial\mathrm{Y}}{\partial\theta}\right)+\frac{1}{\sin^2\theta}\frac{\partial^2\mathrm{Y}}{\partial\phi^2}\right]+\frac{2m}{\hbar^2}\left(\mathrm{E}+\frac{e^2}{r}\right)=0,$$

and this may be broken up into the two equations:

$$\frac{1}{\mathrm{R}}\frac{d}{dr}\left(r^2\frac{d\mathrm{R}}{dr}\right)+\frac{2m}{\hbar^2}(\mathrm{E}r^2+e^2r)=\mathrm{C}$$

$$\frac{1}{\sin\theta}\frac{\partial}{\partial\theta}\left(\sin\theta\frac{\partial\mathrm{Y}}{\partial\theta}\right)+\frac{1}{\sin^2\theta}\frac{\partial^2\mathrm{Y}}{\partial\phi^2}+\mathrm{CY}=0.$$

The latter is the well-known equation of surface harmonics,[2] and the requirements of one-valuedness, finiteness and continuity, make it necessary that C should have the value $n(n+1)$, where n is zero or a positive whole number. The surface-harmonic $\mathrm{Y}(\theta, \phi)$ is then a sum of terms of the form

$$\mathrm{P}_n^{\mu}(\cos\theta)e^{\pm i\mu\phi}$$

where μ is one of the numbers $0, 1, 2, \ldots n$, and where $\mathrm{P}_n^{\mu}(\cos\theta)$

[1] The problem was first solved by Schrödinger, *Ann. d. Phys.* lxxix (1926), p. 361; lxxx (1926), p. 437. For a rigorous proof of the completeness of the set of proper functions, cf. T. H. Gronwall, *Annals of Math.* xxxii (1931), p. 47. It will of course be understood that the solution given above was improved later when relativity, electron-spin, etc. were taken into account.

[2] cf. Whittaker and Watson, *Modern Analysis*, § 18·31

is the Associated Legendre Function.[1] The differential equation for R now becomes

$$\frac{d}{dr}\left(r^2\frac{dR}{dr}\right)+\frac{2m}{\hbar^2}(Er^2+e^2r)R-n(n+1)R=0.$$

From the elementary Bohr theory of the hydrogen atom, we know that for the energy-levels whose differences give rise to the Balmer lines, the total energy E is negative. Suppose this to be the case : writing k for $(e^2/2\hbar)\sqrt{(-2m/E)}$ and z for $(2/\hbar)\sqrt{(-2mE)}r$, the equation becomes

$$\frac{d^2R}{dz^2}+\frac{2}{z}\frac{dR}{dz}+\left\{-\frac{1}{4}+\frac{k}{z}-\frac{n(n+1)}{z^2}\right\}R=0$$

of which the solution that remains finite as $r\to\infty$ is

$$\frac{1}{z}W_{k,\,n+\frac{1}{2}}(z)$$

where $W_{k,\,n+\frac{1}{2}}(z)$ is the confluent hypergeometric function.[2] Thus ψ must be a constant multiple of

$$\frac{1}{r}W_{k,\,n+\frac{1}{2}}(z)P_n^{\mu}(\cos\theta)e^{\pm i\mu\phi}.$$

It is, however, necessary also that ψ should be finite at $r=0$. This condition requires that the asymptotic expansion of $W_{k,\,n+\frac{1}{2}}(z)$,[3] namely,

$$W_{k,\,n+\frac{1}{2}}(z)\sim e^{-\frac{1}{2}z}z^k\left[1+\frac{(n+\frac{1}{2})^2-(k-\frac{1}{2})^2}{1\,!\,z}\right.$$
$$\left.+\frac{\{(n+\frac{1}{2})^2-(k-\frac{1}{2})^2\}\{(n+\frac{1}{2})^2-(k-\frac{3}{2})^2\}}{2\,!\,z^2}+\ldots\right]$$

should terminate ; which evidently can happen only if

$$\pm(n+\tfrac{1}{2})=k-\tfrac{1}{2},\quad\text{or}\quad k-\tfrac{3}{2},\quad\text{or}\quad k-\tfrac{5}{2},\ \ldots,$$

that is, k *must be a whole number* : and on account of the factor z^k, it must be a *positive* whole number : thus

$$\frac{e^2}{2\hbar}\sqrt{\left(-\frac{2m}{E}\right)}$$

must be a positive whole number : so

$$E=-\frac{me^4}{2\hbar^2k^2}\qquad\text{where } k=1,\ 2,\ 3,\ 4,\ \ldots.$$

[1] cf. Whittaker and Watson, *Modern Analysis*, § 15·5
[2] cf. Whittaker and Watson, *Modern Analysis*, § 16·1
[3] cf. Whittaker and Watson, *Modern Analysis*, § 16·3

These values of E *are precisely the energy levels of the stationary orbits in Bohr's theory ; to each value of* E *there corresponds a finite number of particular solutions ;* and thus we obtain *the line spectrum of the hydrogen atom.* It is evident from this equation for E that k *must be identified with the total quantum number.*

The question may now be raised as to the physical significance of the wave function ψ. Schrödinger at first, in a paper [1] received 18 March 1926, supposed that if ψ^* denotes the complex quantity conjugate to ψ, then the space-density of electricity is given by the real part of $\psi\ \partial\psi^*/\partial t$. In a paper [2] received on 10 May, however, he corrected this to $\psi\psi^*$, basing his new result on the fact that the integral of $\psi\psi^*$ taken over all space is, like the charge, constant in time. This interpretation of ψ was, however, soon again modified. The notion of waves which do not transmit energy or momentum, but which determine probability, had become familiar to theoretical physicists from the Bohr-Kramers-Slater theory of 1924 : and in a paper [3] received 25 June, and one received 21 July,[4] both dealing with the treatment of collisions by wave-mechanics, Max Born adopted this conception, and proposed that $\psi\psi^*$ should be interpreted in terms of probability ; to be precise, that $\psi\psi^*\ dx\ dy\ dz$ should be taken to be the *probability* that an electron is in the infinitesimal volume-element $dx\ dy\ dz$. This interpretation was soon universally accepted.

It is convenient to *normalise* the wave-functions by the condition

$$\int \psi_n \psi_n^* d\tau = 1$$

where $d\tau$ denotes the element of volume and the integration is taken over all space. As an example of normalisation, consider the fundamental state of the hydrogen atom, for which $k=1$, $n=0$, $\mu=0$. The wave-function is now some multiple of $(1/r)W_{1,\frac{1}{2}}(z)$, and since $W_{1,\frac{1}{2}}(z)=e^{-\frac{1}{2}z}z$, this gives for the wave-function a multiple of $e^{-\frac{1}{2}z}$, or $e^{-(r/a)}$ where a is the radius of the first circular orbit in Bohr's theory of the hydrogen atom. The wave-function may therefore be written

$$\psi = Ce^{-\frac{r}{a}} \quad \text{where C is a constant.}$$

The normalisation condition is

$$C^2\int_0^\infty \int_0^\pi \int_0^{2\pi} e^{-\frac{2r}{a}} r^2 \sin\theta\ dr d\theta d\phi = 1$$

which gives

$$\pi a^3 C^2 = 1$$

[1] *Ann. d. Phys.*(4) lxxix (1926), p. 734, equation (36)
[2] *Ann. d. Phys.*(4) lxxx, p. 437, note on p. 476
[3] *ZS. f. P.* xxxvii (1926), p. 863
[4] *ZS. f. P.* xxxviii (1926), p. 803

and therefore *the normalised wave-function of the fundamental state of the hydrogen atom is*

$$\psi = \pi^{-\frac{1}{2}} a^{-\frac{3}{2}} e^{-\frac{r}{a}}.$$

Of course this is to be multiplied by the time-factor $e^{-(i/\hbar)E_1 t}$, where E_1 is the energy of the state.

Similarly the normalised wave-function of the hydrogen atom in the state $k=2$, $n=0$, $\mu=0$, is

$$\psi = 2^{-\frac{3}{2}} \pi^{-\frac{1}{2}} a^{-\frac{3}{2}} e^{-\frac{r}{2a}} \left(\frac{r}{2a} - 1\right) e^{-\frac{i}{\hbar}E_2 t}$$

where E_2 is the energy of the state ; and the normalised wave-function of the hydrogen atom in the state $k=2$, $n=1$, $\mu=0$, is

$$\psi = (32\pi a^5)^{-\frac{1}{2}} e^{-\frac{r}{2a}} r \cos\theta \, e^{-\frac{i}{\hbar}E_3 t}$$

where E_3 is the energy of the state.

Now suppose that to two different proper-values of the total energy, say E_n and E_m, there correspond respectively solutions ψ_n and ψ_m of the wave equation, so that

$$\nabla^2 \psi_n + \frac{2m}{\hbar^2}(E_n - V)\psi_n = 0$$

$$\nabla^2 \psi_m + \frac{2m}{\hbar^2}(E_m - V)\psi_m = 0.$$

Multiplying these equations by ψ_m and ψ_n respectively, subtracting and integrating over all space, we have

$$\int (\psi_m \nabla^2 \psi_n - \psi_n \nabla^2 \psi_m) d\tau + \frac{2m}{\hbar^2}(E_n - E_m)\int \psi_n \psi_m d\tau = 0.$$

The first integral may by Green's theorem be transformed into a surface-integral taken over the surface at infinity, which vanishes ; and thus we have

$$\int \psi_n \psi_m d\tau = 0,$$

that is, *two wave-functions* ψ_m *and* ψ_n, *corresponding to different values of* E, *are orthogonal to each other.*

Since in this equation we can replace ψ_m by its complex-conjugate $\psi_m{}^*$, we can combine it with the normalisation-equation into the equation

$$\int \psi_n \psi_m{}^* d\tau = \delta_m{}^n$$

where $\delta_m{}^n = 1$ or 0 according as m and n are the same or different numbers.[1]

In the case of the hydrogen atom, a single proper-value of the energy, say

$$E = -\frac{me^4}{2\hbar^2 k^2}$$

where k is a definite whole number, corresponds to many wave-functions

$$\frac{1}{r} W_{k,\,n+\frac{1}{2}}(z) P_n{}^\mu(\cos\theta) e^{\pm i\mu\phi},$$

n and μ being able to take different whole-number values. This phenomenon—the correspondence of several different stationary states to the same proper-value of the energy—is called *degeneracy*, a term already introduced in Chapter III. It is known from the general theory of partial differential equations that we can always choose linear combinations of the wave-functions which belong to the same value of the energy, in such a way that these combinations are orthogonal among themselves[2] (and, of course, orthogonal to the wave-functions which belong to all other proper-values of the energy). The degeneracy of the hydrogen atom, so far as it is due to the fact that many different values of n correspond to the same value of k, can be removed, exactly as in Sommerfeld's theory of 1915, by taking account of the relativist increase of mass with velocity: while the degeneracy, so far as it is due to the choice of different values of μ, can be removed by applying a magnetic field, as in Sommerfeld and Debye's theory of 1916. It is evident that n is essentially the azimuthal quantum number, and that μ is the magnetic quantum number.

Now consider the case when the total energy E of the electron in the hydrogen atom is positive. $\sqrt{(-E)}$ is now imaginary, and all the confluent hypergeometric functions which are solutions of the differential equation for rR remain finite as $r\to\infty$, so R tends to zero as $r\to\infty$. Moreover, at least one solution of the differential equation exists which is finite at $r = 0$. Thus *every positive value of* E *is a proper-value*, to which correspond wave-functions possessing azimuthal and magnetic quantum numbers, in the same way as the discrete wave-functions. Physically, this case corresponds to the complete ionisation of the hydrogen atom, or to the reverse process, i.e. the capture of free electrons.

[1] If the proper-values of Schrödinger's wave-equation have also a continuous spectrum, the above equations persist in a modified form, for which, see H. Weyl, *Math. Ann.* lxviii (1910), p. 220, and *Gött. Nach.* (1910), p. 442; E. Fues, *Ann. d. Phys.* lxxxi (1926), p. 281.

[2] There is a certain degree of arbitrariness in the choice, the arbitrariness being represented by an orthogonal transformation which may be performed on the wave-functions of the set.

In a further paper,[1] Schrödinger extended his theory by finding the wave-equation for problems more general than those considered hitherto. Still restricting ourselves for simplicity to the motion of a single particle, suppose that its Hamiltonian Function $H(q_1, q_2, q_3, p_1, p_2, p_3)$ is the sum of (1) a kinetic energy represented by a general quadratic form T in the momenta (p_1, p_2, p_3), with coefficients which depend on the co-ordinates (q_1, q_2, q_3), and (2) a potential energy V which depends only on (q_1, q_2, q_3). Then the wave-equation may be derived by a process similar to that followed already in the case of the particle referred to Cartesian co-ordinates. It will be evident from what was there proved that the wave-function must be of the form

$$\psi = e^{\frac{i}{\hbar}\{-Et+S(x,y,z)\}}$$

where $S(x, y, z)$ is Hamilton's Characteristic Function, satisfying the equation

$$H\left(q_1, q_2, q_3, \frac{\partial S}{\partial q_1}, \frac{\partial S}{\partial q_2}, \frac{\partial S}{\partial q_3}\right) = E.$$

Differentiating the expression for ψ, we have

$$\frac{\partial^2 \psi}{\partial q_r^2} = -\frac{1}{\hbar^2}\left(\frac{\partial S}{\partial q_r}\right)^2 \psi + \frac{i}{\hbar}\frac{\partial^2 S}{\partial q_r^2}\psi.$$

As before, the second term on the right is very small compared with the first, and may be neglected. Thus

$$T\left(q_1, q_2, q_3, \frac{\hbar}{i}\frac{\partial}{\partial q_1}, \frac{\hbar}{i}\frac{\partial}{\partial q_2}, \frac{\hbar}{i}\frac{\partial}{\partial q_3}\right)\psi = T\left(q_1, q_2, q_3, \frac{\partial S}{\partial q_1}, \frac{\partial S}{\partial q_2}, \frac{\partial S}{\partial q_3}\right)\psi,$$

and therefore

$$H\left(q_1, q_2, q_3, \frac{\hbar}{i}\frac{\partial}{\partial q_1}, \frac{\hbar}{i}\frac{\partial}{\partial q_2}, \frac{\hbar}{i}\frac{\partial}{\partial q_3}\right)\psi = H\left(q_1, q_2, q_3, \frac{\partial S}{\partial q_1}, \frac{\partial S}{\partial q_2}, \frac{\partial S}{\partial q_3}\right)\psi$$

or as we may write it

$$H\left(q, \frac{\hbar}{i}\frac{\partial}{\partial q}\right)\psi = E\psi.$$

This is the extension of the wave-equation, applicable to the more general form of the Hamiltonian Function.

Since

$$\frac{\partial \psi}{\partial t} = -\frac{i}{\hbar}E\psi,$$

[1] *Ann. d. Phys.* (4) lxxix (1926), p. 734

we have

$$H\left(q, \frac{\hbar}{i}\frac{\partial}{\partial q}\right)\psi = i\hbar\frac{\partial\psi}{\partial t}.$$

This equation, which does not involve E explicitly, may be called the *general wave-equation.*[1] It is clearly the equation that would be obtained by replacing p by $\hbar/i\ \partial/\partial q$ and E by $-\hbar/i\ \partial/\partial t$ in the equation of energy, and then operating on ψ.

It would seem from the foregoing derivations that Schrödinger's wave-equation is connected with Hamilton's Principal and Characteristic Functions by equations which are not exact, but only approximate, involving the neglect of powers of $\hbar$. It was shown[2] many years after the discovery of the wave-equation that this conclusion is incorrect: Schrödinger's equation is *rigorously equivalent* to Hamilton's partial differential equation for the Principal Function, provided the symbols are understood in a certain way.

Considering for simplicity a conservative system with one degree of freedom, let the co-ordinate at the instant t be q, and let Q be the value of the co-ordinate at a previous instant T. Then, as Hamilton showed, there exists in classical dynamics a *Principal Function* $W(q, Q, t-T)$, which has the properties

$$\frac{\partial W(q, Q, t-T)}{\partial q} = p$$

$$\frac{\partial W(q, Q, t-T)}{\partial Q} = -P$$

$$\frac{\partial W(q, Q, t-T)}{\partial t} = -H$$

where p and P are the values of the momentum at the instants t and T respectively, and H is the Hamiltonian Function.

Now consider the quantum-mechanical problem which is specified by the Hamiltonian $H(q, p)$. As explained in Chapter VIII, the variables q and p are no longer ordinary algebraic quantities, but are non-commuting variables satisfying the commutation-rule

$$pq - qp = \frac{\hbar}{i}.$$

The quantity Q, which represents the co-ordinate at the instant T, also does not commute with q or p. A function of q and Q is said to be *well-ordered*, when it is arranged (as of course it can be, by

[1] With regard to the general character of Schrödinger's theory, Heisenberg [*ZS. f. P.* xxxviii (1926), p. 411, *at* p. 412] said 'So far as I can see, Schrödinger's procedure does not represent a consistent wave-theory of matter in de Broglie's sense. The necessity for waves in space of f dimensions (for a system with f degrees of freedom), and the dependence of the wave-velocity on the mutual potential energy of particles, indicates a loan from the conceptions of the corpuscular theory.'

[2] E. T. Whittaker, *Proc. R.S. Edin.* lxi (1940), p. 1

use of the commutation-rules) as a sum of terms, each of the form $f(q)g(Q)$, the factors being in this order.[1] Then it can be shown that there exists a well-ordered function $U(q, Q, t-T)$, which formally satisfies exactly the same equations as Hamilton's Principal Function, namely,[2]

$$\frac{\partial U}{\partial q}=p, \qquad \frac{\partial U}{\partial Q}=-P, \qquad \frac{\partial U}{\partial t}=-H.$$

U may be called the *quantum-mechanical Principal Function.* Although it satisfies the same equations as Hamilton's Principal Function, its expression in terms of the variables q, Q, $t-T$, is quite different from the expression of the classical function W: the reason being that the above equations for U are true only on the understanding that all the quantities occurring in them are well-ordered in q and Q: but on substituting the well-ordered expression for p in the Hamiltonian $H(q, p)$, we shall need to invert the order of factors in many terms, by use of the commutation-rules, in order to reduce H to a well-ordered function of q and Q; and this introduces new terms which do not occur in the classical equations. This explains why, although the equations of quantum mechanics are formally identical with those of classical mechanics, the solutions in the two cases are altogether different.

Now introduce a function $R(q, Q, t-T)$, which is obtained by taking the well-ordered function U and replacing the non-commuting variables q and Q by ordinary algebraic quantities q and Q. It may be called the *Third Principal Function.* By what has been said, it is quite different from the classical Principal Function W belonging to the same Hamiltonian. Then if we write

$$\psi(q, Q, t-T)=e^{\frac{i}{\hbar}R(q, Q, t-T)}$$

it may be shown that $\psi\,(q, Q, t-T)$ *satisfies Schrödinger's differential equation for the wave-function belonging to the Hamiltonian* $H(q, p)$. Thus *the relation between Principal Function and Schrödinger's wave-function is rigorous, not requiring the neglect of any powers of* $\hbar$, *provided the Principal Function is understood to be the Third Principal Function* R, *and not Hamilton's classical Principal Function* W.

We must now consider how the proper-values and wave-functions of Schrödinger's equation are to be determined. The following method for determining them, at least approximately, was given in 1926 by G. Wentzel,[3] H. M. Kramers[4] and L. Brillouin.[5]

[1] This conception is due to Jordan, *ZS. f. P.* xxxviii (1926), p. 513

[2] The first two of these equations were given substantially by Dirac, *Phys. Zeits. Sowjetunion,* iii (1933), p. 64.

[3] *ZS. f. P.* xxxviii (1926), p. 518

[4] *ZS. f. P.* xxxix (1926), p. 828

[5] *Comptes Rendus,* clxxxiii (1926), p. 24; *J. de phys. et le rad.* vii (1926), p. 353. The method was to a great extent anticipated by H. Jeffreys, *Proc. L.M.S.* (2) xxiii (1925), p. 428; cf. also R. E. Langer, *Bull. Amer. M.S.* xl (1934), p. 545, who indicated a correction in Jeffreys's paper.

Consider a one-dimensional oscillatory motion defined by a Hamiltonian

$$H = \frac{p^2}{2m} + V(q).$$

The Schrödinger equation for the wave-function ψ is

$$\frac{\hbar^2}{2m}\frac{d^2\psi}{dq^2} + (E - V)\psi = 0$$

where E is the proper-value of the energy. Suppose that the range in which the particle would oscillate according to classical dynamics, namely, that defined by $E - V = 0$, is $q_1 \leqq q \leqq q_2$, so that q_1 and q_2 are roots of the equation $V(q) = E$; within this range put

$$\left\{\frac{2m}{\hbar^2}(E - V)\right\}^{\frac{1}{2}} = \zeta(q),$$

so the wave-equation becomes

$$\frac{d^2\psi}{dq^2} + \left\{\zeta(q)\right\}^2\psi = 0.$$

If V, and therefore ζ, were constant, the solution would be of the form

$$\psi = \text{Constant} \times \sin(\zeta q + \text{Constant}):$$

this suggests that even when ζ is variable, we should try within the range $q_1 \leqq q \leqq q_2$ to represent ψ by a sine-curve of slowly varying amplitude and wave-length, say

$$\psi = A(q) \sin S(q).$$

Substituting in the wave-equation, we have

$$(A'' - AS'^2 + A\zeta^2) \sin S + (2A'S' + AS'') \cos S = 0.$$

Let us try to satisfy this equation by imposing on A and S the two conditions

$$A'' - AS'^2 + A\zeta^2 = 0, \qquad 2A'S' + AS'' = 0.$$

Now suppose that A′ varies so slowly that $|A''|$ is very small compared to $\zeta^2|A|$ (it is at this stage that the approximation enters): then we may neglect the first term in the former of these equations, which now becomes

$$S'^2 = \zeta^2$$

giving

$$S = \int_{q_1}^{q} \zeta dq + \alpha, \quad \text{where } \alpha \text{ is a constant.}$$

The second condition

$$2A'S' + AS'' = 0$$

gives at once

$$A = c(S')^{-\frac{1}{2}} \quad \text{where } c \text{ is a constant}$$

or

$$A = c(\zeta)^{-\frac{1}{2}}.$$

The approximate expression for ψ is therefore

$$\psi = c(\zeta)^{-\frac{1}{2}} \sin \left(\int_{q_1}^{q} \zeta dq + \alpha \right).$$

Attention must be given, however, to the behaviour of ψ at the points for which $q = q_1$ and $q = q_2$, say P and Q respectively, since at these points ζ vanishes, and A″ cannot be neglected in comparison with $A\zeta^2$. A closer consideration of this difficulty[1] shows that the function ψ must have at P the phase $\pi/4$, so the approximate expression of ψ in the range PQ is

$$\psi = c(\zeta)^{-\frac{1}{2}} \sin \left(\int_{q}^{q} \zeta dq + \frac{\pi}{4} \right).$$

This is the Wentzel-Kramers-Brillouin approximate solution of Schrödinger's equation.

Similarly the function ψ must have at Q the phase $3\pi/4$: so when $q = q_2$,

$$\int_{q_1}^{q} \zeta dq + \frac{\pi}{4} = n\pi + \frac{3\pi}{4}$$

where n is a whole number, representing the number of nodes comprised between P and Q; that is to say,

$$\int_{q_1}^{q_2} \zeta dq = (n + \tfrac{1}{2})\pi.$$

This equation determines the proper-values E *of the energy, in the Wentzel-Kramers-Brillouin approximation.*

Between P and Q the graph of ψ oscillates like a sine-curve, whereas to the left of P and to the right of Q it decreases exponentially.

The Wentzel-Kramers-Brillouin approximation throws light on the connection between wave-mechanics and the formula of quantification enunciated by Wilson and Sommerfeld in the older quantum

[1] For which cf. E. Persico, *N. Cimento*, xv (1938), p. 133

theory, namely that $\int pdq$, where the integration is taken round an orbit, is a multiple of Planck's constant h.

For the momentum, calculated according to classical dynamics from the equation

$$\frac{p^2}{2m} + V(q) = E$$

is

$$p = \pm\{2m(E - V)\}^{\frac{1}{2}} = \pm\hbar\zeta(q)$$

where the upper sign must be taken for the semi-oscillation from q_1 to q_2, and the lower sign from q_2 to q_1. Thus

$$\int pdq \text{ round the orbit} = 2\int_{q_1}^{q_2} pdq = 2\hbar\int_{q_1}^{q_2}\zeta(q)\,dq,$$

and thus the Wentzel-Kramers-Brillouin condition becomes

$$\frac{1}{2\hbar}\int pdq = (n + \tfrac{1}{2})\pi$$

or

$$\int pdq = (n + \tfrac{1}{2})h,$$

which is the Wilson-Sommerfeld condition, completed by the term $\frac{1}{2}$ which appears in the more accurate theory.

So far we have connected Schrödinger's wave-equation only with the stationary states of the atom, to which correspond proper-values of the total energy. We shall now consider more general states.

In Bohr's theory a stationary state meant a particular kind of orbital motion, so that an atom could be in only one stationary state at one time. In Schrödinger's theory, on the other hand, the stationary states correspond to different solutions of a linear partial differential equation, and therefore the various stationary states can be *superposed* just as overtones can be superposed on the fundamental tone of a violin string. We have to consider what is the physical interpretation of this superposition.

Suppose that plane-polarised light whose vibrations are in a direction α passes through a Nicol prism, which resolves it into vibrations in directions β and γ respectively parallel and perpendicular to the plane of polarisation of the Nicol, and permits only the former to pass through. Fixing our attention on a single photon, which is initially polarised in the direction α, we can regard this state as a *superposition* of two states, namely, that of polarisation in the direction β and polarisation in the direction γ.[1] We can speak of the

[1] The superposition here spoken of must not be confused with superposition in classical mechanics; in classical mechanics, the superposition of a certain state of vibration on itself gives a vibration of twice the amplitude, but in quantum-mechanics it gives merely the same state of vibration.

probability of the photon being in either of the states β and γ, these probabilities being such as to account for the observed intensity of light polarised in the direction β, which emerges from the Nicol.

This connection between superposition and probability will now be extended to wave-functions. Consider two stationary states of an atom, for which the energy has the proper-values E_1 and E_2 respectively. Let ψ_1 and ψ_2 be the corresponding normalised solutions of the wave-equation, with their appropriate time-factors $e^{-(iE_1/\hbar)t}$ and $e^{-(iE_2/\hbar)t}$, and let $\psi_3 = c_1\psi_1 + c_2\psi_2$ where c_1 and c_2 are arbitrary complex numbers. Then the state or physical situation represented by ψ_3 is said to be formed by the *superposition* of the states represented by ψ_1 and ψ_2. ψ_3 of course satisfies the general wave-equation

$$H\left(q, \frac{\hbar}{i}\frac{\partial}{\partial q}\right)\psi = i\hbar\frac{\partial\psi}{\partial t}.$$

Let us find the condition that must be satisfied by c_1 and c_2 if ψ_3 is normalised. We have

$$1 = \int \psi_3\psi_3^* d\tau = \int (c_1\psi_1 + c_2\psi_2)\,(c_1^*\psi_1^* + c_2^*\psi_2^*)d\tau$$

$$= c_1c_1^*\int\psi_1\psi_1^* d\tau + c_2c_2^*\int\psi_2\psi_2^* d\tau$$

$$= |c_1|^2 + |c_2|^2.$$

This equation can be interpreted to mean that *there is an uncertainty as to the value which would be found by a measurement of the energy of the atom, either of the values* E_1 *or* E_2 *being possible: and their respective probabilities are* $|c_1|^2$ *and* $|c_2|^2$.

More generally, if $\psi_0, \psi_1, \psi_2, \ldots$ are the normalised wave-functions (with their appropriate time-factors) belonging respectively to the proper-values $E_0, E_1, E_2, \ldots$ (supposed for simplicity to be all different) of the energy of the atom, and if a normalised solution ψ of the general wave-equation is expanded in the form

$$\psi = c_0\psi_0 + c_1\psi_1 + c_2\psi_2 + \ldots,$$

then *in the physical situation defined by the wave-function* ψ, *the probability that a measurement of the energy will yield the value* E_n is $|c_n|^2$. On account of the relation $\int\psi_n\psi_m^* d\tau = \delta_m^n$, it is seen at once that the value of the coefficient c_n is $\int\psi\psi_n^* d\tau$.

The equations which we have found lead to a certain connection [1] with the classical electromagnetic theory of the emission of radiation. Suppose that we consider an atom in a stationary state, so that the wave-function ψ involves the time through a factor (say) $e^{2\pi i\nu t}$.

[1] E. Fermi, *Rend. Lincei*, v (May 1927), p. 795

Then ψ^* will have the time-factor $e^{-2\pi i\nu t}$, so in the product $\psi\psi^*$ these time-factors will destroy each other : that is to say, the distribution of electric charge in the atom, and therefore its electric moment, will not vary with the time : and therefore *according to the classical theory*, the atom, when it is in a stationary state, will not emit radiation.

Suppose next, however, that the atom is not in a pure stationary state, but is in a state which is represented by the superposition of two stationary states, for which the wave-function has the time-factors $e^{2\pi i\nu_1 t}$ and $e^{2\pi i\nu_2 t}$ respectively, so that

$$\psi = Ae^{2\pi i\nu_1 t} + Be^{2\pi i\nu_2 t}$$

where A and B do not involve the time. Then we have

$$\psi\psi^* = (Ae^{2\pi i\nu_1 t} + Be^{2\pi i\nu_2 t})(A^*e^{-2\pi i\nu_1 t} + B^*e^{-2\pi i\nu_2 t})$$

$$= AA^* + BB^* + AB^*e^{2\pi i(\nu_1-\nu_2)t} + A^*Be^{-2\pi i(\nu_1-\nu_2)t},$$

and hence the electric moment of the atom, which depends on an integral involving t only in the combination $\psi\psi^*$, will be periodic, with frequency $(\nu_1 - \nu_2)$, and consequently the atom, *according to the classical theory*, will emit radiation of frequency $(\nu_1 - \nu_2)$. This radiation will continue until the consequent exhaustion of energy has again reduced the atom to a single pure stationary state.

If E_1 and E_2 are the energies associated with the two stationary states, then $E_1 = h\nu_1$ and $E_2 = h\nu_2$, so the radiation is of frequency $(1/h)$ $(E_1 - E_2)$, just as in Bohr's theory, and this result has now been obtained by what are essentially classical methods—the classical theory of the solutions of partial differential equations—without doing violence to the electromagnetic theory of light.

Schrödinger did not, at the outset of his researches, suspect any connection between his theory and the theory of matrix-mechanics.[1] He now, however,[2] showed that the two theories are actually equivalent. In the first place, the commutation-rules of matrix-mechanics, which in the case of systems with one degree of freedom reduce to

$$qp - pq = i\hbar,$$

become obvious identities if we write $p = (\hbar/i)\ \partial/\partial q$, since

$$q\left(\frac{\hbar}{i}\frac{\partial}{\partial q}\right)\psi - \frac{\hbar}{i}\frac{\partial}{\partial q}(q\psi) = i\hbar\psi.$$

To any physical quantity $\zeta(q, p)$ we can correlate a differential

[1] He says so in *Ann. d. Phys.*(4) lxxix (1926), p. 734 ; 'I naturally knew about his [Heisenberg's] theory, but was discouraged by what appeared to me as very difficult methods of transcendental algebra.'

[2] loc. cit. : cf. also C. Eckart, *Phys. Rev.* xxviii (1926), p. 711

operator $\zeta(q, \hbar/i.d/dq)$. Let us consider the operation of this quantity on a wave-function $\psi(q)$. As we have seen, it is possible to expand $\psi(q)$ as a series of normalised orthogonal wave-functions, in the form

$$\psi(q) = c_1\psi_1 + c_2\psi_2 + \ldots$$

where
$$c_n = \int \psi(q)\psi_n{}^*(q)dq.$$

If $\zeta(q, \hbar/i\ d/dq)\psi(q)$ has a corresponding expression

$$\zeta\left(q, \frac{\hbar}{i}\frac{d}{dq}\right)\psi(q) = c_1{}'\psi_1 + c_2{}'\psi_2 + \ldots,$$

then evidently we must have

$$c_n{}' = \int \psi_n{}^*(q)\zeta\left(q, \frac{\hbar}{i}\frac{d}{dq}\right)\psi(q)dq$$

$$= \int \psi_n{}^*(q)\zeta\left(q, \frac{\hbar}{i}\frac{d}{dq}\right)(c_1\psi_1 + c_2\psi_2 + \ldots)dq$$

$$= e_{n1}c_1 + e_{n2}c_2 + e_{n3}c_3 + \ldots$$

where
$$e_{nr} = \int \psi_n{}^*(q)\zeta\left(q, \frac{\hbar}{i}\frac{d}{dq}\right)\psi_r(q)dq.$$

But this equation shows that the column-vector $(c_1{}', c_2{}', c_3{}', \ldots)$ is derived from the column-vector $(c_1, c_2, c_3, \ldots)$ by operating on it with a matrix whose element in the nth row and r^{th} column is e_{nr}. Thus *the physical quantity* $\zeta(q, p)$, *or the operator* $\zeta(q, \hbar/i\ d/dq)$, *may be correlated to the matrix whose element in the* n^{th} *row and* r^{th} *column* is

$$e_{nr} = \int \psi_n{}^*(q)\zeta\left(q, \frac{\hbar}{i}\frac{d}{dq}\right)\psi_r(q)dq,$$

in the sense that the performance of the operator on any wave-function $\psi(q)$ *is equivalent to the operation of the matrix* (e_{nr}) *on the column-vector of the coefficients* c_r, *which express* $\psi(q)$ *in terms of the normalised orthogonal wave-functions* $\psi_1(q), \psi_2(q), \ldots$.

We observe that the matrix thus found for the physical quantity $\zeta(q, p)$ is specially associated with the set of normalised orthogonal wave-functions $\psi_n(q)$ which belong to a particular Schrödinger's equation

$$H\left(q, \frac{\hbar}{i}\frac{\partial}{\partial q}\right)\psi = i\hbar\frac{\partial\psi}{\partial t}.$$

We may call this set of wave-functions the *basis* of the matrix.

In order to establish completely the identity of this correlation (between physical quantities and matrices) with matrix-mechanics, it is necessary to prove some other mathematical theorems, e.g. that the result of operating with the product of two operators $\zeta(q, \hbar/i\ d/dq)$ and $\eta(q, \hbar/i\ d/dq)$ on a wave-function is the same as the result of operating on the column of coefficients of the wave-function with the matrix-product of the matrices corresponding to ζ and η. The proofs will be omitted here.

Suppose now that the basis of the matrix representing $\zeta(q, p)$ is the set of wave-functions belonging to the Hamiltonian function $\zeta(q, p)$, so it is the set of wave-functions of the Schrödinger equation

$$\zeta\left(q, \frac{\hbar}{i}\frac{\partial}{\partial q}\right)\psi = i\hbar\frac{\partial\psi}{\partial t}.$$

Then the matrix-elements are given by

$$e_{nr} = \int\psi_n^*(q)\zeta\left(q, \frac{\hbar}{i}\frac{d}{dq}\right)\psi_r(q)dq$$

$$= \int\psi_n^*(q)E_r\psi_r(q)dq$$

where E_r is the proper-value of the energy corresponding to the wave-function $\psi_r(q)$. Thus

$$e_{nr} = E_r\delta_r^n$$

so *the matrix is now a diagonal matrix whose elements are the proper-values of the energy for the Schrödinger equation*

$$\zeta\left(q, \frac{\hbar}{i}\frac{\partial}{\partial q}\right)\psi = i\hbar\frac{\partial\psi}{\partial t}.$$

Thus *the physical quantity $\zeta(q, p)$, when expressed as a matrix in terms of this basis, is a diagonal matrix whose elements are its proper-values.*

Therefore the problem as formulated in matrix-mechanics, namely, *to reduce the matrix for $\zeta(q, p)$ to a diagonal matrix*, is solved when we have found a solution of the problem as formulated in wave-mechanics, namely, *to find the proper values of* E *and the corresponding wave-functions for the Schrödinger equation*

$$\zeta\left(q, \frac{\hbar}{i}\frac{\partial}{\partial q}\right)\psi = E\psi.$$

The basis constituted by the wave-functions of this equation enables us, by the above formulae, to calculate the matrices that represent q and p. Thus *matrix-mechanics and wave-mechanics are equivalent.*

Let us now investigate more closely the physical meaning of the matrix-elements. Consider the electric moment $\mathbf{M}_{PQ}$ of the classical oscillator which would emit the radiation associated with a transition from a stationary state P to a stationary state Q of the atom: let $\mathbf{M}$ denote $\sum e\mathbf{r}$, the electric moment of the atom, expressed in terms of the co-ordinates of the electrons: and let ψ_P and ψ_Q be the wave-functions of the states P and Q. We may expect that $\mathbf{M}_{PQ}$ will depend in some way on $\mathbf{M}$, ψ_P and ψ_Q. Now we know that the time-factor in $\mathbf{M}_{PQ}$ is $e^{-2\pi i\nu_{PQ}t}$, where $\nu_{PQ} = (1/h)\ (E_P - E_Q)$, the time-factor in ψ_P is $e^{-iE_Pt/\hbar}$, the time-factor in ψ_Q is $e^{-iE_Qt/\hbar}$, and $\mathbf{M}$ has no time-factor. We therefore expect that the expression for $\mathbf{M}_{PQ}$ will involve the product $\psi_P\mathbf{M}\psi_Q{}^*$, where $\psi_Q{}^*$ is the complex-conjugate of ψ_Q. The explicit expression of the co-ordinates has to be removed from this, which can evidently be done by integrating over space. Thus we obtain the expression

$$\int \psi_P \mathbf{M} \psi_Q{}^* d\tau,$$

and *this matrix-element we shall identify with* $\mathbf{M}_{PQ}$, *the electric moment of the classical oscillator which would emit the radiation associated with the transition* P→Q. This identification is justified by comparison with the results of experiments.

Let us now illustrate the equivalence of matrix-mechanics and wave-mechanics by considering the harmonic oscillator in one dimension, for which the Hamiltonian function is

$$H = \frac{1}{2m}(p^2 + m^2\omega^2 q^2)$$

where m and ω are constants. The Schrödinger wave-equation is

$$\hbar^2 \frac{d^2\psi}{dq^2} + (2mE - m^2\omega^2 q^2)\psi = 0$$

where E is the total energy of the motion.

Writing $q = (\hbar/2m\omega)^{\frac{1}{2}} z$, this may be written

$$\frac{d^2\psi}{dz^2} + \left(\frac{E}{\hbar\omega} - \tfrac{1}{4}z^2\right)\psi = 0$$

which is the well-known differential equation of the parabolic-cylinder functions.[1] It has a solution which is finite for all real values of z only when

$$\frac{E}{\hbar\omega} = n + \tfrac{1}{2}$$

[1] cf. Whittaker and Watson, *Modern Analysis*, § 16·5

where n is a whole number,[1] and the solution is then a constant multiple of the parabolic-cylinder function $D_n(z)$, which is a polynomial multiplied by $e^{-\frac{1}{4}z^2}$. Thus *the proper-values of* H *are*

$$E = (n + \tfrac{1}{2})\hbar\omega \quad \textit{where} \quad n = 0, 1, 2, 3, \ldots$$

and the corresponding wave-functions are

$$\psi_n(q) = \lambda D_n\left(q\sqrt{\frac{2m\omega}{\hbar}}\right)$$

where λ is a constant to be determined by the normalising condition

$$\int_{-\infty}^{\infty} |\psi_n(q)|^2 dq = 1.$$

Since

$$\int_{-\infty}^{\infty} \{D_n(z)\}^2 dz = (2\pi)^{\frac{1}{2}} n!,$$

this gives

$$\lambda = \left(\frac{m\omega}{\pi\hbar}\right)^{\frac{1}{4}} (n\,!)^{-\frac{1}{2}} e^{i\alpha_n}$$

where α_n is real ; so the normalised wave-functions are

$$\psi_n(q) = \left(\frac{m\omega}{\pi\hbar}\right)^{\frac{1}{4}} (n\,!)^{-\frac{1}{2}} D_n\left(q\sqrt{\frac{2m\omega}{\hbar}}\right) e^{i\alpha_n}.$$

The definition of the element in the lth row and n^{th} column of the matrix representing the co-ordinate q, in terms of the wave-functions, is

$$\int_{-\infty}^{\infty} q\psi_l^*(q)\psi_n(q)dq.$$

Using the known properties of the parabolic-cylinder functions

$$zD_n(z) = D_{n+1}(z) + nD_{n-1}(z)$$

and

$$\int_{-\infty}^{\infty} D_m(z)D_n(z)dz = (2\pi)^{\frac{1}{2}} n\,!\,\delta_n^m,$$

[1] The Wilson-Sommerfeld quantum condition would lead us to expect the Action to be a whole multiple of h, that is, energy × period to be a multiple of h, or E equal to a multiple of $\hbar\omega$. The occurrence of $(n + \frac{1}{2})$ instead of n is characteristic of the quantum-mechanical solution.

the value of the matrix-element can readily be calculated, and we obtain (inserting the time-factor)

$$q=\left(\frac{\hbar}{2m\omega}\right)^{\frac{1}{2}}\begin{pmatrix} \cdot & e^{-i(\omega t+\beta_1)} & \cdot & \cdot \\ e^{i(\omega t+\beta_1)} & \cdot & 2^{\frac{1}{2}}e^{-i(\omega t+\beta_2)} & \cdot \\ \cdot & 2^{\frac{1}{2}}e^{i(\omega t+\beta_2)} & \cdot & \cdot \\ \cdot & \cdot & \cdot & \cdot \end{pmatrix}$$

in agreement with the value found in chapter VIII.

Similarly the element in l^{th} row and n^{th} column of the matrix representing the momentum p is expressed in terms of the wave-functions by

$$\frac{\hbar}{i}\int_{-\infty}^{\infty}\psi_l^*(q)\frac{d}{dq}\psi_n(q)dq$$

which when evaluated gives the same value as was found in Chapter VIII.

It was shown by C. Eckart [1] by use of the wave-equation that in general problems, an element of the matrix representing a co-ordinate can be expressed as a series of the form

$$\chi(m, n)=\chi_0(m, n)+h\chi_1(m, n)+h^2\chi_2(m, n)+\ldots$$

whereas $n\to\infty$, $\chi_0(m, n)$ tends to the coefficient of the $(m-n)^{th}$ harmonic in the Fourier expansion of the co-ordinate as determined by the classical theory of the motion. This is in accord with Bohr's correspondence principle.

In a later paper [2] he illustrated this theorem by calculating the matrix-elements of the radius vector for the hydrogen atom, and showing that their limiting values, as $h\to 0$ and the quantum numbers tend to infinity, coincide with the Fourier coefficients of the classical motion. In particular, the diagonal terms of the matrix tend to the constant term of the Fourier expansion.

In a paper entitled 'The Continuous Transition from Micro- to Macro-mechanics,' [3] Schrödinger showed how to construct for the harmonic oscillator a *wave-packet*, i.e. a group of wave-functions of high-quantum number and small quantum-number-differences such that the electric density is very nearly concentrated at a single point. He proved that the differential equation of the wave-functions of the harmonic oscillator, namely,

$$\frac{1}{2m}\left(-\hbar^2\frac{\partial^2\psi}{\partial q^2}+m^2\omega^2q^2\right)=i\hbar\frac{\partial\psi}{\partial t}$$

is satisfied by

$$\psi(t)=\exp\left(-\tfrac{1}{2}i\omega t+ke^{-i\omega t}q\sqrt{\left(\frac{2m\omega}{\hbar}\right)}-\frac{m\omega q^2}{2\hbar}-\frac{k^2}{2}-\frac{k^2}{2}e^{-2i\omega t}\right)$$

[1] *Proc. Nat. Ac. Sci.* xii (1926), p. 684; cf. P. Debye, *Phys. ZS.* xxviii (1927), p. 170
[2] *ZS. f. P.* xlviii (1928), p. 295
[3] *Naturwiss*, xxviii (1926), p. 664

when k is any constant. This gives

$$\psi\psi^* = \exp\left[-\frac{\{q - (2\hbar/m\omega)^{\frac{1}{2}} k \cos \omega t\}^2}{\hbar/m\omega} \right].$$

The probability that at the instant t the co-ordinate q lies in the range from q to $q+dq$ is $\psi\psi^* dq$: and clearly if $\hbar$ is very small (as it is) and k is very large, while $\hbar k^2\omega$ is finite, say equal to E, so $(2\hbar/m\omega)^{\frac{1}{2}}k$ becomes $(2E/m\omega^2)^{\frac{1}{2}}$, then $\psi\psi^*$ is negligibly small except when the numerator of the argument of the exponential is approximately zero; that is, the wave-function represents a wave-packet whose position at time t is given by

$$q = \sqrt{\left(\frac{2E}{m\omega^2}\right)} \cos \omega t,$$

which is precisely the equation determining the position of the particle in the classical problem when the energy is E.

The *quantum-mechanical theory of collisions* was founded in two papers of 1926 by Max Born.[1] A classical treatment of the problem had been given in 1911 by Rutherford,[2] whose investigation of the scattering of particles by a Coulomb field will first be described.

Consider the scattering of a narrow beam of α-rays by a sheet of metal foil on which the beam impinges at right angles. The scattered particles afterwards strike a screen of zinc sulphide, and the number of scintillations on each square millimetre of the screen is observed.

Let the number of atoms per unit volume in the foil be n, and let q be the thickness of the foil, so there are nq atoms per unit cross-section of the beam. The area of that part of this unit cross-section which is within a distance B of the centre of an atom is therefore $\pi nq B^2$, and out of σ incident particles the number that are scattered at distances between B and $B+dB$ is $2\pi\sigma nq B dB$. Now according to classical dynamics the path of a particle is a hyperbola having the centre of the atom as its external focus: and since the perpendicular from the focus on an asymptote of the hyperbola is equal to the minor semi-axis, it follows that B is the minor semi-axis of the hyperbola. The angle θ through which the particle is scattered is equal to the external angle between the asymptotes of the hyperbola, so

$$\cot \tfrac{1}{2}\theta = \frac{B}{A}$$

where A is the major semi-axis. The number of particles scattered in the annulus between angles θ and $\theta + d\theta$ is therefore

$$\pi\sigma nq A^2 \cot \tfrac{1}{2}\theta \operatorname{cosec}^2 \tfrac{1}{2}\theta \, d\theta.$$

[1] *ZS. f. P.* xxxvii (1926), p. 863; xxxviii (1926), p. 803
[2] *Phil. Mag.* xxi (1911), p. 669

These fall on the screen, which is supposed to be at a distance l beyond the foil, on an annulus of area

$$d(\pi l^2 \tan^2\theta) \quad \text{or} \quad 2\pi l^2 \tan\theta \sec^2\theta \, d\theta.$$

The number of scintillations observed per unit area is therefore

$$\frac{\sigma n q A^2 \cos^3\theta \cos\frac{1}{2}\theta}{2l^2 \sin\theta \sin^3\frac{1}{2}\theta} \quad \text{or} \quad \frac{\sigma n q A^2 \cos^3\theta}{4l^2 \sin^4\frac{1}{2}\theta}.$$

Writing r = distance of the scintillation from the atom $= l/\cos\theta$, we see that the number of scintillations per unit area perpendicular to r is

$$\frac{\sigma n q A^2}{4r^2 \sin^4\frac{1}{2}\theta}.$$

Now if E be the charge at the centre, e the charge and m the mass of the incident particle, and v_0 the velocity at infinity, the usual formulae of hyperbolic motion give $A = eE/mv_0^2$: so the number of scintillations per unit area perpendicular to v is

$$\frac{\sigma n q e^2 E^2}{4m^2 v_0^4 \, r^2 \sin^4\frac{1}{2}\theta}.$$

Thus if the beam of incident particles is of such intensity that one particle crosses unit area in unit time, and the beam falls on a single scattering centre, then the probability that in unit time a particle should be scattered into the element of solid angle $\sin\theta \, d\theta d\phi$ (whose θ, ϕ are spherical-polar co-ordinates, the polar axis being the direction of the incident particle) is

$$\frac{e^2 E^2}{4m^2 v_0^4 \sin^4\frac{1}{2}\theta} \sin\theta \, d\theta d\phi.$$

This formula, given in Rutherford's paper, was verified experimentally. It afforded a means of determining the charge E on the nucleus of an atom, and led to the conclusion that the nuclear charge is equal to the electronic charge multiplied by the atomic number.

The quantum-mechanical treatment of collision problems originated in the two papers of Max Born already referred to. Consider the scattering of a beam of particles by a fixed centre of force, the potential energy of one particle at a distance r from the scattering centre being $V(r)$. Schrödinger's wave-equation for a particle in presence of the scattering centre is

$$\nabla^2\psi + \left\{k^2 - \frac{2m}{\hbar^2}V(r)\right\}\psi = 0$$

where $k=mv_0/\hbar$, m being the mass and v_0 the initial velocity of the incident particles. The general solution of the equation

$$\nabla^2\psi+k^2\psi=f(x, y, z)$$

where f is a given function, is known to be

$$\psi=C(x, y, z)-\frac{1}{4\pi}\int\frac{e^{ik\rho}}{\rho}f(x', y', z')d\tau'$$

where $C(x, y, z)$ is the general solution of the equation $\nabla^2\psi+k^2\psi=0$, and where $\rho^2=(x'-x)^2+(y'-y)^2+(z'-z)^2$, and $d\tau'$ is the volume-element $dx'dy'dz'$, the integration being extended over all space. So a solution of the wave-equation for the particle is a solution of the integral-equation

$$\psi(x, y, z)=C(x, y, z)-\frac{m}{2\pi\hbar^2}\int\frac{e^{ik\rho}}{\rho}V(r')\psi(x', y', z')d\tau'. \qquad (1)$$

Now if the incident beam is directed parallel to the z-axis, it may be represented by a wave-function, which involves z only: since this wave-function must satisfy the Schrödinger equation

$$\nabla^2\psi+k^2\psi=0,$$

it must satisfy $(d^2\psi/dz^2)+k^2\psi=0$, so it must be a linear combination of e^{ikz} and e^{-ikz}: a stream moving in the positive direction along the axis of z, with electron-density unity, will be represented by the term e^{ikz} alone. The incident beam is thus represented as a plane monochromatic ψ-wave, whose wave-length is inversely proportional to the momentum of the particles.

The complete wave-function ψ of equation (1) must represent this incident beam together with the scattered wave, which is a wave diverging from the scattering centre, of the form

$$\frac{e^{ikr}}{r}g(\theta),$$

where (r, θ, ϕ) are spherical-polar co-ordinates, with the scattering centre as origin, so that θ is the angle of scattering. Evidently the second term in (1) represents the scattered wave, and therefore $C(x, y, z)$ must represent the incident wave. Thus (1) becomes

$$\psi(x, y, z)=e^{ikz}-\frac{m}{2\pi\hbar^2}\int\frac{e^{ik\rho}}{\rho}V(r')\psi(x', y', z')d\tau'. \qquad (2)$$

Let us find the asymptotic form of the second term on the right at great distances from the origin. When r is very large we have approximately

$$\rho=r-r'\cos(\widehat{rr'}),$$

so (2) becomes

$$\psi(x, y, z) = e^{ikz} - \frac{m}{2\pi\hbar^2}\frac{e^{ikr}}{r}\int e^{-ikr'\cos(\widehat{rr'})}V(r')\psi(x', y', z')d\tau'.$$

When the particles of the incident beam are travelling so fast that the scattering is not very great, the function $\psi(x', y', z')$ in the integral may be replaced by the wave-function of the incident wave alone, so the last equation becomes

$$\psi(x, y, z) = e^{ikz} + \frac{e^{ikr}}{r}g(\theta)$$

where

$$g(\theta) = -\frac{m}{2\pi\hbar^2}\int e^{ikr'(\cos\widehat{r'z} - \cos\widehat{r'r})}V(r')d\tau'.$$

Making use of Gegenbauer's formula,[1] this integral becomes

$$g(\theta) = -\frac{2m}{\hbar^2}\int_0^\infty \frac{\sin p}{p}V(r)r^2dr, \tag{3}$$

where $p = 2kr\sin\frac{1}{2}\theta$.

Now denoting the electric density $e\psi\psi^*$ by ρ, and denoting the quantity

$$\frac{e\hbar}{2mi}(\psi^* \operatorname{grad}\psi - \psi\operatorname{grad}\psi^*)$$

by $\mathbf{s}$, it can be shown from the wave-equation that

$$\frac{\partial\rho}{\partial t} + \operatorname{div}\mathbf{s} = 0$$

and this equation suggests that $\mathbf{s}$ may be interpreted as the *electric current*[2]; and therefore the number of particles which in unit time pass through a unit cross-section is the component, normal to the cross-section, of the vector

$$\mathbf{S} = \frac{\hbar}{2mi}(\psi^* \operatorname{grad}\psi - \psi\operatorname{grad}\psi^*)$$

so

$$S_x = \frac{\hbar}{2mi}\left(\psi^*\frac{\partial\psi}{\partial x} - \psi\frac{\partial\psi^*}{\partial x}\right) \text{ etc.}$$

For the incident plane wave, $\psi = e^{ikz}$, so $S_z = k\hbar/m = v_0$, so the number

[1] G. N. Watson, *Bessel functions*, p. 378

[2] Schrödinger, *Ann. d. Phys.* lxxxi (1926), p. 109. Born, *ZS. f. P.* lxxxviii (1926), p. 803; xl (1926), p. 167. Gordon, *ZS. f. P.* xl (1926), p. 117

of particles which in unit time pass through a unit cross-section at right angles to the beam is v_0, the velocity of the particles, as it must be. For the scattered wave we have

$$\psi = \frac{e^{ikr}}{r} g(\theta)$$

so

$$S_r = \frac{k\hbar}{m}\frac{|g(\theta)|^2}{r^2} = v_0 \frac{|g(\theta)|^2}{r^2}.$$

Thus, *with an incident beam in which one particle crosses unit area transverse to the path in unit time, the number of particles scattered into the solid angle* $\sin\theta\, d\theta d\phi$ *in unit time is* $|g(\theta)|^2 \sin\theta\, d\theta d\phi$, *where* $g(\theta)$ *has the value given by* (3).

Of course in this treatment, which is called the *Born approximation*, the interaction $V(r)$ has been treated as a small perturbation, and only the first approximation has been retained.[1] Born's formula was applied by Wentzel[2] to the scattering of a beam of electrically-charged particles by an electrically-charged centre. Suppose the charges of the centre and of a particle are Ze and $Z'e$, where e is the electronic charge ; we suppose the Coulomb force of the centre to be modified by 'shielding,' so we can write

$$V(r) = \frac{e^2 ZZ'}{r} e^{-\frac{r}{a}}$$

where the factor $e^{-\frac{r}{a}}$ represents the shielding, a being the *effective radius* of the atom. Evaluating (3) on this assumption, we find

$$g(\theta) = -\frac{2me^2 ZZ'}{\hbar^2 \left\{ (2k \sin \tfrac{1}{2}\theta)^2 + \frac{1}{a^2} \right\}}.$$

In the usual experiments, $1/a^2$ is small compared with $(2k \sin \frac{1}{2}\theta)^2$, so may be omitted. Thus

$$g(\theta) = -\frac{me^2 ZZ'}{2\hbar^2 k^2 \sin^2 \frac{1}{2}\theta}$$

$$= -\frac{e^2 ZZ'}{2mv_0^2 \sin^2 \frac{1}{2}\theta},$$

and this is precisely the formula obtained classically by Rutherford. Agreement with experiment confirmed the interpretation of $\psi\psi^*$ that had been used in the derivation.

[1] The wave-function ψ of the quantum-mechanical treatment of the problem was derived analytically from the classical solution (by considering the Action) by W. Gordon, *ZS. f. P.* xlviii (1928), p. 180. [2] *ZS. f. P.* xl (1926), p. 590

Born's method was applied in 1927 and following years to investigate the collision of an incident beam of electrons with the atoms of a gas, the atoms being possibly raised to excited states, and the electrons scattered : this work, however, falls outside the limits of the present volume.[1]

On 10 May 1926 Schrödinger contributed another paper to the *Annalen der Physik*,[2] in which he set forth a general method for treating perturbations (i.e. solving problems which are very closely related to problems that have already been solved), and applied his theory to investigate the Stark effect in the Balmer lines of hydrogen. The method was essentially the same as that given by Born, Heisenberg and Jordan, a few months earlier, or indeed as that used long before by Lord Rayleigh[3] in discussing the vibrations of a string whose density has small inhomogeneities. The wave-equation for the unperturbed system is

$$\left\{H\left(\frac{\hbar}{i}\frac{\partial}{\partial q}, q\right) - E\right\}\psi = 0.$$

Let the proper-values of E, and the corresponding normalised wave-functions ψ (which are supposed to be known) be $E_1, E_2, E_3, \ldots$; $\psi_1(q), \psi_2(q), \psi_3(q), \ldots$. Let the perturbation be represented by the adjunction of a small term $\lambda r\psi$ to the left-hand side of the wave-equation, where λ denotes a small constant and r is a known function of the q's, so that the wave-equation for the perturbed system is

$$\left\{H\left(\frac{\hbar}{i}\frac{\partial}{\partial q}, q\right) + \lambda r - E\right\}\psi = 0.$$

Let the new proper-values of the energy be $E'_1, E'_2, \ldots$, where

$$E'_s = E_s + \lambda\epsilon_s + \text{terms involving higher powers of } \lambda,$$

and let the new wave-functions be $\psi'_1, \psi'_2, \ldots$, where

$$\psi'_s = \psi_s + \lambda v_s + \text{terms involving higher powers of } \lambda.$$

Substituting in the wave-equation, and retaining only the first power of λ, we have

$$\{H - E_s + \lambda(r - \epsilon_s)\}(\psi_s + \lambda v_s) = 0.$$

Since $(H - E_s)\psi_s = 0$, this gives

$$(H - E_s)v_s = -(r - \epsilon_s)\psi_s.$$

[1] Though Born himself obtained important results in *Gött. Nach.* 1926, p. 146.
[2] *Ann. d. Phys.*(4) lxxx (1926), p. 437
[3] *Theory of Sound*, 2nd edn. (London, 1894), i, p. 115

Now in Rayleigh's problem of the vibrating string, if there is resonance between the applied force and a proper vibration of the unperturbed system, the oscillation increases without limit, and no finite solution exists: it can be shown that corresponding to this in the present case, in order that there may be a finite solution, the right-hand side of the last equation must be orthogonal to the wave-function which is the solution ψ_s of the equation obtained by equating the left-hand side to zero; so we must have

$$\int (r-\epsilon_s)\psi_s\psi_s^*dq=0$$

or, since $\int \psi_s\psi_s^*dq=1$,

$$\epsilon_s=\int r\psi_s\psi_s^*dq.$$

This equation gives the new proper-value of the energy as

$$E'_s=E_s+\lambda\int r\psi_s\psi_s^*dq.$$

We now proceed to find the corresponding wave-function, the new part of which is given by the equation

$$(H-E_s)v_s=-(r-\epsilon_s)\psi_s.$$

The integration of this presents no difficulty, and gives for the perturbed system the wave-function

$$\psi'_s=\psi_s+\lambda\sum_k{}'\frac{\int r\psi_k^*\psi_s dq}{E_s-E_k}\psi_k$$

where the prime above the $\sum$ means that the term corresponding to $k=s$ is to be omitted.

Schrödinger then considered the case of degeneracy, when to a single proper-value E_s of the energy there correspond several wave-functions: in the perturbed system, the degeneracy will in general be wholly or partly removed, so that a single original energy-level gives rise to a number of energy-levels close together, and spectral lines are split, as in the Stark and Zeeman effects.

In applying this theory to the investigation of the Stark effect in the Balmer lines of hydrogen, we suppose that there is an electric field of intensity F in the positive z-direction, so the wave-equation is

$$\nabla^2\psi+\frac{2m}{\hbar^2}\left(E+\frac{e^2}{r}-eFz\right)\psi=0.$$

Schrödinger followed Epstein[1] in introducing space-parabolic co-ordinates λ, μ, ϕ by the equations

$$x=(\lambda\mu)^{\frac{1}{2}}\cos\phi, \qquad y=(\lambda\mu)^{\frac{1}{2}}\sin\phi, \qquad z=\tfrac{1}{2}(\lambda-\mu)$$

for which the volume-element $dxdydz$ is $\frac{1}{4}(\lambda+\mu)d\lambda d\mu d\phi$. The wave-equation now becomes

$$\frac{\partial}{\partial\lambda}\left(\lambda\frac{\partial\psi}{\partial\lambda}\right)+\frac{\partial}{\partial\mu}\left(\mu\frac{\partial\psi}{\partial\mu}\right)+\frac{1}{4}\left(\frac{1}{\lambda}+\frac{1}{\mu}\right)\frac{\partial^2\psi}{\partial\phi^2}$$

$$+\frac{m}{2\hbar^2}\{E(\lambda+\mu)+2e^2-\tfrac{1}{2}eF(\lambda^2-\mu^2)\}\psi=0.$$

To solve this equation we write

$$\psi=\Lambda M\Phi$$

where Λ is a function of λ only, M is a function of μ only and Φ is a function of ϕ only. The equation for Φ is

$$\frac{d^2\Phi}{d\phi^2}=-n^2\Phi$$

where n is a constant which (in order that Φ and its derivatives should be single-valued and continuous functions of ϕ) must be a whole number, so $n=0, 1, 2, 3, \ldots$. The equations for Λ and M are both of the form

$$\frac{d}{d\xi}\left(\xi\frac{d\Lambda}{d\xi}\right)+\left(D\xi^2+A\xi+2B-\frac{n^2}{4\xi}\right)\Lambda=0.$$

The term $D\xi^2$ represents the Stark-effect perturbation. So for the unperturbed system

$$\frac{d}{d\xi}\left(\xi\frac{d\Lambda}{d\xi}\right)+\left(A\xi+2B-\frac{n^2}{4\xi}\right)\Lambda=0.$$

Put $\Lambda=\xi^{-\frac{1}{2}}u$, $2\xi\sqrt{(-A)}=\eta$, and $B(-A)^{-\frac{1}{2}}=p$. The equation becomes

$$\frac{d^2u}{d\eta^2}+\left(-\frac{1}{4}+\frac{p}{\eta}+\frac{1-n^2}{4\eta^2}\right)u=0$$

the solution of which is the confluent hypergeometric function

$$u=W_{p,\frac{1}{2}n}(\eta),$$

[1] cf. p. 121 *supra*

so in order that the solution may be finite everywhere we must have

$$p=\tfrac{1}{2}(n+1)+k \qquad (k=0,\ 1,\ 2,\ \ldots),$$

and the corresponding solutions are

$$u_k(\eta)=W_{k+(n+1)/2,\ n/2}(\eta).$$

There are two values of k, corresponding to the equations for Λ and M respectively : call them k_1 and k_2. Then n, k_1 and k_2 are the three quantum numbers of the unperturbed motion. Denoting the values of B in the Λ- and M-equations respectively by B_1 and B_2, we have

$$B_1=\left(\frac{n+1}{2}+k_1\right)\sqrt{(-A)}$$

$$B_2=\left(\frac{n+1}{2}+k_2\right)\sqrt{(-A)}$$

so

$$B_1+B_2=(n+1+k_1+k_2)\sqrt{(-A)}.$$

But

$$B_1+B_2=\frac{me^2}{2\hbar^2} \quad \text{and} \quad A=\frac{mE}{2\hbar^2}$$

so the energy in the unperturbed motion is

$$E=-\frac{me^4}{2\hbar^2(n+1+k_1+k_2)^2}$$

an equation from which it is evident that $(n+1+k_1+k_2)$ is the principal quantum number.

If we now carry out the process described above for finding the energy-levels in the perturbed motion, we obtain

$$E=-\frac{me^4}{2\hbar^2(n+1+k_1+k_2)^2}-\frac{3\hbar^2F(k_2-k_1)(n+1+k_1+k_2)}{2me}$$

which is precisely the formula found by Epstein[1] for the energy-levels in the Stark effect of the Balmer lines of hydrogen.

Schrödinger proceeded to discuss the intensity of the Stark components. In order to find the intensity of a line, it is necessary to calculate the matrix-element relating to the corresponding transition. This matrix-element can be determined, as we have seen, from the wave-functions : and it was found to vindicate the selection and polarisation rules given by Epstein. Some of the theoretical results on intensities were confirmed experimentally in the following year by J. S. Foster.[2]

[1] cf. p. 121 *supra*

[2] *Proc. R.S.*(A), cxiv (1927), p. 47 ; cxvii (1927), p. 137

The terms proportional to the square of the applied electric force in the Stark effect were calculated in the following year by G. Wentzel[1] and I. Waller.[2] The Stark effect was investigated in 1926 also by Pauli[3] from the standpoint of Heisenberg's quantum-mechanics.

Schrödinger's treatment, like that of Schwarzschild and Epstein, ignored the fine-structure of the hydrogen lines: the theory was completed in this respect not long afterwards by R. Schlapp,[4] whose work was based on the new explanation of the fine-structure which had been provided by the theory of electron-spin.

Another phenomenon which is to be treated by the quantum-mechanical theory of perturbations is the Zeeman effect.[5] If in a first treatment of the subject we ignore electron-spin, and consider simply the motion of an electron attracted by a fixed nucleus of charge e and under the influence of a magnetic field of intensity H parallel to the axis of z, then the Lagrangean function is

$$\tfrac{1}{2}m\left\{\left(\frac{dx}{dt}\right)^2+\left(\frac{dy}{dt}\right)^2+\left(\frac{dz}{dt}\right)^2\right\}-\frac{e\mathrm{H}}{2c}\left(x\frac{dy}{dt}-y\frac{dx}{dt}\right)+\frac{e^2}{r}$$

and therefore the Hamiltonian function is[6]

$$\frac{1}{2m}\left\{\left(p_x-\frac{e\mathrm{H}}{2c}y\right)^2+\left(p_y+\frac{e\mathrm{H}}{2c}x\right)^2+p_z^2\right\}-\frac{e^2}{r}$$

(measuring e in electrostatic and H in electromagnetic units, as usual) or

$$\frac{1}{2m}\left\{p_x^2+p_y^2+p_z^2+\frac{e\mathrm{H}}{c}(xp_y-yp_x)+\frac{e^2\mathrm{H}^2}{4c^2}(x^2+y^2)\right\}-\frac{e^2}{r}.$$

The corresponding wave-equation, obtained by replacing p_x by $(\hbar/i)\,(\partial/\partial x)$ etc., is

$$\nabla^2\psi+\frac{2m}{\hbar^2}\left(\mathrm{E}+\frac{e^2}{r}\right)\psi+\frac{ie\mathrm{H}}{c\hbar}\left(x\frac{\partial\psi}{\partial y}-y\frac{\partial\psi}{\partial x}\right)-\frac{e^2\mathrm{H}^2}{4c^2\hbar^2}(x^2+y^2)\psi=0.$$

Introducing spherical-polar co-ordinates $(r,\ \theta,\ \phi)$, with the z-axis as polar axis, this becomes

$$\frac{1}{r^2}\frac{\partial}{\partial r}\left(r^2\frac{\partial\psi}{\partial r}\right)+\frac{1}{r^2\sin\theta}\frac{\partial}{\partial\theta}\left(\sin\theta\frac{\partial\psi}{\partial\theta}\right)+\frac{1}{r^2\sin^2\theta}\frac{\partial^2\psi}{\partial\phi^2}+\frac{2m}{\hbar^2}\left(\mathrm{E}+\frac{e}{r^2}\right)\psi$$
$$+\frac{ie\mathrm{H}}{c\hbar}\frac{\partial\psi}{\partial\phi}-\frac{e^2\mathrm{H}^2}{4c^2\hbar^2}r^2\sin^2\theta\,\psi=0.$$

[1] *ZS. f. P.* xxxviii (1927), p. 518 [2] *ZS. f. P.* xxxviii (1927), p. 635

[3] *ZS. f. P.* xxxvi (1926), p. 336; cf. also C. Lanczos, *ZS. f. P.* lxii (1930), p. 518; lxv (1930), p. 431; lxviii (1931), p. 204

[4] *Proc. R.S.*(A), cxix (1928), p. 313 [5] cf. Schrödinger, *Phys. Rev.* xxviii (1926), p. 1049

[6] cf. V. Fock, *ZS. f. P.* xxxviii (1926), p. 242

We shall ignore the term[1] quadratic in the magnetic intensity H: and we shall regard the linear term in H as a perturbation: so the wave-equation of the unperturbed system is the ordinary wave-equation of the hydrogen atom, which has already been considered: the wave-function is a constant multiple of

$$\frac{1}{r}W_{k,n+\frac{1}{2}}(z)P_n^{\mu}(\cos\theta)e^{\pm i\mu\phi}$$

and the energy-levels are

$$E_k = -\frac{me^4}{2\hbar^2k^2} \quad \text{where } k=1, 2, 3, 4, \ldots.$$

Now in the above wave-equation the perturbation-term implies the replacement of E by $E-\lambda r$ where

$$\lambda r = -\frac{ie\hbar H}{2mc}\frac{\partial}{\partial\phi}:$$

so the term $\lambda\epsilon_k$ to be added to E_k on account of the perturbation is (by the general theory)

$$\lambda\epsilon_k = \int\left\{-\frac{i\hbar eH}{2mc}(\pm i\mu)\right\}\psi_k\psi_k^*dq$$

when ψ_k is supposed to be normalised, so $\int\psi_k\psi_k^*dq=1$. Thus *the displacement of the energy-level in the Zeeman effect is*

$$\lambda\epsilon_k = \pm\frac{e\mu\hbar H}{2mc}$$

where μ *is the magnetic quantum number*: which is precisely the value found in Lorentz's classical theory.[2]

The above derivation accounts only for the normal Zeeman effect; as might be expected, since we know[3] that the anomalous Zeeman effect requires for its explanation the assumption of electron-spin. The problem thus presented was solved by W. Heisenberg and P. Jordan[4] by matrix-mechanical methods in 1926, and by C. G. Darwin[5] by wave-mechanics in 1927.

Darwin's model consists of a charged spinning spherical body moving in a central field of force. Denoting the charge by $-e$,

[1] The effect of the quadratic term was considered by O. Halpern and Th. Sexl, *Ann. d. Phys.*(5) iii (1929), p. 565.
[2] Vol. I, p. 412; cf. P. S. Epstein, *Proc. N.A.S.* xii (1926), p. 634; A. E. Ruark, *Phys. Rev.* xxxi (1928), p. 533
[3] p. 136 *supra*
[4] *ZS. f. P.* xxxvii (1926), p. 263
[5] *Proc. R.S.*(A), cxv (1927), p. 1

the mass by m, the position of the centre by (x, y, z) and the potential energy at distance r by $V(r)$, then the motion of revolution contributes terms

$$\tfrac{1}{2}m\left\{\left(\frac{dx}{dt}\right)^2+\left(\frac{dy}{dt}\right)^2+\left(\frac{dz}{dt}\right)^2\right\}+eV(r)$$

to the Lagrangean function. There is also a contribution

$$\tfrac{1}{2}I(\omega_x^2+\omega_y^2+\omega_z^2)$$

where I is the moment of inertia and $(\omega_x, \omega_y, \omega_z)$ the components of spin about (x, y, z). The orbital motion in the magnetic field H along z gives a contribution

$$-\frac{eH}{2c}\left(x\frac{dy}{dt}-y\frac{dx}{dt}\right).$$

The spin gives

$$-\frac{eHI\omega_z}{mc}$$

since e/mc is the ratio of magnetic moment to angular momentum. Lastly there is the interaction of the spin and the motion. The electric force has components $-(x/r)V'$ etc., so we obtain a term

$$-\frac{eIV'}{mc^2r}\left\{\omega_x\left(y\frac{dz}{dt}-z\frac{dy}{dt}\right)+\omega_y\left(z\frac{dx}{dt}-x\frac{dz}{dt}\right)+\omega_z\left(x\frac{dy}{dt}-y\frac{dx}{dt}\right)\right\}.$$

All these terms were taken together and converted into Hamiltonian form, and the Schrödinger wave-equation was then deduced. From this equation by use of spherical harmonic analysis the proper-values and wave-functions were calculated and Landé's g-formula was obtained.[1]

Schrödinger followed up his work on perturbations and the Stark effect by another paper[2] in which he extended the perturbation theory to perturbations that explicitly involve the time, and succeeded in obtaining by wave-mechanical methods the Kramers-Heisenberg formula for scattering.[3] Let q represent the co-ordinates which specify the state of the atom at the instant t, and let $H_0(q, p)$ represent its energy, so the wave-function for the unperturbed atom is given by the equation

$$i\hbar\frac{\partial\psi}{\partial t}=H_0\left(q, \frac{\hbar}{i}\frac{\partial}{\partial q}\right)\psi.$$

[1] cf. K. Darwin, *Proc. R.S.*(A), cxviii (1928), p. 264
[2] *Ann. d. Phys.*(4) lxxxi (1926), p. 109; cf. O. Klein, *ZS. f. P.* xli (1927), p. 407
[3] cf. p. 206 *supra*

Let the atom be irradiated by light whose wave-length is so great compared with atomic dimensions that we can regard the electric vector of the light as constant over the atom : let this electric vector be

$$\mathbf{E} = \mathbf{V}^* e^{2\pi i\nu t} + \mathbf{V} e^{-2\pi i\nu t}$$

where $\mathbf{V}$ is a complex vector. The additional energy of the atom due to this perturbation is $-(\mathbf{M}\,.\,\mathbf{E})$, where $\mathbf{M}$ denotes the electric moment $\sum e\mathbf{r}$ of the atom: so the wave-function of the perturbed atom is determined by the equation

$$i\hbar\frac{\partial\psi}{\partial t} = \left\{\mathrm{H}_0\left(q, \frac{\hbar}{i}\frac{\partial}{\partial q}\right) - (\mathbf{M}\,.\,\mathbf{E})\right\}\psi.$$

The wave-function, when the atom is in the state P, may be denoted by

$$\psi_{\mathrm{P}}{}^0 = e^{\frac{\mathrm{H_P}t}{i\hbar}}\psi_{\mathrm{P}}(q)$$

and a general solution may be represented by a series of these wave-functions. Supposing that the atom is initially in the state P, let us solve the wave-equation of the perturbed atom by writing

$$\psi = \psi_{\mathrm{P}}{}^0 + \psi_{\mathrm{P}}{}^1, \text{ where } \psi_{\mathrm{P}}{}^1 \text{ is small.}$$

Substituting in the differential equation, and neglecting small quantities of the second order, we have

$$\mathrm{H}_0\left(q, \frac{\hbar}{i}\frac{\partial}{\partial q}\right)\psi_{\mathrm{P}}{}^1 - i\hbar\frac{\partial\psi_{\mathrm{P}}{}^1}{\partial t} = (\mathbf{M}\,.\,\mathbf{E})\psi_{\mathrm{P}}{}^0.$$

We solve this by substituting in it

$$\psi_{\mathrm{P}}{}^1 = \psi_{\mathrm{P}+}e^{\frac{\mathrm{H_P}+h\nu}{i\hbar}t} + \psi_{\mathrm{P}-}e^{\frac{\mathrm{H_P}-h\nu}{i\hbar}t}$$

and equating terms which have the same time-factor : the resulting equation for $\psi_{\mathrm{P}}{}^+$ is

$$\mathrm{H}_0\left(q, \frac{\hbar}{i}\frac{\partial}{\partial q}\right)\psi_{\mathrm{P}}{}^+ - (\mathrm{H_P} + h\nu)\psi_{\mathrm{P}}{}^+ = (\mathbf{M}\,.\,\mathbf{V})\psi_{\mathrm{P}}.$$

Now expand $\psi_{\mathrm{P}}{}^+$ and $\mathbf{M}\psi_{\mathrm{P}}$ as series of the wave-functions ψ_{R}, say

$$\psi_{\mathrm{P}}{}^+ = \sum_{\mathrm{R}} a_{\mathrm{PR}}\psi_{\mathrm{R}}$$

and

$$\mathbf{M}\psi_{\mathrm{P}} = \sum_{\mathrm{R}} \mathbf{M}_{\mathrm{PR}}\psi_{\mathrm{R}} \quad \text{where} \quad \mathbf{M}_{\mathrm{PR}} = \int \psi_{\mathrm{R}}{}^*\mathbf{M}\psi_{\mathrm{P}}d\tau,$$

so $\mathbf{M}_{PR}$ is the element in the (matrix) electric moment of the atom, that corresponds to a transition from the level R to the level P. Thus

$$\sum_R a_{PR}\left\{H_0\left(q, \frac{\hbar}{i}\frac{\partial}{\partial q}\right) - H_P - h\nu\right\}\psi_R = \sum_R (\mathbf{M}_{PR}\,.\,\mathbf{V})\psi_R$$

or

$$\sum_R a_{PR}(H_R - H_P - h\nu)\psi_R = \sum_R (\mathbf{M}_{PR}\,.\,\mathbf{V})\psi_R$$

and therefore

$$a_{PR} = \frac{(\mathbf{M}_{PR}\,.\,\mathbf{V})}{h(\nu_{RP} - \nu)}.$$

Similarly the coefficient of ψ_R in the expansion of ψ_P^- is determined : and thus

$$\psi_P^1 = \frac{1}{h}\sum_R \left\{\frac{(\mathbf{M}_{PR}\,.\,\mathbf{V})}{\nu_{RP} - \nu}e^{\frac{H_P + h\nu}{i\hbar}t} + \frac{(\mathbf{M}_{PR}\,.\,\mathbf{V}^*)}{\nu_{RP} + \nu}e^{\frac{H_P - h\nu}{i\hbar}t}\right\}\psi_R.$$

This equation gives the perturbed value of the wave-function. The electric moment of the classical oscillator which would emit the radiation emitted in the scattering process of an atom in the state P, associated with the transition to the state Q, is the real part of

$$\int \psi_Q^* \mathbf{M} \psi_P d\tau, \quad \text{or in this case} \quad \int (\psi_Q^0 + \psi_Q^1)^* \mathbf{M} (\psi_P^0 + \psi_P^1) d\tau.$$

Neglecting terms of the second order, this is found to be

$$\mathbf{M}_{PQ}e^{-2\pi i\nu_{PQ}t} + \frac{1}{h}\sum_R \left\{\frac{\mathbf{M}_{RQ}(\mathbf{M}_{PR}\,.\,\mathbf{V})}{\nu_{RP} - \nu} + \frac{\mathbf{M}_{PR}(\mathbf{M}_{RQ}\,.\,\mathbf{V})}{\nu_{RQ} + \nu}\right\}e^{-2\pi i(\nu_{PQ} + \nu)t}$$

$$+ \frac{1}{h}\sum_R \left\{\frac{\mathbf{M}_{RQ}(\mathbf{M}_{PR}\,.\,\mathbf{V}^*)}{\nu_{RP} + \nu} + \frac{\mathbf{M}_{PR}(\mathbf{M}_{RQ}\,.\,\mathbf{V}^*)}{\nu_{RQ} - \nu}\right\}e^{-2\pi i(\nu_{PQ} - \nu)t}.$$

The term $\mathbf{M}_{PQ}e^{-2\pi i\nu_{PQ}t}$ represents the spontaneous emission associated with the transition P→Q, and *the other terms are precisely those found by Kramers and Heisenberg for the electric moment corresponding to the scattered radiation associated with the transition* P→Q.

The Compton effect was investigated quantum-mechanically in 1926 by Dirac[1] by means of his symbolic representation of matrix-mechanics, and by W. Gordon[2] by wave-mechanics. Gordon found that the quantum-mechanical frequency and intensity are the

[1] *Proc. R.S.*(A), cxi (1926), p. 405

[2] *ZS. f. P.* xl (1926), p. 117 ; cf. G. Breit, *Phys. Rev.* xxvii (1926), p. 362 ; O. Klein, *ZS. f. P.* xli (1927), p. 407, *at* p. 436 ; G. Wentzel, *ZS. f. P.* xliii (1927), pp. 1, 779 ; G. Beck, *ZS. f. P.* xliii (1927), p. 658

geometric means of the corresponding classical quantities at the beginning and end of the process. Breit[1] remarked that according to Gordon, the radiation in the case of the Compton effect may be regarded as due to a $\psi\psi^*$ wave moving with the velocity of light. Similarly Breit showed[2] that if the motion of the centre of gravity of an atom is taken into account, emission takes place only by means of unidirectional quanta due to $\psi\psi^*$ waves moving with the velocity of light.

In 1927 Schrödinger[3] showed that by means of the conception of de Broglie waves, the Compton effect may be linked up with some other investigations of a quite different character. He began by recalling an investigation of Léon Brillouin[4] on the way in which (according to classical physics) an elastic wave, in a transparent medium, affects the propagation of light. Let there be an elastic compressional or longitudinal wave (i.e. a sound-wave) of wave-length Λ, which is propagated in a transparent medium. It reflects at its wave-front light-rays which traverse the medium; but the reflected ray has negligible intensity except when there is between the wave-lengths and the angles the relation

$$\lambda = 2\Lambda \cos i$$

where i is the angle of incidence and λ is the wave-length of the light in the medium (so it is $K^{-\frac{1}{2}}$ times the wave-length of the light *in vacuo*, where K is the dielectric constant). This is precisely the equation which had been found in 1913 by W. L. Bragg as the condition that X-rays of wave-length λ should be reflected by the parallel planes rich in atoms in a crystal, when i was the angle of incidence and Λ was the distance between the parallel planes in the crystal.[5] In Brillouin's theorem it is supposed that the velocity of propagation of the elastic wave is small compared with the velocity of light: more accurately, the formula must be modified as in the case of reflection at a *moving* mirror.[6]

Schrödinger now showed that the Compton effect can be assimilated to the Brillouin effect, if the electron of the Compton effect in its initial state of rest is replaced by a 'wave of electrical density' and also in its final state is replaced by another wave. The two waves form by their interference a system of intensity-maxime located in parallel planes, which correspond to Bragg's planes in a crystal. The analytical formulae are, as he showed, identical with those found by Compton in the particle-interpretation. Thus *a wave-explanation is obtained for the Compton effect.* Other cases are known in which a transition of a particle from one state of motion

[1] *Proc. N.A.S.* xiv (1928), p. 553
[2] *J. Opt. Soc. Amer.* xiv (1927), p. 374
[3] *Ann. d. Phys.* lxxxii (1927), p. 257
[4] *Ann. d. Phys.* xvii (1921), p. 88
[5] cf. p. 20 *supra*
[6] Brillouin's problem was treated quantum-mechanically in *Phys. ZS.* xxv (1924), p. 89 by Schrödinger, who found that quantum theory led to the same formula as classical physics.

to another may be translated into the superposition of two interfering matter-waves.

The year 1926 saw the publication of some important papers on the problem of atoms with more than one electron. It had been known for some years that the energy-levels or discrete stationary states, that account for the spectrum of neutral helium (which has two electrons), constitute two systems,[1] such that the levels belonging to one system (the *para*-system) do not in general combine with the levels of the other system (the *ortho*-system) in order to yield spectral lines. This fact led spectroscopists at one time to conjecture the existence of two different chemical constituents in helium, to which the names *parhelium* and *orthohelium* were assigned. The para-level system consists of singlets and the ortho-system consists of triplets.

Heisenberg[2] and Dirac[3] now investigated the general quantum-mechanical theory of a system containing several identical particles, e.g. electrons. If the positions of two of the electrons are interchanged, the new state of the atom is physically indistinguishable from the original one. In this case we should expect the wave-functions to be either symmetrical or skew in the co-ordinates of the electrons (including the co-ordinate which represents spin). Now a skew wave-function vanishes identically when two of the electrons are in states defined by the same quantum numbers : this means that in a solution of the problem specified by skew wave-functions there can be no stationary states for which two or more electrons have the same set of quantum numbers, which is precisely Pauli's exclusion principle. A solution with symmetrical wave-functions, on the other hand, allows any number of electrons to have the same set of quantum numbers, so this solution cannot be the correct one for the problem of several electrons in one atom.

In a second paper,[4] Heisenberg applied the general theory to the case of the helium atom. There are energy-levels corresponding to wave-functions which are symmetric in the space-co-ordinates and skew in the spins of the electrons (i.e. the spins of the two electrons are antiparallel) : these may be identified with the para-system. There are also wave-functions which are skew in the space-co-ordinates and symmetric in the spins (i.e. the spins of the two electrons are parallel) : these are associated with the ortho-system. In both systems the wave-functions involving both co-ordinates and spin change sign when the two electrons are interchanged. Now if a wave-function is symmetrical (or skew) at one instant, it must remain symmetrical (or skew) at all subsequent instants, and therefore the changes in the system cannot affect the symmetry (or

[1] For studies of the helium atom and the related ions Li^+ and Be^{++} from the standpoint of the earlier quantum theory, cf. A. Landé, *Phys. ZS.* xx (1919), p. 228 ; H. A. Kramers, *ZS. f. P.* xiii (1923), p. 312 ; J. H. van Vleck, *Phys. Rev.* xxi (1923), p. 372 ; M. Born and W. Heisenberg, *ZS. f. P.* xxvi (1924), p. 216.

[2] *ZS. f. P.* xxxviii (1926), p. 411

[3] *Proc. R.S.*(A), cxii (1926), p. 661

[4] *ZS. f. P.* xxxix (1926), p. 499

skewness), so the para-levels and ortho-levels cannot have transitions into each other ; which explains their observed property.

Heisenberg's first paper was notable for the discovery of the property of like particles known as *exchange interaction*, which had an important place in the physical researches of the years immediately succeeding : an account of these must be reserved for the next volume.

Index of Authors Cited

Subject Index